Engineering Disasters – Lessons to be Learned

University of Plymouth Library

Subject to status this item may be renewed
via your Voyager account

http://voyager.plymouth.ac.uk

Exeter tel: (01392) 475049
Exmouth tel: (01395) 255331
Plymouth tel: (01752) 232323

Engineering Disasters – Lessons to be Learned

Don Lawson

Professional Engineering Publishing

Professional Engineering Publishing Limited
London, UK

First published 2005

© D S Lawson

ISBN 1 86058 459 4

A CIP catalogue record for this book is available from the British Library.

Printed and bound by The Cromwell Press, Trowbridge, Wiltshire, UK.

Contents

Robert Stephenson's Recommendation

Robert Stephenson* in 1856 said:

> ... *for nothing was so instructive to the younger Members of the Profession, as records of accidents in large works, and of the means employed in repairing the damage. A faithful account of those accidents, and of the means by which the consequences were met, was really more valuable than a description of the most successful works. The older Engineers derived their most useful store of experience from the observation of those casualties which had occurred to their own works, and it was most important that they should be faithfully recorded in the archives of the Institution.*

*The railway pioneers George Stephenson and his son Robert were among the 56 engineers and mechanics who started the Institution of Mechanical Engineers. George and Robert Stephenson were the first and second Presidents of IMechE for the years 1847–1853.

Preamble

Plagiarize, Let no one else's work evade your eyes.

Engineers should adopt these words. Not for stealing ideas but to be well informed. They come from *Lobachevsky*, a song by Tom Lehrer, a Harvard mathematics lecturer.

One view of designers is that they start with a blank piece of paper, or these days a blank CAD screen, and proceed to create a new product out of their heads. Only a superhuman person could produce a successful product that way. Design engineering is a structured process using both art and science to create new or improved products – building on experience, bad as well as good.

Engineers do not want their product to fail unexpectedly or to injure anyone. Engineering textbooks do not dwell on failure, yet that is where they start. The load on a structure or mechanism is assessed against the load to failure. Then one knows, or thinks one knows, the safety factor being applied.

Humans design, build, operate, use, maintain, and can wreck engineering products. Humans are fallible. Engineers have to take into account all the potential failures of people – including engineers – as well as failures of equipment and materials. The aim is to reduce the chance of failure to an acceptable level while enjoying all the benefits.

Unfortunately, the very word disaster conjures up negative connotations. It is natural to try to forget failures and steer away from any product or project that has been labelled a failure. We have to reject these emotions and see what we can learn.

The idea for this book came from Whyte's *Engineering Progress Through Trouble* (**1**). Bob Whyte's book has been a constant source of reference since the time he gave me a copy.

Reference
(**1**) **Whyte, R.R.** (1975) *Engineering Progress Through Trouble*, IMechE, ISBN 0 85298 183 X.

Acknowledgements

The Internet search engines are great for research, but libraries are still vital. I would like to thank the staff of the following libraries for their help: IMechE, ICE, RINA, McMaster University, and the British Library.

This book covers a wide range of engineering topics, and I appreciate the assistance from the many experts who have helped me with their comments, or given me permission to quote their words. They include Dr Addison Bain, Brian Clementson, Dr Jerry Cuttler, Dieter Hofmann, Prof. Trevor Kletz, John Lee, Richard Lockett, Eric Magel, Dr Dan Meneley, Dr Sean Leen, William Middleton, Dr Ron Mitchel, Miguel Palomares, Prof. Henry Petroski, Academician N.N. Ponomarev-Stepnoy, Tony Roche, Roger Ridsdill-Smith, Prof. Rod Smith, Dr Victor Snell, and Terry Young.

In addition, the staff of PEP have been most helpful, particularly Sheril Leich who has steered me through this project.

And I would like to dedicate this book to two engineers with whom I have worked and from whom I have learned so much – Everett Long and Gordon Brooks.

Don Lawson
Oakville, Canada

The role of the engineer

The modern use of the word 'engineer' started with the industrial revolution. As the industrial revolution grew, engineers met to exchange ideas. They copied the style of scientific societies, which originated in Europe in the seventeenth century. In 1771, John Smeaton founded the Smeatonian Society of Civil Engineers in the UK. By invitation, like-minded engineers met for dinner. One of the earliest engineering societies open to all qualified engineers, was the Institution of Civil Engineering, which began in a London coffee house in 1818. It was called Civil Engineering to differentiate it from Military Engineering. The Institution of Civil Engineers' Royal Charter of 1828 defines engineering as:

> *The art of directing the great sources of power in nature for the use and convenience of man.*

Institutions were established in the United Kingdom for various branches of engineering. The Institution of Mechanical Engineers was founded in 1847, the Royal Aeronautical Society in 1866, and the Institution of Electrical Engineers in 1871. By the mid-nineteenth century, engineering societies had been established in the United States and most European countries.

The design engineer's tasks are summarized in Fig. 1. The skill of the design engineer is to stay away from failure on the one hand, and on the other be economic in the use of materials and labour.

Figure 1 is for design, but a similar chart can be drawn for maintenance, operations, or management. Management has to design the organization, and assess its effectiveness, against the constraints of effective use of resources and avoiding failure.

The engineer, in whatever role in the profession, needs to know the boundaries of failure. This requires knowledge and learning. The views of Robert Stephenson, quoted in the frontispiece, are as relevant today as when he said them in 1856.

What is a disaster?

Disaster equates to tragedy and catastrophe. The emotional perception of risk results in major engineering failures being labelled disasters.

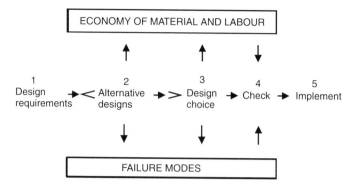

Fig. 1   The design process

Failure is when a person or object fails to perform to expectations. In this broad definition, failure covers the range from disasters to where a product does not perform as well as expected.

Why should I study disasters and failures?

To learn, and to avoid repeating them. Learning from failures and mistakes of others is cost-effective and less painful than experiencing them oneself. And the public expects the profession to learn from its mistakes and not repeat them. With an increasingly litigious society, nobody wants to be caught in the witness box thinking *I wish I had known that.*

Are all failures bad?

Engineering is not precise, and materials have a range of properties. Avoiding all failures might be expensive. A good example is telegraph and power poles. There is a great benefit in standardizing. But there is also a wide range of loading due to different terrain and weather conditions. It would be uneconomic to make all poles stand up to the worst statistical loading. The right engineering solution is to balance the cost of replacing a few failures with economy over the whole population. This is a conscious decision to allow for a few failures.

A learning profession

We are fortunate in seeing fewer failures as engineering knowledge builds. The failures we now see tend to be more complex than earlier ones. The pressure is on investigators to find a quick and simple cause. Investigations get lots of media attention. But the real lessons are often only apparent in quiet analysis after the media hubris has subsided. Engineers can afford to wait and see what lessons can be learned. Lessons tend to be timeless. Hopefully the same types of failure will not get repeated, and lessons from one failure, or near miss, can prevent others – even in different industries.

Most engineering societies have committed their members to Continuing Professional Development (CPD). CPD is an attitude of mind that observes, questions, and learns. Studying failures, and learning from them, is a key part of CPD.

This book

The aim is to show that we can always learn something from disasters and failures.

Part 1 describes selected disasters and failures from a range of industries. Initially, I chose a series of failures that aroused my interest. I wanted to see whether there were common themes from this random selection. During research, it became apparent that more should be written on each example to draw out the lessons. I re-learned the lesson of John Dryden (1678):

> *Errors, like straws, upon the surface flow:*
> *He who would search for the pearls must dive below.*

It takes more words to do justice to the lustre of the pearls. Consequently, the original list of examples was cut in half.

Each example starts with a background description of the failure followed by **comments** and **lessons learned**. The background descriptions come from a variety of sources, including Proceedings of Engineering Institutions, and Accident and Inquiry Reports. The descriptions are not a complete and definitive analysis of each situation. Fuller descriptions can be found in the references. The aim is to draw out interesting and informative aspects from which lessons can be learned.

The **comments** and **lessons learned** are my own. By creating a summary, I have made selections that inevitably introduced bias. No doubt readers will draw their own conclusions, and if they are different from mine, and a lesson is learned, then so be it. The aim is to trigger the thought process that leads to the question *What can I learn from this?* and *Could something like this happen to my work?*

One exception to the cases in Part 1 is the inclusion of the summary of the report in Section 1.7 Lessons from the US Space Program, which logically fits between the descriptions of the two space shuttle accidents.

In studying failures, I came across information and quotations that were worth repeating. A selection of these is included in **Part 2**.

Parts 1 and 2 raise a lot of thoughts and questions. In **Part 3**, I have looked for common threads and conclusions based on the examples in Part 1. The similarity of these conclusions with the writings of others in Part 2 shows that they may be generic. If Part 1 had included all the original list of cases, a quick review suggests that they would have reinforced the conclusions.

PART 1

The *Hindenburg* Disaster – Hydrogen Myth?

In 1937, the airship ***Hindenburg*** burst into flames while preparing to land. The accident was perceived to be the result of using hydrogen. This created a widespread belief that hydrogen is dangerous.

Fig. 1.1.1   The *Hindenburg* on fire. Reproduced by permission of the National Air and Space Museum, ©Smithsonian Institution (SI 73–8701)

1.1.1   The disaster

The ***Hindenburg*** (LZ129), owned by the Deutche Zeppelin Reederei (DZR) of Berlin, left Frankfurt at 8:15 p.m. on 3 May, 1937, for a scheduled flight to Lakehurst, New Jersey. She had been delayed by unfavourable weather and arrived on the evening of 6 May. The weather had been stormy with thunder in the area but was now clearing.

The ***Hindenburg*** circled the field and turned towards the mooring mast. The forward anchor ropes were dropped to the ground at 7:21 p.m. At 7:25 p.m., at a height of around 200 ft and a distance of 600 ft from the mast, a flame appeared near the tail. The flames spread forward rapidly and within 34 s, she was a blazing wreck on the ground.

A crew of 61, including trainees, were on board – 22 died. The passenger accommodation was only half full with 36 passengers – 13 passengers and one member of the ground crew died.

1.1.2   Airship history

By 1900, airships had evolved from balloons through semi-rigid airships, to airships using a rigid structure. The pioneer of the rigid airship was Count von Zeppelin, an aristocrat and cavalry officer. When he was 25, he was sent to America to observe the Civil War for the King of Württemberg. During his visit, he ascended in an army observation balloon, and met Abraham Lincoln. Back in Germany, he left the army, having upset the Kaiser. Alarmed at the developments in France with non-rigid airships, he decided to pursue his concept for a rigid airship. France was Germany's traditional enemy and, with his military background, he was motivated to compete.

The Count saw the development of the first effective internal combustion engine by Gottlieb Daimler as the key to successful flight. The development for producing aluminium at a reasonable price also helped. In 1896, the Count addressed the prestigious Union of German Engineers. Instead of dwelling on the military aspects, he presented his airship for international air transport, carrying mail, and exploring the Arctic and other remote terrain. The King of Württemberg attended, and both he and the Union supported the Count's concept.

The first Zeppelin (Luftschiff Zeppelin #1 or LZ1) was 416 ft long and made her maiden flight over Lake Constance in July 1900. This flight ended with LZ1 unceremoniously entangled with debris in the water.

During the First World War, Zeppelins were used as scouts for the German Navy, and, in a limited extent, to drop bombs over England. They were developed to fly as high as 24 000 ft to avoid fighter aircraft.

Germany was prohibited from building airships for its own use until 1925. However, the LZ126 airship was ordered by the US Navy and built at Friedrichshafen by the Luftschiffbau Zeppelin GmbH. (LZG) as war reparation. This was the first airship to have the outer skin coated with a dope mixed with aluminium powder to tauten the skin, to make it waterproof, and to reflect the heat from the sun. The building ban did not stop LZG from concluding a licence agreement in 1923 with the Goodyear Tire and Rubber Co. A new venture, Goodyear-Zeppelin, was established to make Zeppelins in the United States using German technology and seconded staff.

Once the ban on building large airships was lifted, the LZG raised public funds by lectures and contributions to build LZ127, later named the ***Graf Zeppelin***. This was the largest and best, capable of speeds over 70 mile/h. She was launched in 1928 and in the same year crossed the Atlantic. The following year, she made the first passenger flight around the world. Both flights were partly sponsored by Hearst Newspapers, who gave the flights great publicity.

The success of the ***Graf Zeppelin*** led to plans for a larger commercial airship, LZ128. The crash of the British R101 in France in October 1930 shocked the airship community and halted the design of LZ128. After a delay of three years, design started on a new Zeppelin, LZ129, to be named the ***Hindenburg***.

Funding the construction of the ***Hindenburg*** was difficult for LZG with the depression following the Wall Street Crash of 1929. A loan was finally secured from the Government of the State of Württemberg. The State Government offset the loan with the Central Government. The Nazi party now controlled the Central Government, and Air Minister Göring intervened. He created the Deutsche Zeppelin Reederei (DZR), which was half funded by Lufthansa, to take control of flight operations. LZG was left with airship construction.

By the mid-1930s, Germany was dominant in airships. The UK effort had ended with the disaster of the R101. The US interest stopped with the crashes of the ***Acron*** in 1933 and ***Macon*** in 1935. Despite these crashes and the difficult economic climate, Germany pursued the development of Zeppelin. Zeppelins were seen worldwide as a great success – until the ***Hindenburg*** crashed.

1.1.3 Why were airships popular?

Today, the image of the airship is a cumbersome and archaic behemoth. By the 1930s, the large airships could cross the Atlantic in one-third the time taken by a ship. They were ships of the air. The ***Hindenburg***, at 804 ft long, was only 10 per cent shorter than the ***Titanic***, but only 1/1000 its structural weight. In 1936, she had made 55 flights, including ten scheduled flights to New York and seven flights to Rio de Janeiro. The ***Hindenburg*** had carried a total of 2708 passengers. Until the crash, no fare-paying civilian passenger had lost their life on an airship.

Airplane service across the Atlantic was still in the future. Airplanes were uncomfortable, particularly for long journeys. Their piston engines created vibrations and were not reliable. They were noisy and smelt of engine oil.

1.1.4 The impact of world events and the political climate

In reparation from the First World War, the United Kingdom, the United States, France, and Italy obtained airships and airship technology from Germany. The United States assigned the airships that they received to the navy. As a result, all the airship-operating experience and the bases were with the Navy. When the Zeppelin Company wanted to fly to New York, they had to make arrangements with the US Navy to use their base at Lakehurst, New Jersey.

The political atmosphere of the 1930s was charged. The Nazi Government used the Zeppelins for propaganda within Germany, flying over the Berlin Olympic Games as well as the infamous Nürnberg Rally.

The *Hindenburg* was originally designed to use a combination of hydrogen and helium to reduce fire risks. The United States was the sole source of helium. They would not provide helium to Germany because of the political climate. At least, that has been the accepted position. The research of John Duggan (1) shows that this may be another myth. He has found no record in either Germany or the United States of a request for helium. There was a Helium Act giving the US Bureau of Mines a control over helium. The reasons for not requesting helium could be that:

- it was believed, without asking, that supplies would be denied because of the Helium Act;
- it would be too difficult to establish a network of helium supply and processing at each stop;
- the cost of helium was much higher than hydrogen and the Zeppelin project was short of cash;
- the Nazi Government would not release precious foreign currency to buy the helium; and
- the *Graf Zeppelin* had been operating successfully with hydrogen.

Testimony by Dr Eckener shows that he decided to change from helium to hydrogen for the *Hindenburg* as late as 1935, well into the construction phase. He planned to use helium for future operations, once economic conditions improved.

1.1.5 The key players
Hugo Eckener (1868–1954)

Hugo Eckener was a prized student in psychology, philosophy, history, and economics. His psychology doctorate from Leipzig under the renowned Dr Wundt was titled *Variabilities in Human Perception*. While he, aged 24, was still at Leipzig, he was asked whether he was interested in a post in the new psychology laboratory at the University of Toronto. He expressed interest initially, but decided to spend a brief period in the military.

As a part-time journalist, Dr Eckener reported on Count von Zeppelin's second airship flight over Lake Constance in 1900 as well as the first flight of LZ2 in 1906. LZ2 crashed in a strong gust of wind. Count von Zeppelin read Dr Eckener's articles and decided to meet him. The Count was seen as a crazy man, a mad inventor and an aerial dreamer. Dr Eckener saw the prospect of changing the image of the Count and his airships by an educational and publicity campaign. Eckener became sold on the Zeppelin concept and learnt to fly them.

Dr Eckener's first flight as a Captain, in May 1911, was inauspicious. In gusty conditions, he ended up landing on top of the hangar. From this experience, he decided that the safety of the passengers and crew would be essential for the successful promotion of airships. This would be achieved by training the crew and providing better information on weather conditions.

During the First World War, Dr Eckener trained Zeppelin pilots. After the war, he was running the Zeppelin Company, hands on. Eckener commanded LZ126 on the Atlantic crossing to deliver her to the US Navy in 1924. He negotiated the agreement with Goodyear. He commanded the Graf Zeppelin on her first passenger flight across the Atlantic in 1928 and her round-the-world flight in 1929.

Dr Eckener gained a legendary status for his skills as a Zeppelin commander. He was a keen sailor, and this experience gave him a good understanding of meteorology and the ability to judge changes in the weather. However, his greatest skill was in promoting and marketing the image of Zeppelin. He flew Zeppelins over key landmarks like the Statue of Liberty, the White House, and the Chicago World's Fair, and sent greetings to Heads of State. He met with the Emperor of Japan, as well as three Presidents of the United States and the Presidents of Brazil and Argentina. Despite the control of operations moving to DZR, with construction left with LZG, Dr Eckener was indispensable and effectively ran both companies.

President Coolidge invited Dr Eckener to the White House when he delivered the LZ126. On the same occasion, he met Henry Ford but failed to convince him to develop airships instead of airplanes. When he flew the *Graf Zeppelin* across the Atlantic in 1928, there was a ticker tape parade along the Broadway in his honour, despite the fact that the Atlantic had already been crossed by the British airship R34 in 1911. R34 was the first aircraft of any sort to make the return Atlantic crossing. Dr Eckener met with President Roosevelt to get him to authorize the commercial use of the navy base at Lakehurst.

Following the crash of the *Hindenburg*, Dr Eckener ordered modifications to the LZ130, the *Graf Zeppelin II*. He obtained an export permit from the US government to buy helium. By the time it was ready for delivery, the political climate leading up to World War II prevented shipment. In 1940, the last of the Zeppelins was scrapped on Göring's orders.

Dr Eckener was outspoken against the Nazi regime, and considered political asylum in the United States. He sat out WWII but was bombed out of his house in Friedrichshafen. In 1946, a US Navy Lieutenant sought out Dr Eckener at his home and took him a box of cigars from Vice Admiral Rosendahl. In 1947, at the instigation of the Goodyear Corp., Eckener visited the United States and worked with Goodyear for seven months on a proposal for a 950 ft, 250-passenger airship (**2**). The idea did not fly and he retired to Friedrichshafen.

Charles E Rosendahl (1892–1977)

Commander Charles E. Rosendahl was the Naval Officer in charge of Lakehurst at the time of the *Hindenburg* crash. He had 19 years of experience with airships, and later retired in 1946 with the rank of Vice Admiral. He had commanded the *Los Angeles*, which was originally the LZ126, the airship delivered by Dr Eckener.

During 1928, he visited the United Kingdom and Germany to study the construction of airships. He participated in the trials of the *Graf Zeppelin*. Rosendahl was on the first westward flight across the Atlantic in the *Graf Zeppelin* and, at the invitation of Eckener, was aboard when she circumnavigated the globe in 1929. He made several flights in the *Hindenburg* as a Naval Observer.

1.1.6   The US investigation

The *Hindenburg* disaster had to be investigated, but by whom? The airship was German, landing in the United States and on a Navy base. Should the investigation be technical, under commercial law, or should a criminal investigation be held?

The US Secretary of Commerce, in an order dated 7 May, designated South Trimble Jr, a solicitor at the Department of Commerce, to investigate *the facts, conditions, and circumstances of the accident and report.* Commander Rosendahl was one of the technical advisors. The Committee began hearings on 10 May and reported on 15 August 1937 (**3**).

1.1.7   The Department of Commerce Report

The Commerce Report on the Hindenburg disaster lists the data for the ***Hindenburg*** as having $7\,063\,000\,\text{ft}^3$ of hydrogen, giving a maximum lift of $472\,940\,\text{lb}$. The pressure was from $\frac{1}{2}$ in to 1 in water pressure. The nominal weight was $430\,950\,\text{lb}$, including $143\,650\,\text{lb}$ of diesel fuel. The maximum speed was 84 mile/h and cruising speed was 75 mile/h. The outer skin was a mixture of cotton and linen coated with a dope containing fine aluminium powder, to protect against ultraviolet heating from the sun. The inner surface of the upper part was coated with red iron paint. The hydrogen was contained in 16 gas cells. Either ballast water or hydrogen was released to change height. Gas shafts between the gas cells took hydrogen released from the manoeuvring valves to the top of the ship for ventilation. Four power pods each had a diesel of 1100 HP driving a four-bladed, 19 ft 9 in propeller.

The flammable limits of a mixture of hydrogen and air were quoted as probably between 4.5 and 62 per cent hydrogen. The mixture is explosive in the range 15–45 per cent hydrogen.

The Report summarizes the findings as:

- *sabotage – no evidence;*
- *accidental causes:*
  - *presence of combustible mixture of hydrogen and air by:*
    - *diffusion/osmosis – unlikely;*
    - *failure of valve mechanism – no evidence;*
    - *decreased ventilation – although the ship was nearly stationary, there was a breeze and hydrogen disperses upwards quickly;*
  - *broken propeller – damaged in the crash, not the cause;*
  - *fracture of hull wire – shear wires had broken in other ships without causing a fire;*
  - *structural failure – no evidence;*
  - *ignition of mixture due to:*
    - *mechanical friction – insufficient evidence;*
    - *chemicals – no evidence of spontaneous combustion;*

- *thermodynamic – sparks from the engine exhaust were not close to the start of the fire;*
- *electrical – from power supply, instrumentation or wireless – remote;*
- *electrostatic – in older types of cell fabric, a spark could be produced by tearing the fabric. Since virtually all the cells were consumed by the fire, no tests were made on the cell fabric. Did the landing ropes serve as conductors? National Bureau of Standards tests showed that the airship would be 90 per cent discharged within 0.6 s to 170 s of the ropes touching the ground. The ropes had been touching the ground for four minutes before the flame appeared. Brush discharge due to the meteorological conditions of a charged damp atmosphere – possible.*

The report concluded as follows.

The cause of the accident was the ignition of a mixture of free hydrogen and air. Based on the evidence, a leak at or in the vicinity of cells four and five caused a combustible mixture of hydrogen and air to form in the upper stern part of the ship in considerable quantity: the first appearance of an open flame was on the top of the ship and a relatively short distance forward of the upper vertical fin. The theory that a brush discharge ignited such mixture appears most probable.

1.1.8   The role of the FBI
Throughout the investigations, the FBI told everyone that they were not conducting a formal investigation, but would support others. Their file provides some interesting background as well as showing the modus operandi of the FBI.

The modus operandi of the FBI
The FBI file (**4**) shows how their agents dutifully recorded everything, and kept their colleagues and the Director fully informed. These written reports primarily present the facts and also give a feel for the politics of the situation.

The files show that the FBI was inundated with data and leads. Many leads claimed sabotage. All the leads were diligently followed until the FBI was convinced that they were either false or that they led nowhere.

Closing the file
By the end of 1937, the two FBI agents involved wrote notes to the record, suggesting that the *Hindenburg* case was closed. They

received letters from J. Edgar Hoover, the Director, telling them: *first tidy up your file documentation so that it will be available in the event inquiries are subsequently made relative to this case.*

1.1.9   The German investigation

Following the accident, Air Minister Göring appointed an investigating committee headed by Dr Eckener. The committee consisted of Dr Ludwig Dürr, Professor Dieckmann, and four others. Dr Dürr had been on the design team for all Zeppelins, and the Chief Designer of all but the first.

The German investigation concluded that the most likely cause was a hydrogen leak, maybe caused by a broken tensioning wire puncturing a gas cell. The escaping hydrogen was then ignited by a static discharge.

An onlooker to the disaster, Mark Heald, an academic, saw an electrostatic display, St Elmo's fire, over the **Hindenburg** before she caught fire. This was not reported in the US hearings. A specialist in static electricity, J.A. Hering from The Netherlands, wrote to Professor Dieckmann saying that St Elmo's fire would not have had the temperature to ignite hydrogen.

Professor Dieckmann conducted tests on the dope from the **Graf Zeppelin** and could not get it to explode, but he could when he tested dope from the **Hindenburg**. He postulated the conditions necessary for static electricity to cause ignition as:

- *electrical isolation of the cover from the structure;*
- *gaps between the cover and the structure;*
- *sufficient conductivity in the ropes touching the ground;*
- *the height of the airship to create enough potential difference; and*
- *a sufficient gap to generate a spark.*

All the conditions would have to be present to ignite hydrogen, and all were present at the time of the disaster. He recommended that LZ130 have a conductive outer cover, well earthed to the structure, and to use impregnated landing ropes to give a slow release of static electricity.

The German tests on the cover and dope took place after the US investigation had concluded.

1.1.10   New developments in the 1990s

The story of the **Hindenburg** took a new twist in the 1990s with public presentations by Dr Addison Bain, a former manager of

Hydrogen Programs at the NASA Kennedy Space Center. Dr Bain's interest in the *Hindenburg* started in the late 1960s. He studied the evidence from both photographs and testimony of eyewitnesses, and found contradictory information. Eyewitnesses reported that the *Hindenburg* burnt from the tail with a bright flame tinged yellow or red. Hydrogen burns with a near-colourless flame. If any colour is present, it is blue. The falling pieces were burning in the same way as the main body of the airship. There were similar flames on the covering of the elevator fins, although they did not contain hydrogen. The airship retained her trim and kept altitude for several seconds while well on fire. If the hydrogen had been burning, surely the airship would not be staying airborne?

No crew member had reported smelling gas. Hydrogen has no smell, but garlic was added to make it easier to detect leaks. Captain Pruss insisted that his instruments on the bridge did not show any leakage from the gas cells.

Addison Bain's tests

In 1994, Dr Bain managed to get samples of the outer covering of the *Hindenburg* and carried out tests. The outer covering of cellulose acetate butyrate dope and fine aluminium powder was similar to the rocket fuel he was familiar with at NASA. He found that the aluminium and dope had bled through the fabric in places and had combined with the iron oxide on the inner surface. The resulting mixture is similar to a thermite fuse mixture used to achieve high temperatures in welding.

Dr Bain tested samples of similarly doped fabric for their potential to be ignited by an electrostatic discharge. When an arc struck down onto the fabric samples, it only resulted in local damage. Airships struck by lightning had shown similar localized damage. When the arc was parallel to the surface of the fabric, the electrical energy was sufficient to ignite the sample, which was quickly consumed by fire.

Addison Bain's conclusions

The testimony of Herman Lau began to make sense. He was a crewman who had been near gasbags four and five at the time of the accident. While the witnesses on the ground reported the first flames appearing on the port side, Lau reported seeing a glow like a *Japanese lantern* on the starboard side. He saw no fire at first but a bright reflection through the inside of the cell. Cell number four disappeared because of the heat. Lau had not smelt any gas before the fire.

Dr Bain agrees with the Commerce Inquiry Report that the cause of ignition was static electricity. However, he maintains that the airship would have burned even if no hydrogen had been present, because of the flammable nature of the outer fabric coating. He first presented his findings in the Smithsonian Air and Space Magazine (**5**). See also *Afterglow of a Myth* (**6**).

Changes made by the Zeppelin Company

Addison Bain was given access to the historic records of the Zeppelin Company and found that they had made changes to the sister ship the ***Graf Zeppelin II***, after the ***Hindenburg*** disaster. A fire-retarding calcium compound was added to the dope. Bronze was substituted for aluminium. Although heavier, it was far less combustible and a better conductor of electricity. Other changes were made to reduce the likelihood of electrical potential differences building up between different parts of the structure and the cover. Finally, Bain found a handwritten letter from Otto Beyersdorff, an independent electrical engineer retained by the Zeppelin Company. Beyersdorff wrote: *The actual cause of the fire was the extremely easy flammability of the covering material brought about by discharges of an electrostatic nature.*

1.1.11   Is this the end of the story?

In 2000, Norman Peake, writing in the Journal of the Airship Heritage Trust – *Dirigible* (**7**), challenged Addison Bain's conclusions. He points out that hydrogen-filled barrage balloons shot down in WWII burnt with an orange flame, and also that no lightening strike was seen to hit the ***Hindenburg***. In addition, he says that not all the covering burnt – or at least did not burn very quickly.

In a rebuttal article in a subsequent edition of *Dirigible* (**8**), Richard Van Treuren dismissed Peake's criticisms. He said that the colour of the flames in the barrage balloon fires was due to the cover material burning. Van Treuren reports that Dr Dieckmann found that the dope on the ***Hindenburg*** had no electric conductivity. The structure had been painted with a primer to prevent corrosion. This primer was non-conductive and flammable. Van Treuren lists 11 helium-filled US blimps that were all lost by fire. He wrote the book – '***Hindenburg***: *The Wrong Paint, Hydrogen: The Right Fuel'*.

The photographs and film of the crash show that the covering was all burnt within half a minute. This does not support Peake's criticism.

1.1.12   Some loose ends

Some loose ends remain, and may have some bearing on the accident. Harold Dick of the Goodyear Corp. took part in many Zeppelin flights and tests. He noted that there was always a smell of diesel fuel in the stern of the *Hindenburg* and thought that the fuel line joints should be tightened.

Even in the early trial flights, some fluttering of the outer cover near the aft engines had been observed. This was possibly due to vibration in that area from the engines. The structure was also said to become loose after long flights.

In a 2001 film for the TV Discovery Channel, *The Hindenburg Disaster – Probable Causes,* the staff of the Aeronautics Department at Cranfield postulated that a fatigue failure in the structure might have initiated the accident. They say that there was little understanding of fatigue at the time the structure was designed.

The *Hindenburg* was being brought in for a high landing at Lakehurst. The US Navy crew liked to winch down airships whilst the German approach was to fly down to the mooring. In the thundery weather, the high landing created a greater electrical potential difference between the airship and the ground than in a normal landing.

Dr Eckener blamed Captain Pruss and considered that he should have waited for better weather conditions. Eckener wrote a Zeppelin operating manual that said that hydrogen gas should not be vented in thundery conditions. Pruss had vented hydrogen four times while coming into land, the last time just six minutes before the fire. Pruss also made a tight turn over the field that may have strained the structure and broken tensioning wires. Some suggest that Rosendahl contributed by pressing Pruss to make an early landing.

COMMENTS

After many decades, the story of the *Hindenburg* takes a new twist. The analysis of Addison Bain is very credible. Why were some of the facts that are now obvious not spotted during the 1937 investigation? The clear message from Addison Bain's work is: *Don't blindly accept what doesn't feel right*. The Commerce Report examined many potential causes, but its conclusion left one feeling – *Yes, but …* Dr Bain's answer has that indefinable quality that leads one to say, *I can buy that.*

The photograph, Fig. 1.1.1, is worth studying. Dr Bain estimates that it was taken 17.2 s after ignition. John Dugan (1) says that 15 s following the first sign of fire there was a muffled explosion.

Figure 1.1.1 shows two forward water tanks falling from the airship, presumably shaken from their quick release mountings by the explosion. If the flames seen in the photograph were from the hydrogen in the rear of the airship, then they are large enough to have consumed all the hydrogen in that section. By law of gravity, one would expect to see the tail on the ground by that time. A stone would only take 3.5 s to fall to the ground from that height. Even a parachutist would have reached the ground! The only conclusion is that the hydrogen was still supporting the tail in the early stages of the fire.

From Bain's work, the cause of the crash was unlikely to have been initiated by the ignition of a hydrogen leak. The fire was probably initiated by a static electrical discharge. The development of the fire was due to the flammability of the outer skin, which then burnt the hydrogen in the gas cells now surrounded by a fierce fire. The diesel fuel added to the flames.

The question of a structural failure, leading to the disaster either by the breaking of a shear wire or fatigue, does not seem consistent with the photographs that show a normal landing until the fire started. Crewman Lau, who was in the tail close to the source of the fire did not see or hear any structural failure. The structure only appeared to fail when the airship broke her back after the fire was well established.

Still in question is the source of the initiating spark. Dr Dieckmann pointed out that the low conductivity of the outer cover, the insulation of the anti-corrosion paint on the structure, and the poor electrical bonding between the various components provided an environment for the generation of sparks. While an electrostatic glow was observed over the *Hindenburg*, no direct lightning strike was seen. It seems likely that the charged atmosphere and high landing procedure were sufficient to generate enough electrical potential differences to produce the initiating spark.

At last, we now have a credible explanation for the *Hindenburg* disaster – and hydrogen was not the root cause. It was a Frankenstein version of Benjamin Franklin's experiment with a kite in a thunderstorm. Only this time the kite was over 800 ft long, coated with rocket fuel surrounding 7 million cubic feet of hydrogen, and with a few tons of diesel fuel thrown in!

How dangerous is hydrogen?

Hydrogen self-ignites at a temperature of 585°C versus gasoline at 228°C to 501°C, depending on the additives. Flammable and explosive limits of various gases in air are dependent on the geometry and other environmental parameters, but typical figures are below.

	Flammable limit (%)	Explosive limit (%)
Hydrogen	4	18
Gasoline	1	1.1
Natural gas	5.3	6.3

From a safety point of view, hydrogen is similar to natural gas. However, it burns quickly and emits very little heat. It has the advantage, due to rapid dispersion, of not forming pools, unlike natural gas and gasoline. This is because hydrogen is 15 times lighter than air and disperses upwards at about 20 m/s. It is very difficult to detonate hydrogen in open air.

Dealing with the 'red herrings'
There will always be a lot of information to sort in a high-profile inquiry. There will be crank letters, false leads, and misleading information. Somehow, these have to be accepted, reviewed, and put to rest without distracting from the main investigation. All leads have to be considered until they can be closed.

The value of careful observation and 'ballpark' calculations
The work of Addison Bain shows the value of careful observation. Observe and record, then ask *why* rather than being constrained by just looking for signs that fit a theory. Simple calculations can be very revealing and give an understanding of the situation. The US investigation concentrated on eyewitness accounts and expert comments, with no experimental tests being requested to check the facts.

The Zeppelin era
The Count's first Zeppelin flew in 1900, before the Wright brothers got off the ground. As the decades passed, both the Count and Dr Eckener recognized that aircraft would overtake airships for intercontinental travel. The difficult political events surrounding the Zeppelin development took valuable time – time that aircraft used to develop their competitive capability.

Zeppelins covered the period from 1900 to 1940. They were immersed in war, political unrest, and harsh economic times. It was not the best environment in which to develop expensive experimental projects. Yet the image that Dr Eckener was able to create was amazing. He stirred the imagination of the public. He was a market-ing and public relations guru. Maybe his early psychological research into human behaviour helped, but he was a dominating personality and had an authoritarian management style. Engineering was a mys-tery to him, and he had little time for engineering matters. His Chief Designer, Dr Dürr was a reclusive individual, and we might assume that there was little communication of the engineering decisions and risks. At the same time, virtually all the work of the Zeppelin Com-pany was developmental. The continuous pattern of design changes during the construction of the *Hindenburg* does not show clarity of technical concept and specification.

Dr Dürr had the benefit of the strong structural design background in Germany at that time. Professor Otto Mohr, A. and M. Ritter, and Heinz Müller-Breslau were all internationally known for their contribution to structural analysis **(9)**. On the other hand, the attitude of the company to doping of the outer cover appears indifferent. The specification called for tests of water-tightness, flexibility, and flame resistance. There is no record of the tests having been completed. Changes were made, and flights took place, before the final coat was applied.

It is ironic that such strong measures were taken to ensure that no passenger carried matches or cigarette lighters, while the fire resis-tance and resistance to electrostatic discharges of the outer cover were treated so casually. There is no record of why the change was made from the apparently acceptable dope used on the *Graf Zeppelin* to the incendiary dope on the *Hindenburg*. This provided a latent fault waiting for the combination of circumstances that came together at Lakehurst, seven years later.

1.1.13   Lessons learned
The broad management scene
Reviews of the *Hindenburg* accident concentrate on the technical aspects, but there are management lessons.

- Try and carry out complex engineering development in a con-ducive climate with adequate funding.
- Ensure that the top manager knows and understands the risks in the detailed engineering decisions.
- Know when a project is in a development stage and do not believe it is mature until clearly demonstrated.
- In the development phase, beware of being overconfident and believing that past experience will apply to the new changes.

- Make sure that there is a comprehensive plan up front and do not adopt a casual ad hoc approach to development.
- Beware of a dominating and authoritarian executive who may not fully recognize the risks being taken in the details.

Designers should recognize how all their materials interact

The individual design decisions within the Zeppelin Company were effective in their narrow sphere but missed the overall impact on risk. Were the designers aware of the potentially dangerous properties of the materials they were selecting? The doping kept the skin taught, the aluminium powder reflected the sun's heat, but the combination was explosive.

- Make sure that you know all the relevant properties of the materials being used.
- Do not make changes unless you know the full impact of those changes, and that they meet the overall specification.
- Check the effect of individual design decisions on the design as a whole.
- Keep asking *What if. . . ?*
- When there is flammable material about, whether solid, liquid, or gas, beware of electrostatic sparks. This requires a very detailed assessment. Reference (**10**) provides useful guidance.
- Ensure that at least someone is able to comprehend the technical aspects of the whole project.

Careful observation is invaluable

- Look, observe, question, and analyse until the answer feels right.

Simple calculations and tests can be very revealing

- Ballpark calculations can give a good feel for the credibility, or otherwise, of a theory.
- Simple tests can give an insight into a phenomenon.

Inquiries do not always get the whole answer

The political and commercial environment can influence inquiries. This might explain the different focus and conclusions between the American and German inquiries.

- Knowledgeable witnesses are human and can sometimes miss the obvious.

What can engineers learn from the FBI?

The FBI has to deal with a mass of data, often conflicting – and so do engineers. The FBI files show the effectiveness of good reporting and record keeping. J. Edgar Hoover demanded this from his staff, as well as replying courteously to everyone who communicated with him. Record keeping can be a chore, but it is worth the effort. When done succinctly, it helps to keep all levels in the organization well informed.

- Keep good, accurate, and succinct notes.
- Distribute the notes to those who need to know and to the record file.
- Always reply, with courtesy, to all communications that are sent to you.

Lessons from the myth

- Always be ready to question any apparently well-established position, particularly if you have any doubts.
- Myths are easier to create than dispel.

References

 (1) **Duggan, J.** (2002) *LZ129 Hindenburg – The Complete Story*, Zeppelin Study Group, UK, ISBN 0 95141 148 9.
 (2) **Botting, D.** (2001) *Dr Eckener's Dream Machine*, Harper Collins, London, ISBN 0 00653 225 X.
 (3) *Air Commerce Bull.*, Bureau of Air Commerce, **9**(2), 1937.
 (4) FBI file 62–48190 available at http://foia.fbi.gov/hindburg.htm.
 (5) Smithsonian National Air & Space Magazine, April/May 1997.
 (6) **Bain, A.** and **Schmidtchen, U.** (2000) *Afterglow of a Myth: Why and How the Hindenburg Burnt*, DWV, January, www.dwv-info.de.
 (7) **Peake, N.** (2000) What happened to the Hindenburg? *Dirigible*, **XI**(2), 14–15.
 (8) **Van Treuren, R.G.** (2000) What happened to the Hindenburg ... Indeed!! *Dirigible*, **XI**(3), 4–6.
 (9) **Timoshenko, S.P.** (1953) *History of Strength of Materials*, Dover Publications, New York, ISBN 0 48661 187 6.
(10) **Glor, M.** (2001) *Gases, Electrostatic Ignition Hazards Associated with Flammable Substances in the Form of Vapors, Mists and Dusts*, Swiss Institute for the Promotion of Safety and Security, Basel, www.vtt.fi/aut/rm/projects/stata/sem2-99/glor.pdf.

1.2.1 What's wrong?

The Railway Age started in England and was a key part of the Industrial Revolution. Now, 150 years later, the public looks with despair, frustration, and, at times, with anger at the state of the British railway system. Their list of concerns is long: no leadership, poor political decisions, abysmal management, frequent accidents creating concern on whether it is safe to travel by train, bad timekeeping, squalid trains, lack of service, high costs, and engineering failures being just a start. Is it as bad as this, or is this media hype?

In this section, the general problems of the industry are explored. In the following sections, two of the technical issues are studied – signals passed at danger (SPADs), and the wheel/rail interface. A review of the history puts the issues into context.

1.2.2 The early history of British railways

It was George Stephenson's vision, building on the pioneering work of Trevithick, Murray, and Blenkinsop that led to steam engines being used for locomotion rather than stationary power sources. The Liverpool and Manchester Railway introduced the concept of transporting passengers as well as goods. They introduced tracks for each direction, stations, timetables, and respectable rolling stock. The cost and speed advantages of trains soon killed off the canals, long-distance horse-drawn coach traffic, and coastal shipping of materials.

The commercial opportunities offered by railways, coupled with the laissez-faire approach taken by the government, led to an explosion of railways. Put together an entrepreneur, an engineer, and a lawyer and you had the seed for another Railway Company. Railways had to expropriate land for the track, and required an Act of Parliament for each railway. In 1846, at the height of the railway mania, 273 Railway Bills received Royal Assent. By the start of the First World War, 1279 Acts of Parliament had been passed authorizing railways.

Like most manias, there were many casualties, with inadequate business planning, inadequate financing, and greed, all taking their toll. Many made fortunes including Robert Stephenson, Isambard Kingdom Brunel, and George Hudson. Robert Stephenson became the first millionaire engineer, and the MP for Whitby. Brunel made

and lost many fortunes. By 1848, George Hudson, a railway promoter, controlled nearly a third of the railway system. He was the
Mayor of York three times, and the MP for Sunderland. He ended up
in the debtors' prison, accused of defrauding his shareholders.

Amalgamating railways became popular, and the maximum number of railway companies in operation hovered around 150 in the
latter half of the nineteenth century. The London and North Western
Railway had the largest capitalization of any company at the turn of
the century.

Parliament had plenty of opportunity to provide guidance. They
never took the lead. The Duke of Wellington, once the popular soldier
and hero, became a politician (Tory) and PM who polarized public
opinion. He was worried that railways would allow hooligans and
criminals to roam the country. He expressed concern over misman-
agement of the railways, and the potential for monopolies to develop.
He even floated the idea of state ownership. William Gladstone (Lib-
eral) proposed that all future railways be state owned. His idea did
not survive the debate. Parliament discussed nationalization again in
1873 but decided to leave the industry in private hands; no doubt,
the fact that 120 Railway Directors sat in Parliament influenced the
outcome of the debate.

1.2.3   Railways in the first half of the twentieth century

The advent of cars and trucks brought an end to the dominance of
the railways.

Railways played essential roles in the First and Second World Wars.
In both wars, the government took over the running of the railways,
but spent nothing on maintenance. As a result, the privately owned
railways struggled after each war.

In the period between the wars, the four private railway companies did a good job of promoting themselves. This was the era of
luxury trains, such as the *Flying Scotsman* and the *Golden Arrow*.
The railway companies promoted holiday travel for the masses to
visit the seaside. Commuting by train was encouraged, leading to the
growth of dormitory communities along the rail routes. Despite these
promotions, the railway companies lacked investment and had poor
financial returns due to competition from the roads.

1.2.4   Safety, risk, and regulation

Safety was on the agenda as early as the 1829 Rainhill Trials. The
Conditions for the Trials stipulated the right to put the boilers to a

water pressure test of three times the working pressure. They also required that *there must be two safety valves, one of which must be completely out of reach or control of the Engine-man, and neither of which must be fastened down while the engine is running.*

The opening of the Liverpool and Manchester Railway on 15 September 1830 saw the first fatality. William Huskisson, an MP for Liverpool and a former President of the Board of Trade, had his leg crushed by the **Rocket** when he was standing between the tracks during a break in the ceremonies for the engines to take on water. He died that evening. This early and prominent fatality did not slow the development of railways.

While the government took a laissez-faire approach to the commercial side of railways, they took a more decisive approach to safety regulation. A series of Regulation of Railway Acts started in 1840. The most notable was the 1889 Act, prompted by the disaster in Armagh.

On 16 June 1889, an inexperienced driver took an overloaded excursion train out of Armagh station. The train stalled on the hill, and the driver and guard uncoupled the rear half of the train. The brakes failed, and it rolled back and crashed into the next train. Over 80 were killed, many of them young children.

The public was outraged and demanded action by the government. The Regulation of Railways Act 1889 came into force 79 days after the accident. In the industry, it became known as the *lock, block,* and *brake* Act. *Lock* because the levers in the signal box had to be interlocked so that they were consistent with the points. *Block* because the railway routes had to be divided into blocks, each under the control of a signalman. Only one train at a time was to be allowed in a block. Previously, trains had been dispatched at timed intervals. *Brake* because all trains must have an automatic brake system for emergencies. The aim was a *fail-safe* system, using either vacuum or air pressure.

1.2.5   Nationalization 1947

The Attlee (Labour) government nationalized the railways in 1947. They were structured along the regional lines of the private companies. Each region had its own manager and Board of Directors.

The original shareholders received a generous payment in the form of Government Bonds. British Railways (BR) took over 632 000 staff and 7000 horses for working the railway yards, as well as worn-out equipment.

The government imposed low fares and freight charges, which left BR with $\frac{3}{4}$ of its revenue used for wages, before any other costs could be considered. Strikes by railway workers and increased competition from the roads brought mounting losses. The answer was the 1955 Modernisation Plan. This was a 15-year plan to replace steam engines with high-speed diesels, electrify key lines, and generally update the entire infrastructure. In addition, this dealt with the pressing environmental issue, and reduced smog.

The hoped-for benefits of the Modernisation Plan did not materialize. In 1959, Ernest Marples, Minister of Transport (Conservative), appointed Dr Richard Beeching to Chairman of the British Railways Board. Dr Beeching had been a technical director of ICI. His brief was to make British Railways profitable. His recommendations were, basically, to close all unprofitable business. This resulted in cutting the rail system to about half, and abandoning the rural areas.

In 1968, Barbara Castle, Minister of Transport (Labour), introduced a Transport Act, which for the first time raised the concept that the railways provided a social benefit not recognized in the financial accounts. The 1968 Transport Act forgave BR's historic debt. It also ushered in a new modernization era with more electrification, higher speeds, and InterCity trains. BR reorganized along business lines that could stand alone, each with profit centres, instead of the regional structure inherited from the start of nationalization.

The railway culture

The history of the railways is a tale of confused direction and ever-changing government views. But inside the industry, there was a strong culture. A hierarchy evolved where station staff, footplate men, signalmen, and track workers, each had their role. There was mutual respect and communication between these groups. The railway industry developed effective apprenticeship schemes for technicians and graduate engineers.

The HM Railways Inspectorate (HMRI) reviewed all accidents; and the lessons learnt were incorporated into the system. The industry was effectively self-monitoring. The first guidelines for safety were introduced in 1858.

Long-term employment in the industry was the norm. There was a gradual building and consolidation of knowledge. Experience was passed from one generation of railway workers to the next. There was pride in being part of the railway industry.

Technical developments

British railways have always been a great place for engineers. Despite low levels of investment, there was never a lack of interesting projects. In the 1920s and 1930s, the companies looked to their engineers to develop improved products giving a commercial advantage. Engine development went from record-breaking steam engines to diesel and electric traction. There were steam-turbine-driven locomotives, even a coal-burning gas-turbine engine. An experimental diesel engine was built in 1912. When diesel drive became the preferred system, there were over 50 prototype designs tested.

In 1903, the Mersey Railway was the first steam railway to be electrified. Since then, ac, dc, various voltages, third rail, and pantograph have all been tried. Bogies, braking systems, and coach designs, all have interesting development histories (1).

These engineering developments are interesting, but were they appropriate for the business of British Rail? Would it have been better to concentrate the limited funds on fewer and standardized designs?

BR encapsulated a lot of its experience in standards, practices, and handbooks. These covered a wide range of topics from driver training to categorizing defects in rails. Some were incorporated into British Standards such as BS11, which standardized rail dimensions.

The end of British Rail

British Rail's life ended with the Railways Act of 1993. British Rail had existed for about 50 years. The public saw it as a typical government-run enterprise, not too efficient and bureaucratic. But the trains ran, and the public relied on the railways. The railways were part of the national scene.

1.2.6   Privatization

With the Conservative Government of Margaret Thatcher, BR was subject to increasing financial discipline. Peripheral activities such as hotels, ferry services, and workshops were sold off. In three years, the business increased 18 per cent as measured by the number of train miles. The cost per passenger mile dropped by about one-third. The subsidies to BR, on a GDP basis, were around 30 per cent of the average subsidy in Europe.

During Thatcher's second term, there was a flood of privatizations in a variety of industries. She did not raise the issue of privatizing the railways, although the idea did bubble to the surface. Cecil Parkinson, her Minister of Transport, said in Parliament in 1990 that the government was *determined to privatize British Rail.*

Railway Act 1993

Following the election of John Major's Conservative government in 1992, the privatization of the railways was committed. The plan was to break the industry into many pieces, separate the infrastructure from operations, franchise passenger services, and control the Railtrack monopoly with a Rail Regulator.

Politicians, civil servants, and consultants worked feverishly to implement the plan. The Treasury had a hyperactive privatization unit looking for the next project. The Treasury saw the opportunity to reduce the subsidy to the rail industry. In the run-up to privatization, the government spent £450 million on consultants – many of whom had little knowledge of the industry (2).

The Board of BR, while being reluctant in promoting privatization, favoured a BRplc model. This is the model that had been used for most of the earlier privatization of other industries. Their views do not appear to have been seriously considered.

An argument was tabled that the EU bureaucrats in Brussels required separation of track and train operations. This was misleading, as Brussels only wanted separate accounts to be prepared. This was to handle the issue that occurred at several places in Europe where trains from different companies ran on the same track.

The Railway Act 1993 finally received Royal Assent on 5 November, following almost a year of debate on the details, rather than the principles. Over 500 amendments were made to the Bill during its passage through Parliament. This Act formally approved the separation of the infrastructure from the franchised operating companies.

Creating the new structure

During 1993, both a Franchising Director and Rail Regulator were appointed.

The infrastructure was carved out of BR as a separate company, Railtrack, which was formally vested as a legal entity on 1 April 1994. Railtrack was not initially part of privatization. It was seen as a monopoly in need of government funding. They inherited 23 000 miles of track, 1100 signal boxes, 2500 stations, and the remaining property assets of the old BR. Railtrack had around 10 000 staff of whom $\frac{2}{3}$ were signalmen.

The 25 train operating companies (TOCs) were franchised between December 1995 and April 1997. The freight operations were sold during the same time frame. Most of the freight services went to EWS (English Welsh and Scottish Railway), with international container business going to Freightliners Ltd. The BR maintenance companies (BRISCOs) were sold off separately.

Privatization started slowly because of the concern that a Labour Government might renationalize. This discouraged investors. The Government recognized that the process needed impetus, so they gave priority to the sale of Railtrack and the Rolling Stock Leasing Companies (ROSCOs). Railtrack was floated on the Stock Exchange in May 1996 for £1.9 billion.

The process continues under new Labour

The Labour government of Tony Blair came to power in 1997. Although they had challenged the privatization when in opposition, they decided to live with it once they were in office.

The government published a Railways Bill in 1999. The Bill established the Strategic Rail Authority (SRA) and devolved some powers to Scotland and Wales. The SRA was established to develop long-term thinking for the whole railway system as well as to manage the government subsidies.

How was the industry supposed to work?

The government, through its civil servants and consultants, drew up the contracts for controlling business between the parties. There were many contracts and legal documents. The freight access agreements between the Freight Operating Companies (FOCs) and Railtrack involved 224 separate legal documents (**3**).

The key issue was believed to be the access to, and control of, the track that was a monopoly. This was to be managed by access agreements with associated charges. The passenger train and freight companies (TOCs and FOCs) had to pay for each access to the track. If they did not use the track at the planned time, then a series of penalties were charged. Similarly, Railtrack was penalized if the track was not available for the planned trains.

The government drew up the terms of contract between Railtrack and the infrastructure companies (BRISCOs) for track maintenance and replacement.

The passenger and FOCs would lease their rolling stock from the ROSCOs.

The Franchising Director would award and manage the franchises for the TOCs.

The Rail Regulator would issue licences to operate. The regulator would set track access charges. The regulator was expected to make sure that the industry operated efficiently.

What actually happened?

The Rail Regulator considered the access charges he initially set to be paid to Railtrack as too rich (2). They were reduced by eight per cent the following year and two per cent less than inflation for subsequent years. This led Railtrack to cut corners and gave little incentive to invest in track improvements. Railtrack appealed to the government for financial support. At the same time, it wanted to show a sound private sector image to the stock market. This was achieved by property development and selling off the property assets acquired from BR, and the shares rose from £3.80 when Railtrack was floated to a peak of £17. Railtrack paid dividends and had a policy of increasing dividend payments.

Railtrack brought in the management advisors McKinsey. They recommended that the assets such as track should only be replaced when needed and not at fixed time intervals. This just-in-time approach complemented Railtracks' philosophy of getting the most from the assets – sweat-the-assets.

The track maintenance contractors had long-term fixed price contracts from Railtrack, as drawn up by the government. The initial rates were based on previous BR expenditure, which was already inadequate. The Rail Regulator decided that these would be cut further, by three per cent below inflation for each year. As a result, track maintenance was under-funded.

If track maintenance interfered with the running of trains, then Railtrack had to pay a penalty. This led them to try and force maintenance within train schedules.

A computer system, TRUST (Train Running System) was installed to monitor the use of the track. There are several thousand input points. Some are automatic and some need manual input from signalmen and train and station staff. This system monitors the track use against the timetable, down to the minute. Accountants then use these data to generate the monthly billings of charges, penalties, and incentives between the 29 players.

The BRISCOs had been acquired by civil engineering contractors. They were experienced in managing the financial outcome of difficult contracts with claims for extras and variations. The boundary between maintenance and replacement was ill defined. The maintenance contractor could argue that a questionable track should be replaced. It would then be subject to a different contract with the replacement contractor. Railtrack was ill equipped to manage this engineering interface.

Railtrack did not understand its real role in life. It aimed to manage, for commercial gain, the back-to-back contracts with the train companies using the tracks. The engineering needs were ignored. Railtrack had no comprehensive technical database of its assets.

The Blair Government was under pressure to renationalize. They resisted this, but took a hands-on approach. The reduction in political involvement that was expected to come from privatization failed to materialize.

A string of accidents

On 19 September 1997, a collision occurred at Southall, in which seven people died. The collision was caused by the driver passing a signal at danger. The term SPAD (signal passed at danger) became generally known, and used, by the public. A Public Enquiry was established, led by Professor John Uff (**4**).

On the 5 October 1999, 31 people were killed at Ladbroke Grove in another collision caused by a SPAD. A Royal Commission was appointed under Lord Cullen (**5**).

A derailment at Hatfield on 17 October 2000 caused four deaths (**6**). The cause of the derailment was the fracture of the rail by fatigue. The specific reason was said to be gauge-corner-cracking. Railtrack reacted with a massive programme of rail inspection and track speed limits. This completely disrupted the train operations, causing chaos for the travelling public. The disruptions lasted well into 2001.

In May 2002, the rear coach of a train derailed at Potters Bar because of problems with a switch, and seven people were killed.

The history of the track at Hatfield

On 17 October, the 12:10 train from King's Cross to Leeds derailed on the curve at Hatfield while travelling at around 115 mile/h. The 35 m section of the outer rail was broken into over 200 pieces, Fig. 1.2.1 (**6**).

The cause was quickly found to be a form of fatigue called gauge-corner-cracking. Later, it was generalized to rolling contact fatigue.

The track maintenance contractor, Balfour Beatty, had warned of the potential for gauge-corner-cracking near Hatfield in November 1999 (**2**). It has also been reported that concerns had been raised the

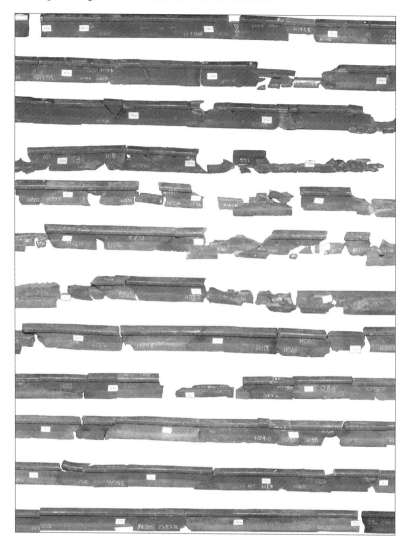

**Fig. 1.2.1   Reconstruction of multiple-fractured rail at Hatfield.
Reproduced courtesy of HSE ©Crown copyright 2000**

previous year (**7**). Balfour Beatty recommended rerailing within six months. Railtrack inspected this rail in February 2000 and decided to replace it. The work of replacing the rail was given to the replacement contractor Jarvis. There was then a series of issues, each minor in itself. The wrong part of the track was surveyed, access times for working on the track were missed. Three attempts were made to deliver the new rail to the site before it was eventually delivered in two lots in April. On one occasion, the wrong type of train brought the rails. The train could not be unloaded under the electric overhead lines.

For four months Railtrack argued with Jarvis about the timing of the work. Finally, it was scheduled for November. The rail failed two weeks before it was going to be changed.

There has been no explanation as to why there was no speed restriction posted on this section of track while it was waiting to be replaced.

It is probable that the rail failed before the derailment. An electric signal is used to detect trains in a section of track. If the track is broken, then the signal cannot be received. This is back-up information for signallers and is not relied upon to indicate track integrity. The signal was missing for the four previous trains (2).

The fallout from the Hatfield crash

After the crash, Railtrack embarked on an urgent inspection programme. Initial checking surfaced 300 sites where gauge-corner-cracking might exist. No site was found to be as bad as Hatfield (8). Further inspection identified around 7000 sites with fatigue cracks. On 6 November, the Health and Safety Executive (HSE) imposed a 20 mile/h speed limit on any line where cracks over 20 mm had been detected.

The net result was that the rail timetable was thrown into chaos. Journey times were doubled or more. Only half of the long-distance trains arrived on time in the last three months of 2000. The financial impact of the chaos caused problems for all the rail companies. It had the greatest impact on Railtrack, which suffered huge losses and looked to the government for help. The government provided some finance, but in October 2001 the Transport Secretary, Stephen Byers, appointed receivers to take over Railtrack.

The Hatfield crash was the straw that broke Railtrack's back.

In July 2003, six people were charged with gross negligence and manslaughter for the Hatfield accident. A further six, including Gerald Corbett, the boss of Railtrack at the time, were summoned for offences under the H&S Act. The legal process continues.

The rail industry in its own words

The government asked Mercer Management Consultants to survey members of the industry, stakeholders, and specialists. In May 2002, they reported and saw four underlying root causes:

- *lack of industry leadership, leading to confused and conflicting priorities;*

- *Railtrack failed as infrastructure manager;*
- *a general lack of consideration of value for money;*
- *the fragmented nature of the industry.*

These root causes resulted from four problem areas:

- *a failure to implement correctly the maintenance and renewal of the track, stemming from a loss of knowledge and expertise, compounded by historic under-investment;*
- *poor investment strategies and planning;*
- *inefficient capacity utilization on a congested network;*
- *onerous and bureaucratic safety concerns, with no single body balancing the benefits and cost of safety regulation.*

Enter Network Rail

In 2002, a new concept, Network Rail was devised to take over the rail infrastructure. Network Rail is a company limited by guarantee. Any profits will be reinvested in the company. The company has members rather than shareholders. It was hoped that there would be around 100 members, and the total is now 116. Amongst the members are the other companies in the rail industry, unions, and public interest groups as well as the SRA. Network Rail works within the strategy set by the SRA. The SRA will provide stand-by loans, although Network Rail has secured £9 billion bridge financing from the banks. They are classified as being in the private sector. However, should they start to use the SRA funds, they could be classed in the public sector (**9**).

Most of Network Rails' Directors have engineering or operations backgrounds. Non-executive Directors provide additional railway experience from both the United Kingdom and internationally.

Network Rail formally completed the acquisition of Railtrack's assets in October 2002. Network Rail paid £500 million of which £300 million was paid by the government. Network Rail has assumed Railtrack's debts of around £7 billion.

The pieces of the UK rail industry

The main players in the UK rail industry are shown in Fig. 1.2.2.

- DOT, the government Department of Transport.
- SRA, Strategic Rail Authority provides the overall direction for the industry. It awards and manages franchises for TOCs and FOCs. It channels government funds into the industry.
- RPC, Rail Passenger Council and Committees. 'The voice of the passenger'. There are eight regional committees coordinated by the Council that also deals with national issues.

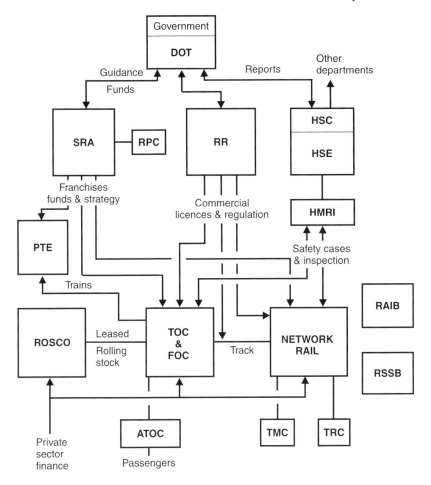

Fig. 1.2.2   The UK rail industry

- RR, Rail Regulator licences Network Rail and the operating companies. Provides economic regulation. Appointed by the Minister and given authority to grant licences, but is independent of government. From July 2004, the single-person role of the Regulator was replaced by a statutory board called the Office of Rail Regulation (ORR).
- HSC*, the Health and Safety Commission. A non-departmental organization responsible to government for administration of the Health and Safety at Work Act. Reports to relevant departments. Oversees and effectively acts as the Board of HSE.
- HSE*, a body of three, appointed by HSC, who implement the work required by HSC with a staff of around 4000.

- HMRI, HM Rail Inspectorate. Transferred from Department of Transport to HSE following the Clapham accident. Accepts and reviews Railway Safety Cases. Approves and inspects new works and rolling stock to meet standards. Investigates selected incidents.
- PTE, seven self-managed Passenger Train Executives established to revitalize suburban transport in and around Manchester, Newcastle, Glasgow, Birmingham, Liverpool, Leeds, and London.
- ROSCOs, three rolling stock companies who lease trains to TOCs and FOCs.
- TOC, Train Operating Companies. Operate the passenger train services. Currently 25, but SRA aims to reduce the number to 20.
- FOC, Freight Operating Companies operate freight trains and services, and own all their locomotives and many wagons. Currently three companies, but EWS dominates.
- ATOC, Association of Train Operating Companies. An unincorporated association. Provides ticketing and enquiry services so that passengers have one point to get tickets and information across the whole system.
- Network Rail. The infrastructure monopoly owning the track, bridges, and tunnels, and controlling the signalling. Network Rail owns the stations, but they are operated by the TOCs except the main London stations operated by Network Rail.
- TMC, Track Maintenance Companies, operated by the BRISCOs for fixed price track maintenance.
- TRC, Track Replacement Companies, operated by BRISCOs.
- RSSB, Rail Safety and Standards Board. An independent body within the industry to set standards for the whole industry. One of the recommendations from Lord Cullen's second report.
- RAIB, Rail Accident Investigation Body. A group to be set up separate from HSE, following the recommendation of Lord Cullen.

*NB: On 15 July 2004, the Secretary of State for Transport announced that the responsibility for railway safety regulation would move from HSE to a new Office of Rail Regulation.

Who owns the pieces?

There are 25 TOCs. Bus companies have bought several of them and created local transport monopolies in some areas of the United Kingdom. Around 70 per cent of the revenue of all the TOCs were controlled by four companies, National Express, First Group, Connex, and Virgin/Stagecoach **(10)**. Connex is now no longer a TOC.

In freight, the US railroad Wisconsin Central was the major investor in EWS, together with investment companies from Boston and Geneva, and the bank Goldman Sachs. Wisconsin has been taken over by the Canadian railway company, CN.

The banks bought out the ROSCOs, providing handsome profits for the original purchasers. The Royal Bank of Scotland, HSBC, and Abbey National own the ROSCOs. While most of the industry is heavily regulated, the ROSCOs are not regulated.

The 13 BRISCOs at privatization were consolidated in traditional civil engineering companies. There were seven companies, although four, Balfour Beatty, Jarvis, First Engineering, and Carillion controlled 80 per cent of the maintenance and replacement market (**10**). In October 2003, Jarvis backed out of this work. Shortly afterwards, Network Rail decided to take over all maintenance.

Future plans
The government's 10-year plan has been adopted in the SRA Strategic Plan. This calls for a 50 per cent increase in passengers and an 80 per cent increase in freight volume by 2010. The aim is to move people and freight off the roads, particularly in congested regions such as around London. The government estimates that £60 billion investment will be needed and plans on half of this coming from the private sector and the rest from public funds.

COMMENTS
The brief comment is '*What a mess*'. There are signs of improvement, and change is likely to continue to happen. The issues can be explored at two levels – the broad market issue and the detailed issues within the industry.

Managing a subsidized monopoly
The private sector free market economy works well when there is competition. Given a competitive product or service and adequate finance, a private sector company can flourish, especially when free of restrictive regulation. But this model does not ensure that good executives run the companies. It just gets rid of them when they fail, either by Board or shareholder decision, by being taken over or bankruptcy.

A rail infrastructure company is by definition a monopoly. Furthermore, in most places around the world, passenger rail traffic is not economic. There is no easy answer on how to efficiently manage a subsidized monopoly. Choosing a competent executive is an obvious first step. The uncertain and stop/go financing provided by the government gives no sound basis for long-term planning. Had the government chosen a model in which they had a significant shareholding, then they could have had a say in the appointments of key executives.

Railtrack's financial business

Railtrack was intent on presenting itself as a successful private sector company. This was a sham. In reality, it was boosting its apparent performance by selling property assets acquired from BR. It was dependent on government subsidy and had its main source of revenue, access charges, set by the Regulator. In Railtrack's defence, it would be near impossible to deliver private sector performance in these circumstances. Railtrack was so intent on showing its private sector success that it even announced an increased dividend after the Hatfield crash.

Railtrack's management

Railtrack's executive did not appear to consider that they needed an understanding of the technology of the business. A quote from 'The Requisite Organization' by Elliot Jaques says it all – *You cannot manage what you do not understand.*

Railtrack's management style was extremely confrontational. With so many players and the whole public as the customer, this did not work.

The immediate reaction to the Hatfield crash by Railtrack was to go into panic mode, possibly led by legal and commercial concerns. An interesting question is whether engineering judgement would have produced a different course of action. Experience with derailments had shown few fatalities. Was it essential to bring the whole network to a near standstill that eventually led to the demise of Railtrack?

The impact of the contracts

The government dictated the contractual relationship between all the parties. The government drew up these terms using civil servants and consultants with little knowledge of the industry. The nature of these contracts resulted in challenges and conflict between the parties. What was needed was collaboration.

The access charge system focussed attention on avoiding penalties rather than running the system to time. If a train were held to meet a connecting train, then it would trigger penalties. In the case of Railtrack, it provided an excuse to avoid the necessary scheduling of maintenance and replacement.

Was it really necessary for the government to draw up all the contractual terms?

Should the track infrastructure have been separated from the running of the trains?

This question has raised a lot of debate. The separation of track and trains has worked successfully in several countries. However, in each case there has been a well-established relationship between the parties built up over years. Both sides recognize the common interest and are pragmatic in reaching decisions they can live with.

In the United Kingdom, the infrastructure has lacked investment and maintenance funds for decades. The contractual arrangements post-privatization resulted in conflict. With these circumstances, taken together with the break-up of the BR technical team, it would have been better to have kept the track infrastructure and running the trains under one management.

Breaking up the knowledge base

BR had a technical culture and a knowledgeable technical team. One aim of the Conservative government in privatization was to reduce the power of the unions. But separating BR into 100+ pieces also broke up the technical team. The industry suffered from the loss of this knowledge and data.

Knowledge is built up in teams and generally resides in the minds of the members of the team. Records help maintain the knowledge. But the records tend to be most easily accessed and understood by the members of the team that created them. A stranger will find them less easy to follow. The break-up of the BR technical team lost much of the knowledge built up over decades.

The impact of regulation

Regulation came from both the Rail Regulator and HSE/HMRI. Did the Rail Regulator's pressure on Railtrack to improve performance conflict with the HSE/HMRI pressure for attention to safety? The success of the whole industry is highly dependent on the sound decisions of the Regulator, who is effectively the Tsar of his domain.

The benefit of standards

Standards, guidelines, and handbooks are good ways to gather knowledge and experience that can easily be applied to current work. In the case of the railways, many of the standards relate to safety and are of interest to the safety regulator. BR had extensive standards that were inherited by Railtrack. However, other parts of the fragmented industry were not happy with Railtrack dictating the standards.

Following the Cullen recommendations after the Southall crash, the Safety and Standards Directorate was hived off into a separate subsidiary of Railtrack, Railway Safety. It is now proposed that a separate industry body, Rail Safety and Standards Board be created and take over the role and staff of Railway Safety. The Rail Regulator may make it a licensing requirement that the key players get involved in, and help fund, RSSB.

Safety perception and reality

Prominent accidents raise public concern. It has been cynically said that the media has fuelled this worry, especially when the accidents occur close to media studios. While all fatal accidents are a concern, the facts are that there has been a continuing improvement in railway safety over the last 30 years. This continued through privatization.

HSE statistics on annual deaths in the United Kingdom for passengers travelling by different forms of transport over the decade ending 2001 are car, 1765; air, 32.5; bus or coach, 17.6; and rail, 10.

During the period 1994–2002, the average number of fatalities for trespassers and suicides on the railway was 264 per year, split roughly equally between the two (11). The total fatalities of passengers and railway staff in train accidents over these eight years were 88, of which 42 were from the three accidents at Southall, Ladbroke Grove, and Hatfield. Fatalities at level crossings have averaged 10.5 per year.

Studying some of the reactions from HSE/HMRI raises the question whether public perception and comments from politicians have influenced decisions more than reality. Where should attention be paid? Would reducing accidents to trespassers and at level crossings save more lives for the money spent than trying to reduce train accidents?

Views from the World Bank

The World Bank had no direct involvement in the privatization of BR. They studied the top-down restructuring and privatization in Japan, New Zealand, Argentina, Sweden, and the United Kingdom. In addition, they studied the bottom-up restructuring in Canada and some railroads in the United States. Their studies (12) were completed in 1994 and have not covered the recent developments in the United Kingdom.

They noted that railways are frequently *nations within a nation* and, as large inert organizations, are difficult to move. Well-managed privately owned railways are in a perpetual state of realignment to capitalize on the changes to their markets. Government ownership impedes this type of change. They systematically use railways as instruments of social policy. They subsidize and reluctantly invest in railways, but sacrifice them to year-to-year financial juggling rather than planning for optimal results. By the time politicians react, the state-owned railways have often lost touch with their customers.

There is no single best approach to restructuring and privatization. While restructuring initially appears straightforward, experience shows that it is difficult to realize. Restructuring is a process as well as an end result – and both need to be well conceived, effective, and well executed. Railway managers tend to believe that each railway is unique – and that experience elsewhere has limited relevance. Many lessons have to be reinvented just to fit local thinking. Also, governments do not recognize that changes from privatization can be just as fundamental for them as for the industry they are privatizing.

The UK case is one extreme of the restructuring spectrum – a strategy that can aptly be termed *unbundling*. The plan was to shatter the old, vertically integrated railway. The process was complex due to simultaneously trying to liberalize, bring in new participants, and fundamentally redefine the position of the government into a regulatory role. The complexity increased with the number of enterprises involved and the number of interfaces. In the unbundling, they noted that *the devil is in the concept – as well as in the details*. The government wanted to rationalize the perceived social and broad economic needs and at the same time have a free market commercial entity. The UK rail privatization has not been a simple process; it has created a great deal of fragmentation and risks have been taken for uncertain rewards.

Their discussion paper concludes that, *even in railways, fundamental change is possible, though not necessarily predictable.*

The World Bank paper, in careful diplomatic language, warns others of the risks of copying the UK model. They appear to favour a phased and learning approach.

Positive signs

Government and industry seem committed to make improvements. There is more investment now than there has been in decades: track is being refurbished, and the Channel Tunnel Link is in place. There is new rolling stock – some 4500 new units have entered service. Results are slowly being seen – there are fewer broken rails, and fewer delays. Research studies are being funded at various universities. A network of researchers has been created with the formation of Rail Research UK. And there is a willingness to learn from experience outside the United Kingdom. However, changes will take time. In March 2003, John Armitt, the CEO of Network Rail and a civil engineer, said that despite spending £2.5 billion/year on renewal and maintenance, it will take another three to five years to remove the backlog of maintenance.

In January 2004, Alistair Darling told MPs that changes were needed to streamline the over-complicated structure, and a fundamental review will take place. The press questioned whether this signals creeping nationalization or a weird hybrid. Railway evolution is continuing.

The opening question, asking whether the railways are as bad as stated, can be answered with a tentative *no*. The public complains, but votes with their feet. Passenger numbers are 30 per cent greater than at privatization and are still growing. The railways apologize for being a few minutes late while it is impossible to time a car journey to that accuracy on the congested roads. Despite the gloom, there are positive signs.

1.2.7   Lessons learned
Wrecking an industry is easier than rebuilding it! Working in a government-funded and regulated environment is not easy.

Creating organizations
- Designing organizations requires the same care as designing equipment.
- Study the proposed boundaries between units. Can they work together effectively?
- Look and learn from history, as well as what others have done worldwide.
- Make sure there is competent and knowledgeable management in place to provide leadership.
- Aim to reduce bureaucracy.
- Consider whether a phased approach may be better than trying to change everything at once.

Ensuring the continuation of knowledge
- Make sure you have the knowledge base necessary for the work.
- Do not break up a technical capability until you are sure you have a satisfactory alternative.

Safety
- The perception of safety and the resultant public/political pressures can create great emotion and distort the proper allocation of resources.
- Always try and understand the realities of risk and safety.
- Try and communicate the real facts to a wider audience so that the right problems get solved.

Recognize the benefits of standards
- Standards can assist in getting the best knowledge and experience in the day-to-day work.
- Develop and use standards.
- Keep standards up to date, easy to read and apply.
- Avoid bureaucracy in applying standards.

References
(1) Roche, A.P. (2001) *IMechE Presidential Address, Engineering – The Force for Change*, 23 May 2001.
(2) Wolmar, C. (2001) *Broken Rails – How Privatisation Wrecked Britain's Railways*, Aurum Press, London, ISBN 1 85410 857 3.
(3) Freeman, R. and **Shaw, J.** (2000) *All Change: British Rail Privatisation*, McGraw-Hill, ISBN 0 07709 679 7.
(4) Uff, J. (2000) *The Southall Rail Accident Report*, HSE Books.
(5) Lord Cullen, *The Ladbroke Grove Rail Inquiry* Part 1, HSE, 2001, Part 2, HSE, 2002.
(6) Train Derailment at Hatfield, 17 October 2000, First HSE Interim Report, 20 October 2000, Second Interim Report, 23 January 2001.
(7) *The Daily Telegraph*, 8 May 2001.
(8) *The Independent*, 2 November 2000.
(9) Setting the record straight on Network Rail, News release by National Statistics, 11 July 2002.
(10) Murray, A. (2001) *Off the Rails*, Verso, London and New York, ISBN 1 85984 640 8.
(11) HMRI Safety Reports.
(12) Kopicki, R. and **Thompson, L.S.** *Best Methods of Railway Restructuring and Privatization*, CFS Discussion Paper #111.

In addition to the above sources, general information has been obtained from:
International Railway Journal. See www.railjournal.com
The Railway Forum. See www.railwayforum.com
Jack, I., *The Crash that Stopped Britain*, Granta Books 2001.

In the spirit of openness with the public on rail matters there is a mass of reports, consultative documents, and minutes of meetings on the Internet. Most of these can be found on the Internet web sites of Network Rail, SRA, ORR, HSE, and HMRI.

The acronym SPAD – signal passed at danger – became well known to the UK public following the accidents at Southall in 1997 and Ladbroke Grove in 1999. But the risk of passing signals at danger has always been a feature of railway life. The public, when driving cars expect to stop at red lights, so they question – *Why can't train drivers obey the signals?* What are the real facts and issues?

1.3.1 Accidents – road versus rail

There are huge differences between road and rail. Road accidents depend on the driver's field of vision, and whether the driver obeyed the signals. The car driver can swerve to avoid a collision. Train drivers get advanced warning of signals and rely on the signalling system to provide adequate separation, rather than their vision. The consequences of SPADs can be higher because the train driver cannot swerve to avoid a collision.

We have become immune to deaths on the road. Rail travel is much safer. However, the rail collisions that do occur get higher attention.

Reducing road deaths is a broad subject with no unique answer. The players are the whole population, with no single organization to blame or improve the situation. On the other hand, the media and regulators believe that there is a singular solution to SPADs, namely, automatic train signalling and control.

Comparing road and rail does not lead to useful answers. Studying the history and experience with signalling systems is more instructive, as well as asking whether automatic signalling systems are cost-effective in saving lives.

1.3.2 History

The Great Western Railways started to install an automatic warning signal (AWS) to give drivers an audible and visible signal in the cab, telling them the status of a distant signal. In 1906, they devised an automatic train control (ATC) system by coupling the AWS to automatic brake application. The system was used for several years before it was replaced by the BR system. In the 1950s, BR installed a trial test of a different form of AWS **(1)**. This system used electro-magnets on the track between the rails to activate relays on the locomotive that showed whether the next signal was clear or at caution or red. There were both audible and visual displays in the cab. The driver had to acknowledge the signal within 5 s, otherwise the brakes would be fully applied. Following a crash that killed 112 people at Harrow and Wealdstone in 1952, where the driver passed several signals at danger, there was public pressure for the introduction of AWS. Progress was slow and by 1958, only 288 miles of track had been fitted. However by 2000, almost the entire UK passenger system had been fitted.

With higher rail utilization in congested commuter areas, the warning signal for single and double yellow signals comes on frequently. Drivers are continuously cancelling these warnings in order to keep the trains moving. The warning is noisy, and failure to cancel applies the brakes. Psychologically, drivers' instantaneous reaction is to cancel the horn, and then reflect why it sounded. By which time they may have passed the signal and forgotten whether it was double yellow (advanced caution) or single yellow (caution). In dense traffic, the repetition of warnings followed by cancellations can be both confusing and hypnotic. Will the warning be followed by an all-clear or further warnings or a full stop? This ambiguity can lead a driver to approach the next signal too fast.

1.3.3   Accidents at Clapham (1988), Southall (1997), and Ladbroke Grove (1999)

Clapham

At Clapham, a train rear-ended one stopped at a red signal. The debris spread onto the adjacent line and was struck by a third train – 35 people died. An enquiry was held under Justice Hidden. The cause was a badly wired signal cubicle. The root cause was sloppy procedure and training.

Hidden's report was highly critical of BR management and made 93 recommendations including the installation of Automatic Train Protection (ATP) nation-wide within five years of making a system selection. It is worth noting that ATP would not have prevented the Clapham accident. BR never accepted that the timeframe for installing ATP was realistic although they had committed two pilot schemes prior to the Hidden Report. Eventually, in 1995, the Government agreed with BR/Railtrack that nation-wide installation of ATP was not a cost-effective decision. Further installation was abandoned. HSC/HSE concurred. BR/Railtrack looked for alternative and cheaper systems. From this work emerged the Train Protection and

Warning System (TPWS). TPWS is a relatively simple and cheap bolt-on system. It can detect a train that is travelling too fast at a caution signal – a speed trap function. It also applies the brakes on a train that has passed a red signal – a train stop function. It can bring a train travelling at up to 70 mile/h to a halt before the next signal.

Southall

The Southall crash killed seven people. An InterCity train from Swansea to Paddington was fitted with AWS, but it had been isolated after giving trouble the previous day. Owing to confusion and poor communication, the fault had not been repaired. Great Western Trains (GWT) had been experiencing AWS faults on their trains at a rate of around ten per week with two isolations/week. The train was also fitted with ATP as part of the original BR pilot scheme. However, the driver did not feel competent to use it, as he had not been fully trained on the system. He switched it off. The driver was experienced, but he was not paying attention. He was tidying his belongings, ready for the end of the journey, and missed a double yellow and then a single yellow warning signal. At the speed he was travelling, the signals would be visible for 16.3 s. He eventually saw the red signal and braked hard, but it was too late to avoid hitting a hopper-wagon train crossing his path.

HSE charged the driver with manslaughter, and GWT with corporate manslaughter. Both charges were eventually dropped. GWT was fined £1.5 million under the Health and Safety at Work Act. Two years after the crash, an enquiry opened under Professor John Uff. The Ladbroke Grove accident occurred during this enquiry.

Ladbroke Grove

A commuter train passed a stop signal and, because of the nature of the track layout, collided head on with an InterCity train. The combined impact speed was around 140 mile/h. The escaping diesel fuel caused a massive fire and 31 people died. The 31-year-old commuter train driver had just finished driver training, having joined the company as a cleaner. He died in the crash. His train was entering a complex rail system with six bi-directional lines. Routing was decided by the Auto-Route-Setting System and so the driver was never sure which track the system would put him on. The red signal had a history of being difficult to see. It had been passed at red seven times in the previous five years and was one of the 22 signals most frequently passed at danger on the whole railway. It is possible that the driver thought that he had a yellow signal. At that time of day it appeared to be lit, because of reflected sunlight. It is not clear whether the AWS gave a warning.

A public enquiry was constituted under Lord Cullen. He delivered a devastating critique of management failures. He saw a *slack and complacent regime* that did not recognize the dire consequences of SPAD. At the enquiry, the solicitors for the bereaved and injured saw *bureaucracy, inertia, and management speak.* Bureaucracy was in the production of voluminous papers, the context of which bore little reality to what had happened. Management speak consisted of generalizations or vague assertions couched in clichés.

Joint study

A joint study was constituted under Uff and Cullen to review the effectiveness of TPWS and consider the future implementation of ATP. They had a report on ATP written by Sir David Davies that had been requested by the Deputy PM (**2**). By this time, a European system of ATP was well into development. The European Traffic Management System (ERTMS) is intended to be the single next generation control and traffic management system, to be used across the whole of Europe.

Uff and Cullen endorsed the introduction of ERTMS, with TPWS to be completely installed as an interim measure. They proposed the fitting of ERTMS by 2010 on high-speed lines and on all main lines by 2015.

The final reports by Uff and Cullen (**3**) are voluminous and include 295 recommendations. The reasons for the three accidents have been summed up by Roger Kemp, the Group Safety Director at ALSTOM (**4a**) as:

> Ladbroke Grove can be attributed to poor infrastructure layout, inadequately trained staff and a management environment that refused to learn from near misses, Southall to poor maintenance, unreliable equipment, inadequate operating procedures and inattentive staff and Clapham Junction to poor management of on-site modifications and testing.

He noted that most accidents do not have a single cause but occur because of a coincidence of lesser events.

1.3.4   What are ATP, ERTMS, ETCS, and GSM-R?

Automatic train protection (ATP) is a concept and a generic term. ERTMS covers a number of technical areas aimed at improving

EU rail competitiveness. One of its sub-areas is the European Train Control System (ETCS). ERTMS includes a Global System for Mobile communication – Railway (GSM-R). The acronyms ERTMS and ETCS are often used interchangeably in the UK rail context. The United Kingdom has been party to the development and specification of ERTMS. The UK Government has committed to the relevant EU Directives – hence the installation of ERTMS/ETCS has become mandatory for new and upgraded signalling systems. ERTMS/ETCS have three levels, each with prescriptive specifications.

ERTMS/ETCS level 1

ERTMS/ETCS level 1 is the most basic. A system of balises at the side of the track provides information to the train-borne computer. This in turn checks the driver's actions and intervenes if required, like a more intelligent form of TPWS. One issue is the reliability of the additional equipment. Not that it is unreliable, but the system now has more equipment that has to work.

The driver enters train length, weight, and braking performance into the on-board computer before the start of each journey. A relatively simple task for a passenger train, but more complex for a freight train dropping off and picking up wagons. Route details and signal information comes from the balise. The computer uses all these data to calculate a safe speed envelope.

The driver takes his information from the signal. ERTMS level 1 checks that the driver stays within that envelope. The on-board computer is unaware of when a signal clears until it passes the next balise and is updated. This leads to trains being controlled to a speed slower than necessary. It can reduce capacity at busy stations and junctions.

ERTMS/ETCS level 2

Level 2 uses a radio system (GSM-R) for driver information. The driver takes his information from the in-cab display and does not need the lineside signals. The balises are still used for confirming the position of trains. Removing the need for trackside signals offers a significant overall cost saving.

With full and continuous in-cab signalling, level 2 allows more capacity at busy stations and junctions. The headway, or time gap between trains, can be optimized. It also allows speeds in excess of 200 km/h. Safety is improved because emergency stops can be given to a train at any time.

Level 2 creates a challenge for the whole railway. The system changes the concept the railways have used to date. The signalling system changes from lineside to radio-controlled on-board computers. Everyone's role will be impacted. It needs an integrated track–train interface, which will cross all the commercial boundaries of the fractured UK railway industry.

Because it is the only means of communicating with the train, the reliability of the radio system has to be high, in excess of 99.9 per cent. Dealing with degraded modes becomes more complex with less lineside signal information to fall back on.

ERTMS/ETCS level 3

Level 3 creates an even more fundamental change than level 2. The Lock, Block, and Brake Act of 1889 introduced the philosophy of protecting trains in a fixed block, defined by the position of signals along the track. Level 3 introduces a moving block, or protective space, that moves with the train. All track-based train detection systems, such as track circuits and axle counters can now be dispensed with. The balises are retained, just to advise the trains of their position.

The current need for insulated rail joints to allow track circuit information to be transmitted is no longer required, and can bring cost savings in installation and maintenance. But this would lose one means of detecting breaks in the rails. Dealing with, and recovering from, major system failures is a major concern with level 3.

1.3.5   The plan forward

The recommendations of the Uff/Cullen Joint Inquiry on train protection and the evolution of ERTMS have effectively merged. TPWS, as an interim step, has been fully installed. An ERTMS Programme Board (EPB) was set up to respond to the UFF/Cullen recommendations. The EPB was co-chaired by SRA and Railway Safety. An ERTMS Programme Team (EPT) was set up in May 2001 with representation from across the industry to produce an industry plan for fitting ERTMS. Their Final Report on the planning phase was issued in April 2002.

As the work of EPT progressed, it became clear that the time frame proposed by the Joint Inquiry was impractical. EPT drew up their own recommendations. They updated the estimates presented by experts at the UFF/Cullen enquiry. In turn, the HSC/HSE commissioned their own report to validate EPT's analysis. While HSC praised the high quality of EPT's work, they recommended to the Minister that an even more strategic approach is probably required.

The HSC made formal recommendations to the Minister on 5 February 2003. They endorsed the EPT recommendation for ERTMS level 2. In addition, the HSC said that they would pursue TRWS+ as an improvement on the interim measures. HSC told the Minister that the Uff/Cullen timetable was 'not robust', as the level 2 development would not be completed until 2008. They endorsed the EPT-proposed schedule for fitting ERTMS level 2 on high-speed lines by 2015 and on all main lines by 2030. Uff/Cullen had recommended that installation should be required by regulation, EPT recommended against regulation except for the EC Interoperability Directives. HSC advised the Minister that it was too soon for them to propose regulation owing to the early stage of development of the system. In his letter to the Minister, the Chairman of HSC said that HSC recognizes *that investment in ERTMS is not only – or even mainly – about delivering safety benefits. It is about a vision that the nation has for its railways in the 21ˢᵗ century. The political and financial dimensions of such strategic decisions mean that they cannot be left to the industry alone. Government must take a lead.*

1.3.6   What has to be done?

The scale of the programme is indicated by some basic numbers:

- between 30 000 and 50 000 signals fitted with additional and complex equipment;
- 8000 locomotive cabs retrofitted;
- 19 000 drivers retrained;
- 30 000 km of track equipped with balises and GSM-R wiring;
- 80 000 industry workers have their work drastically changed.

1.3.7   Some statistical data

There are several hundred SPADs each year, although the total is reducing. The rate has halved during the last decade. Most of them are minor misjudgement of braking distance, or at low speed during shunting. With most SPADs, the train stops within the safety overlap provided – usually 200 yd. Monthly statistics are published by HSE. Over the last 30 years, 2.7 per cent of all collisions and derailments were the direct result of SPADs.

Train driver errors

What failure rate should we expect for drivers passing red signals? Trevor Kletz has made some simple calculations (**5**). He sees that SPADs occur at about one in 10^4 times that a signal is approached at danger. This is at the upper end of human reliability. A slightly different approach by Duffey and Saull (**6**) extrapolated SPAD data over six years to estimate the minimum failure rate achievable. They derived a minimum rate of one SPAD per 20 000 train hours. They note that this is the same as the rate for industrial errors.

Risks to the public

The installation of TPWS has reduced the risk of accidents that could have been prevented by full installation of ATP, by an estimated 65–80 per cent. The recommendations of Uff/Cullen were predicted to prevent 83 fatalities over 40 years. The EPT figures are 74 fatalities saved over 40 years. The HSE assessment is that with TPWS and TPWS+, there is a risk of one APT-preventable accident every 10 years, with an average of four fatalities. They estimate that this will fall to one in 60 years once ERTMS is fitted and reliable. To put these figures in context, there were 3431 deaths on the UK roads in 2002.

1.3.8   The safety case versus commercial costs

One way of getting a feel for the impact of decisions affecting both safety and commercial activities is by a simple chart like Fig. 1.3.1 (**4b**). The safety benefits are compared with commercial costs by placing a monetary value on each life saved. Figure 1.3.1 shows the *Pass* zones where this philosophy gives total benefits. The area under the gradient line is the *extra mile* zone where regulators might insist on measures that are more costly than the agreed cost per fatality. Figure 1.3.1 also shows the limit of SRA funds. If the regulator's request is outside that limit, then new funding is needed – presumably from the Government. ERTMS level 1 is seen to have significant commercial costs, outside SRA's ability to fund and with little safety benefit. But ERTMS level 2 has both safety and commercial benefits. Modal shift takes account of moving traffic from road to rail.

1.3.9   Cost/benefit

A detailed cost/benefit study has been made of the ERTMS installation plan (**7a**). This shows the capital costs at £2.4 billion with a possible additional risk of £1.3 billion. The funding peaks at £2.2 billion in 2015 after which benefits begin to kick in. The benefits break-even with costs by 2030. The expected net-present worth based on a 3.5 per cent discount rate is £3.68 billion. The bulk of the benefits are commercial due to the improved infrastructure and greater carrying capacity – including reducing congestion on the roads by greater use of the railways. The safety benefits were only assessed as three per cent of the total. This is clearly different from Fig. 1.3.1

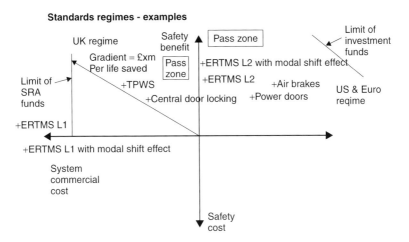

Standards regimes - examples

Fig. 1.3.1   Safety and commercial costs of various decisions. From Ref. (4b)

and would move the point for ERTMS 2 well to the right, outside the limits of the industry to fund.

1.3.10   Experience with TPWS

Although Professor Uff said that the development of TPWS between 1995–1999 was *undistinguished* and that until the Southall accident, progress had been *desultory*, the industry considers that since then it has done a good job. TPWS was installed across the whole railway system in seven years. The HSE's concern over the slow progress led them to recommend that TPWS installation be mandated by Regulation, which was passed in 1999. HSE works by reviewing and approving safety case documentation produced by the many constituents in the industry. The industry believes that regulation creep from this process added to the problems and doubled the cost of installing TPWS with little true safety impact (**8**).

1.3.11   Lessons learned to date

Lessons from BR/GWML ATP pilot scheme

The Great Western pilot scheme covers 89 sets of train-borne equipment and 455 signals, which provides full protection on 153 route miles. It is almost identical in function to the technology of ERTMS level 1 but predates, and does not meet the interoperability specifications. Installation started in 1992 and it has been fully operational for several years. The project has had problems and the lessons learned (**7b**) are summarized as:

- any scheme is easier to control with a single organization, at least there must be a single authority to enforce action;
- avoid changes to proven technology and application, unless absolutely necessary. But system assumptions may have to change as equipment is developed and technologies evolve;
- a change control process must be set up at the start of the project to manage changes. This requires a System Authority to manage, approve, and record changes;
- optimizing new equipment with existing vehicles is an iterative process. Problems may not surface until equipment has been operating for some time;
- know the environment into which complex equipment is being installed. For example, it was found that counters on wheels could experience loads of up to 50 g in each plane. Retrofits also complicate the space and flexibility for mounting additional equipment. Avoid track-mounted equipment whenever possible;
- defect reporting and loss of redundancy need careful attention – particularly for intermittent faults. Defect and trend analysis need to be in place to establish the root cause of problems. Fault reporting needs to be simple, timely, and in a form to allow electronic data analysis and manipulation for easy presentation;
- pay attention to the driver/machine interface;
- good resources, both human and equipment need to be readily available. Early recruiting and training/retraining is essential. Driver training has to be linked to the equipment and system changes;
- time to initialize a train system at the start of a journey has been an issue when the time taken to resolve testing problems is included;
- operating policy on not using degraded systems has to be coupled with providing good logistical support to sort out the problems quickly.

Dutch experience (7c)

The decision to install ATP was taken in 1962 and has been steadily modified since. ERTMS level 2 is being installed on new and high-speed lines. The lessons learned are:

- drivers can find ways to operate/manipulate the system that the designers had never anticipated. So get the designers to travel in the cab from time to time;

- it is not easy to manage a huge installation programme with many players;
- ATP is not clearly endorsed by a commercial business case and therefore it has to be separately financed;
- it is important to keep an on-going communication process with the public.

New York City Subway experience (7d)

The train control system developed on the New York Subway has highlighted issues such as:

- safety has to be ensured in both the normal mode and when equipment has failed or degraded;
- train-borne displays should be developed with the drivers – right from the concept phase.

Dockland Light Railway experience (7e)

The London DLR opened in 1987 and now has 12 km of routes with 16 stations. It uses a moving block ATC system. There has been a steady programme of upgrades and the lessons learned are:

- with a complex software-based system, development timescales are very difficult to predict – even when it is believed that the equipment has been 'proved' elsewhere;
- the most significant threat to system reliability is common mode failures. Failure modes can often remain dormant for years – only surfacing because of a coincidence of uncommon events;
- if you modify a key part of the basic system, consequential problems will show up in other areas;
- wheel tachometers have a short life. They are used for determining the position of the train on the track. Errors are caused by wheel slip, which is aggravated by the weather. The more accurately they tried to know the position of the train on the track, the worse the reliability;
- despite extensive testing, service operation appears to stress the system more than expected, and in different ways. So there has to be careful monitoring of early operation and then corrections made. This experience leads to the recommendation that new systems should be introduced gradually and in manageable steps. Each step should be proved before the next step is taken.

Advice from the EPT (7f)

From their work to date the EPT have the following thoughts:

- programme co-ordination is key;
- plagiarize from the experience of others. Have 'no ego';
- make sure that everything has been done somewhere else first.

COMMENTS

SPAD issues fall into three categories – technical details, driver performance, and the overall organizational and management decisions. Technical solutions for limiting SPADs have been around for a century in various forms. The question is, *how affordable are the solutions for improving safety?*

Blaming the driver for SPADs is too easy. It is more difficult to put the reasons for the failure in perspective. First it has to be recognized that drivers are human. Expecting zero errors is unrealistic. Next, the climate in which the driver operates is determined by the management and company approach.

Management has been blamed in enquiries following SPAD accidents. While some of the criticism appears justified, little has been said about the constraints on management. The industry has been under-funded for decades. In recent times, regardless of whether it is in government or private sector ownership, infrastructure investment funds have to come from the government. The government's decisions on funding are based on their policy for the industry as well as safety policy. Both are set in the context of overall financial constraints. It is not easy to get *seamless* performance among all players.

Over-reaction?

Accidents focus attention and emotion on a topic. The result can often be an over-reaction. The installation of TPWS+ could have given an acceptably low risk of fatalities from SPAD. However, this line of thinking is now overtaken by the commitment to the European Directives and ERTMS level 2.

The expenditure of £2.4 billion on ERTMS installation is predicted to reduce the risk of an ATP-preventable accident, killing on average four people, from one every 10 years to one every 60 years. In other words, one less person killed on the railway every three years. Could the £2.4 billion have saved more lives if used in a different way?

ERTMS is recognized as a good system for high-speed lines, but do occasionally used freight and rural passenger lines need the full high-density radio coverage, speed supervision, and £250 000 worth of signalling in every cab? The Swiss, backed by others, are proposing a limited supervision form of ERTMS. It supervises speed only at high-risk points and leaves the driver in control, unsupervised in low-risk open country. As it is compatible with the full ERTMS cabs, it offers a potentially affordable migration path.

Learning from enquiries
Recommendations from enquiries are made at one point in time – the presentation of the report. Technology moves on and can change the basis of some recommendations. Arguments can be made to move away from the recommendations because of changed circumstances. Some arguments are genuine, while some are foot dragging.

Enquiries are expected to produce many recommendations. Sometimes these are unrelated to the cause of the accident, as was Justice Hidden's ATP recommendation.

Is it reasonable to expect judicial enquiries, taking place over a few months, to be able to instantly judge the best detailed solution? Sometimes they are contradictory. For example, the Great Heck Enquiry recommended the removal of bogie retention straps from class 91 locomotives. This was because the straps pulled the bogies off the rails as the locomotive body lifted up. Otherwise it would have stayed on the rails. Meanwhile, the Sandys Enquiry recommended the strengthening of the very same straps, because the bogies had gone flying off and caused damage.

Enquiries often focus the parties on closing each and every stable door after all the horses have bolted!

Experience with regulations
The experience with installing TPWS appears to be different depending on where you sit – in the rail industry or the HSE. The HSC/HSE seems to have changed course several times over the decades. Nigel Harris, the Managing Editor of the magazine Rail has questioned whether HSE has the necessary in-house expertise to fulfil its role (**8**). He accuses them of rewriting history and concludes *if HSE/HSC is incapable of the mature approach required, Government should replace it. Soon.*

A competent, consistent, and respected regulator is needed. The complexity of introducing ERTMS across the whole rail system requires sound regulation to ensure consistency of standards. In this case, the standards are mostly set in Brussels – the UK Regulator's role is to assure compliance.

Who decides the value of a life?

As part of the work to provide recommendations to the Minister on ATP, the HSE retained consultants to find out what the public wanted. The consultants selected members of the public and involved them in an exchange of information, expert's forum, and a deliberation conference. While the selected members of the public had mixed views on whether train protection (ATP) was a priority for investment, the majority agreed that implementation of ERTMS level 2 was appropriate. There appeared to be no debate on the value of a life saved, which is basic to any decision on the level of investment in safety equipment.

The public are the people who jaywalk amongst the traffic, gamble with safety and have a record of higher risks in the home than at work. They tolerate a higher level of risk on the road than for other forms of transport. Should the HSE poll the public to determine the amount that should be spent on safety? One can argue that the public takes the risk and eventually is the source of payment. Therefore, they should have a say. Can we distinguish between what society is willing to pay in the long term, and the short-term gut reaction of the public? There is no easy answer to this difficult question, which lurks beneath every safety investment decision.

Straightening out the system

SPADs are more dangerous at complex and busy junctions. When there are problems at a road junction, one answer is to change the road system – straighten it out and remove the hazard. Could a similar approach help reduce rail accidents? In Europe, the TGV runs on its own track and avoids interaction with other traffic.

If the United Kingdom had invested in a new north–south high-speed line, it would have cost about the same as the West Coast remodelling. But it would have given much more capacity and a huge safety benefit. This would come from segregating high-speed trains on the new line, and leaving the existing West Coast Main Line (WCML) for freight and local passenger trains. It did not happen, mainly because commercial and safety policies were decided independently.

Who is in charge of the train?

With the proposed changes, is the driver or the system in charge? The current position is clear – the driver is in charge, aided by the warning signal. At the higher levels of ERTMS, the computer takes greater control until a driverless train could be an option. For now, ERTMS is there to support the driver.

Who is in charge of the ERTMS installation?

The HSC asked the Government to take a lead, and they have decided to give the lead to SRA. EPT has now evolved into the National ERTMS Project Team (NEP), working under the instructions and funding of SRA. EPB is now an advisory group to SRA.

The industry is still reeling from being fragmented. It is about to get an engineering project of enormous dimensions imposed on it. At the same time, it is expected to meet the government's plan for a 50 per cent increase in passenger traffic and 80 per cent increase in freight volume. Will there be any lapses in safety or service glitches that will upset the travelling public during the long and complex transition process? The history of the relationship between all the parties does not bode well. On the other hand, most of the past problems have been recognized, and commitments have been made to do better next time.

In summary, the SPAD issue has morphed into the installation of ERTMS under EU Directives. ERTMS is far more than a safety response to SPADs. However, the commercial benefits are well into the future and depend on several factors not easy to predict – like a move away from road transport to the railways, as well as the effectiveness of the management of the installation.

1.3.12   Lessons learned

Knee-jerk responses can be unsafe and expensive, and lead to decisions that do not resolve the problem. Recommendations concentrating on the technical details and driver error can miss the organizational causes of an accident.

There is a lot of experience listed under Section 1.3.11, which needs to be built into the implementation of ERTMS. The lessons are also applicable to most engineering projects.

An issue like SPADs that involves government policy for an industry, government funding, government regulation, government safety policy, and an array of private companies is difficult to resolve. Each party has its own agenda, objectives, priorities, limited resources, and culture. When problems occur, it is too easy to blame someone else. Deciding who is ultimately responsible is elusive. The lesson is for all parties to work seamlessly together, assist each other, and not hide behind whoever is in trouble. Hopefully this will happen with the implementation of ERTMS.

References

(1) Vaughan, A. (2000) *Tracks to Disaster*, Ian Allan Publishing, UK, ISBN 0 71102 731 5.

(2) Davies, D. (2001) *Automatic Train Protection for the Rail Network in Britain: A Study*, Royal Academy of Engineering, London.

(3) Professor John Uff, Q.C. (2000) *The Southall Rail Accident Inquiry Report*, HSE Books; **Rt. Hon. Lord Cullen, P.C.** (2001) *The Ladbroke Grove Rail Inquiry*, Part 1 and 2 Reports, HSE Books; Uff/Cullen (2001) *The Joint Inquiry into Train Protection Systems*, HSE Books, London.

(4) *Delivering Rail Safety in Changing Times, IMechE*, 17 April 2002, London.

> **(4a) Kemp, R.** *New rolling stock – real safety not just a safety case.*

> **(4b) Lockett, R.** *All Change for Europe.*

(5) Kletz, T. (1991) *An Engineer's View of Human Error*, Gulf Publishing Co., ISBN 0 85295 265 1.

(6) Duffey, R.B. and **Saull, J.W.** (2003) *Know the Risk*, Butterworth Heinemann, ISBN 0 75067 596 9.

(7) *Train Protection – The Way Forward?* IMechE, 24 February 2003.

> **(7a) Williams, N.** *Costs/Liabilities – Who Pays and How?*

> **(7b) Wright, N.** *UK ATP Lessons Learned.*

> **(7c) Coenraad, W.** *Experience with Train Protection Systems in the Netherlands.*

> **(7d) Rumsey, A.** *Implementing Communications Based Control at New York City Transit.*

> **(7e) Lockyear, M.** *Case Study: Docklands Light Railway.*

> **(7f) Jones, S.** ERTMS Programme Team (EPT).

(8) Harris, N. *Rail*, **455**, February 19–March 4 2003, 3.

The public can well say *You engineers have been working with wheels and rails for a couple of centuries; isn't it about time you knew all about them?*

So what is the story and status of this basic piece of engineering?

At the point where the wheel meets the rail, there is a large and equal load on each of them. Nominally it is a point contact. With deformation, the actual contact patch under each wheel is a small ellipse – the size of a 5p or 1¢ piece.

1.4.1 The rail as a beam

The rail acts as a beam to support the wheel. The sleepers and ballast act as an elastic foundation. Peter Barlow (1776–1862) in the United Kingdom made early estimates of the stresses. Winkler in Prague made improved calculations, followed by Zimmerman in Berlin in 1888 (**1**).

Rail deflections were measured by Wasiutyński in Russia in 1889 and the calculations were proved. Later, Westinghouse in USA used strain gauges to confirm that the theory of a beam-on-an-elastic-foundation was appropriate for railway tracks. The strain gauge tests also highlighted that the bending and torsional stresses cannot be ignored.

Fieldwork showed that there are significant dynamic loads in addition to the static loads. Crude braking can produce flat spots on the wheels that result in dynamic loads.

With continuously welded rail, there are additional loads dependent on the initial installation and the temperature. The process of manufacture can lock residual stresses into the rail – tensile in as-rolled heat-treated rails and compressive in head-hardened rails.

1.4.2 Local contact stresses

Where the wheel and rail meet, the contact stresses dominate. Hertz, in Germany, first analysed contact stresses in the 1880s. Belajev, in Russia, was the first to apply Hertz's theory to the wheel/rail interface in 1917.

Experiments show that typical surface roughness can theoretically produce local peak asperity pressures, up to eight times the nominal

Hertzian stress (**2a**). In addition, there can be peaks due to grit and sand on the rail.

The contact patch is related to wheel size and loading. In the United Kingdom, there has been an increase in vehicle mass and at the same time a reduction in wheel diameter. As well as the vertical load from the wheel, there are longitudinal loads along the rail due to friction, traction, braking, and wheel slip. Curving and the dynamic motion of the bogies produce lateral forces on the contact patch. The longitudinal and lateral forces combine to give the total shear load on the contact patch.

The stresses on both the wheel and rail at the contact patch are complex and at a microscopic level are in the plastic region. So the material on both wheel and rail are being cycled at very high stress levels. A busy line with around 100 trains per day can accumulate 10^6 wheel passes each year. So fatigue is a potential issue.

1.4.3   Vehicle dynamics

Most locomotive and passenger rolling stock have a pivoted bogie at each end. An exception is the French TGV, where one bogie supports the ends of two coaches. Each bogie usually has two wheel sets. Each wheel set has wheels rigidly attached to each end of an axle. The wheels are flanged and usually have a conical tread profile.

As a bogie goes around a curve, the leading wheel set moves towards the outside of the curve. The conical wheel tread on the outer rail is then at a larger radius on the cone than the wheel on the inner rail. This provides a steering force to the bogie. With a pivoted bogie, this means that the trailing wheel set is forced in the opposite direction.

The conical wheel profile can produce kinematic oscillations of the bogie. Bogies have yaw stiffness, springs, and dampers to assist in providing a good ride as well as reducing the peak load on the rail. Bogie design is a trade-off between stability on a straight track at high speed, and curving performance. As a result, not all wheels can avoid some slipping or creep in a curve.

In a curve, the outer rail is raised higher than the inner rail to provide some offset to the curving forces. The difference in height, the cant, is limited to 6 in, so that the tilting is not too great if the train stops on the curve. The cant can only fully offset the curving forces at one speed. At higher speeds there is a *cant deficiency*, which means that the wheels have to produce a lateral force against the outside rail to balance the centrifugal forces. In turn, this force has to be resisted by the ballast. If the ballast is in poor condition, this will exacerbate the dynamic forces and lead to a 'spiral-of-decline'.

Predicting the dynamics of the vehicle and action of the wheel on the rail is complex and requires modern computer programs (**2b**). These can use quasi-steady-state analysis or be more complex and introduce all the dynamics and non-linearities.

The motion of the wheel on the rail is complex. At the symposium *Wheels on Rails* (**2c**), Malcolm Dobell from London Underground showed a video taken with a camera mounted under a carriage looking at the wheel. This gave stark visual evidence that the dynamics are not simple and that the contact point can wander across the railhead.

1.4.4   Shakedown theory

What happens to the material at the highly stressed contact patch? Carter first studied rolling contact theory in the 1920s (**3**). In the 1960s, Johnson at Cambridge developed an understanding of plastic flow and the shakedown of the surface material (**4**).

When the contact stress exceeds the elastic limit and goes into the plastic range, the local material is constrained by the material around it. This leads to residual stresses. The steel used for rail's strain hardens. Tests on small samples show that the surface hardness can increase by 50 per cent after 40 000 cycles (**2a**). As a result, the contact load can increase to a new *elastic shakedown limit* before it reaches the plastic zone. At higher stresses there is a region where the plastic deformations are matched at each end of the cycle. At even higher stresses, above this *plastic shakedown limit*, the material will ratchet on each load cycle. Eventually, the material will fail by *ductility exhaustion* (Fig. 1.4.1).

The picture is further complicated when there is a horizontal load on the contact patch as well as the vertical load. The 'shakedown diagram' is shown in Fig. 1.4.2. The diagram shows regions where the plastic flow is on the surface, and other regions where it is sub surface.

The quasi-static calculations for various power cars and coaches on a 1500 m radius curve at Hatfield are shown in Fig. 1.4.3 (**2d**). The calculations show that some of the contact points would be in the gauge corner region of the rail. Most vehicles will take the rail over the shakedown limit for standard rail. Also shown is a higher shakedown limit for hardened rail, although some tests suggest that the benefit might not be as large as shown in the figure.

With full dynamic analysis, the points in the figure become quite broad patches. So any perturbation such as minor track variations,

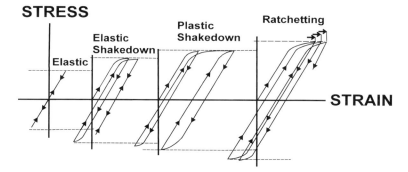

Fig. 1.4.1   Four states of cyclic loading

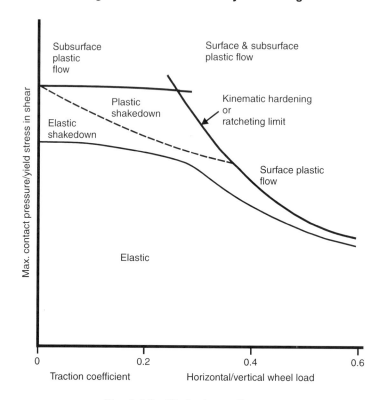

Fig. 1.4.2   Shakedown diagram

out-of-profile wheels, or variations over welds, can produce stresses well into the plastic region.

1.4.5   Crack propagation

Cracks start in the distressed surface layer and grow at a shallow angle of around 15 to 20 degrees, in the direction of travel. These are shear

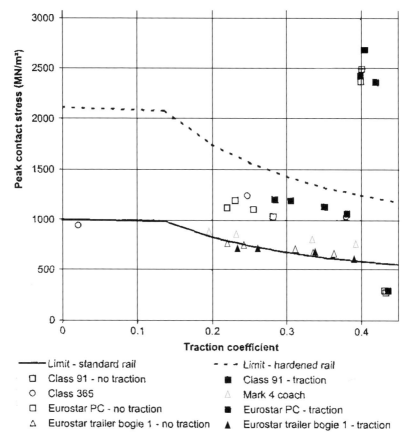

Fig. 1.4.3   Quasi-steady-state rolling contact fatigue conditions

cracks caused by the cyclic ratcheting. Sometimes the cracks then return to the surface, resulting in a piece of rail chipping off – known as a squat. In other situations, after the crack has reached a depth of around 5 to 10 mm, it can turn down at an angle of 70 degrees and eventually lead to fracture of the rail. In some cases, the cracks can branch.

With the lateral loads produced in curves added to the tangential loads, the initial cracks develop at an angle rather than straight across the rail, Fig. 1.4.4.

The different phases of crack growth are shown in Fig. 1.4.5 (**2e, 5**). Starting from the distressed surface layer, the cracks initially grow quickly, but the rate of crack growth drops off quickly as the crack grows.

Then the contact stresses determine the crack growth rate. As the stresses reduce with depth, the crack growth rate reduces. In this stage,

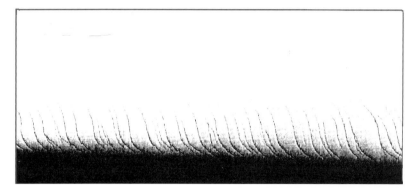

Fig. 1.4.4   Moderate rolling contact fatigue cracks

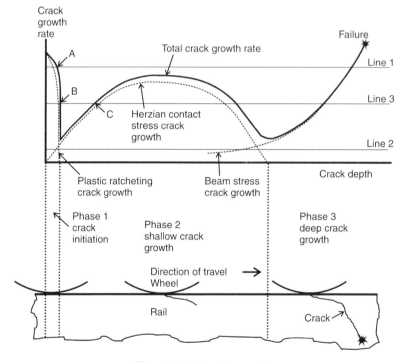

Fig. 1.4.5   Crack growth

any moisture – water or lubricant – can be forced into the crack and enhance the cracking rate at the crack tip in the tensile mode.

The third phase of crack growth is driven by the beam stresses. The crack growth rate is the sum of all these effects. The shape of the curve is dependent on the specific site conditions – radius of track, wheel loads, train speeds and so on.

1.4.6   Fracture mechanics

The development of fracture mechanics over the last 50 years has given some ability to predict crack growth. At crack initiation, Microstructural Fracture Mechanics (MFM) looks at the damage within the grain. As the crack grows to around 1 mm, Elastic Plastic Fracture Mechanics (EPFM) can be used. At defects over this size, Linear Elastic Fracture Mechanics (LEFM) can adequately predict crack growth.

1.4.7   What limits rail life?

The life of the rail is a battle between wear and fatigue. If the wear rate is high, then the fatigue cracks are worn away before they can propagate. The trade-off between wear and fatigue can be seen in Fig. 1.4.5. Line 1 shows the effect of a high wear rate where the fatigue cracks do not get a chance to grow beyond point 'A' before being worn away. Line 2 shows a low wear rate where the fatigue cracks can grow with little effect from wear. Line 3 shows an intermediate wear rate where the cracks can grow to point 'B' but not propagate further. But if for some reason it was able to grow to 'C', then the Hertzian stresses would allow it to grow further.

If the wear is not enough to prevent fatigue, then grinding can remove the fatigue cracks. But grinding can only take place at discrete points in time. Take, for example, grinding to give the equivalent wear of line 3. If the grinding takes place after point 'C', then even achieving the average wear rate of line 3 will not stop the second phase of crack growth.

Kalousek from the National Research Council of Canada (NRCC) has worked for several decades in this area and coined the phrase *magic wear rate* (**6**). The *magic wear rate* is the optimal, where fatigue is arrested by the sum of wear and grinding.

1.4.8   Lubrication

Lubrication can dramatically reduce rail wear. This is particularly true in curves. Lubrication has reduced wear rate by as much as 200 times (**6**). The lubrication can be applied by equipment on the track or on the rolling stock. Both grease and solid stick lubricants are used. Lubrication is particularly effective in the gauge corner in curves. The issue is to reliably get the lubricant where it is needed and nowhere else. Driving wheels need traction without wheel spin. Wheel spin would add to the damage to the rail. Excess lubricant can be a fire hazard in subway systems and tunnels. In the United Kingdom, there is concern that lubrication could accidentally lead to poor braking performance – especially after bad experience with leaves on the track.

Lubrication, besides reducing wear, can move the points on the shakedown diagram towards the left. In some cases, this will move them below the shakedown limit.

1.4.9   Wheel/rail profiles

Over recent decades, there have been many studies around the world at improving the profiles of both wheel and rail. BR Research reported in 1987 on their work to optimize the rail profile to minimize contact stresses (**7**).

In North America, rail grinding has been a feature since the early 1980s (**8, 9**). It is used to remove corrugations and improve the wheel/rail profile. These railroads are mostly labelled *heavy haul* with long freight trains using heavily loaded wheel sets.

NRCC has developed a computer rail profile optimization program (*Pummelling* (**10**)). This program takes account of the fact that the rail has to handle a range of wheel profiles from a variety of rolling stock, in both new and worn states. The program predicts the contact frequency and load intensity across the railhead. The aim is to distribute the contact points on the surface to optimize the use of the railhead material. The programme can evaluate the frictional loads and optimize profiles to minimize locomotive fuel consumption.

The wheel treads have to be checked and reprofiled when worn. Wheel wear life is not very predictable. Some wear badly in 50 000 miles, while others will last 270 000 miles. Sometimes wheels on the same axle will wear differently (**2f**).

Differences of the order of 1 mm in wheel/rail profiles can make significant differences in both wear and fatigue life. Wheel and rail profiles need to be matched for optimum performance.

1.4.10   Metallurgy

The early railways used iron rails. When Henry Bessemer created mild steel, railways quickly turned to using steel rails. The first steel rails were laid by Midland Railway in Derby in 1857. The old iron rails had to be replaced three or four times a year on busy lines, whereas the steel rails lasted over 15 years.

The early steel rails had a pearlitic structure based on the carbon/manganese content. Standard rail is still basically the same material. Improved steel making has reduced the manufacturing defects in rails and resulted in cleaner material The use of electric arc furnaces, vacuum degassers, continuous casting, and controlled cooling have improved the final product.

Harder and tougher steels have been developed. Developments in steels have improved wear, and have resulted in fatigue failures becoming more prominent as the factor limiting life. Specialist steels are more expensive and so their use is often limited to the higher loaded curves.

Selecting the best rail metallurgy for the various situations – straight track, curves, high speed, and heavy haul – is usually a compromise. The relationship of the rail properties to service life is still not fully understood.

Coated rails

Research on coated surfaces is showing promise (**11**). The INFRA-STAR project aims to prevent rolling contact fatigue and reduce squeal noise in curves. This is a collaborative programme between Sheffield University, AEA Technology, and Chalmers University in Sweden, together with industrial partners. Different alloy materials as well as ceramic coatings are being reviewed. Two methods of producing the coated material are being tested – rolling and laser spraying. Laboratory tests help select the best sample materials. Experimental work on a sample rail track started in 2001.

1.4.11   Inspection

Inspection of rail track starts at the basic level of walking the lines looking for problems. Initial cracks are not easy to see and are reported (**2d**) to feel like fish scales. The usual way of inspecting rail cracks is by ultrasonics although eddy current methods are also used. BR had used ultrasonic test carriages run at speeds up to 40 mile/h but stopped using them because they gave too many false readings. As a result, they moved to a manual technique where track workers walked an ultrasonic head along the rail.

As well as inspecting for cracks in the rail, the track has to be inspected to ensure that the geometry is correct and that the fasteners, sleepers, points, and ballast are in good condition.

Railways use an electrical signal through the rails to provide a back-up signal for the presence of a train in a block. If there were a complete break in the rail due to a fatigue crack, then no signal would be present. While there are often false signals, this technique will show about 30 per cent of rail breaks.

1.4.12   Experience on rail systems around the world
UK

At the 3rd International Conference on Contact Mechanics and Wear of Rail/Wheel Systems held at Cambridge in July 1990, a paper was presented by staff from BR Research. They said that *on curved track the outer 'high' rail has been found to be susceptible to rolling contact fatigue resulting in the initiation of cracks from the running surface of the rail. These shallow cracks can develop into squats which can link together resulting in loss of surface material or turn and propagate transversely through the rail.*

The Rail Regulator read in an HSE report that the number of broken rails in 1998/9 had increased to 937. Railtrack had predicted about half that number. The Rail Regulator together with HSE commissioned a study by Transportation Technology Center Inc. (TTCI) to report on the way Railtrack manages its track. TTCI is a subsidiary of the American Association of Railroads. TTCI reported that the increase in rail breaks was not statistically significant when the 30-year average of 767 breaks per year was considered. However, Railtrack was significantly worse than other passenger railways, but similar to North American railways that carry much higher axle loads. The broken rails per track mile for Railtrack were 3 times the French and 1.65 times the German statistics.

TTCI noted that from 1994 to 1999, UK rail traffic had increased by one-quarter and freight, with higher axle loads, had increased by 40 per cent. Concurrent with this, there had been fewer rail replacements and less inspection. Track quality and wheel irregularities had worsened. Dynamic loads can be four to five times the static loads.

TTCI saw gauge corner cracking as a serious emerging problem, treatable at this stage only by grinding. There had been no grinding in the last year of BR and in the first year of Railtrack. TTCI noted that Railtrack now had one track-grinding car in use and several more on order.

TTCI found it difficult to assess the UK situation because Railtrack had few records. They recommended that Railtrack create and maintain a central database of defects, and not rely on their contractors to keep records. They also recommended that Railtrack reintroduce automatic ultrasonic testing, saying that *Railtrack is slow in adopting best practice in ultrasonic inspection.*

The TTCI report did note that Railtrack had started to make improvements and were starting to adopt better management practices. Ironically, their report was completed and tabled just days after the Hatfield crash.

It has been said that the privatized Railtrack *marginalized the role of engineers* (**12**). Until the Hatfield crash, nobody on the Railtrack Board paid attention to engineering issues. In May 2002, Railtrack decided to take direct control of inspection rather than relying on the maintenance subcontractors.

Railtrack reported fewer broken rails in 2001/2002. Of the 539 broken rails, 29 per cent came from inherent mid-rail defects caused by gaseous impurities in the rail material as well as inclusions that eventually led to fatigue failures. Welds result in 24 per cent of the failures and 23 per cent are at the rail ends. Only eight per cent of failures are from rolling contact fatigue although the number may be higher. Some of the reported mid-rail defects may be rolling contact fatigue when examined more closely. The average life of rails is around 22 years. But an example of rolling contact fatigue cracks has been found after only one year in service.

Railtrack has introduced several inspection and measuring tools. In 1999, Railtrack started to install strain gauge systems on selected track (Wheelchex) to monitor the forces from wheels. By 2002, Railtrack had installed Wheelchex at 16 sites. This system generates the statistics on wheel loading and highlights cases of severe overload. Early statistics showed significant overload on some lines (**2h**).

Railtrack developed an eddy current inspection tool *Lizzard* specifically to inspect for gauge corner cracks. After Hatfield, Railtrack decided to buy a fleet of track inspection vehicles. The aim is to replace the men walking the track. The new vehicles use ultrasonics, ground radar, and video.

With the help of consultants, Railtrack started to apply a whole-life rail management model to manage the work on the track. This provides a co-ordinated engineering basis for rail inspection, grinding, and replacement. The model looks at the probability of damage at specific sites using calculations of the dynamics of the vehicles used on the track, together with data collected by inspection.

An industry-wide organization, the Wheel Rail Interface System Authority (WRISA) was set up in May 2001. The objective was to bring together all parties involved and help facilitate interface issues. WRISA is in the process of being disbanded in early 2004. Strategic direction for the vehicle/track interface will then come from the SRA and a new Systems Interface Committee created by RSSB.

Japan

Japan introduced the Shinkansen high-speed train in 1965. Around 1980, the Tokaido Shinkansen line was troubled by what was called

Shinkansen shelling (**13**). Squats were developing and becoming detached from the railhead. A programme of improved inspection and grinding cured the problem. A large, government-funded R&D programme supported this work and covered all aspects of the science, metallurgy, and engineering.

An inspection car travels the lines each night. There has been a very low frequency of broken rails and no fatality due to broken rails.

European joint study

European Railways, including BR, sponsored work at the European Rail Research Institute (ERRI) at Utrecht. Since 1987, rolling contact fatigue has been the subject of one of their research programmes. In their 1990 report, they say that the incidence of rolling contact fatigue is increasing and that it *presents a major threat to the integrity of modern railway systems*. They concluded that there is *no obvious single solution* and that *in service grinding to remove small surface cracks and establish low contact stress rail profiles should be given further attention.*

Germany

In the 1970s many cases of gauge corner fatigue damage occurred on the track of the German Railways, which carries mixed passenger and goods transport. Sometimes, the defects resulted in vertical fractures (**14**).

On 3 June 1998, an ICE train crashed at Eschede, north of Hanover, killing 102 people. The cause was a wheel tyre broken by fatigue. The train used a special wheel with a layer of rubber between the wheel body and the tyre. The broken wheel caught in a switch, derailing that car. The following cars and rear power car ran into a concrete bridge, which crashed onto several of the cars.

France

In 1992, surface-initiated rolling contact fatigue caused 20 per cent of rail breaks on French Railways (**15**). They have had shelling or squats similar to Japan. Inspection and grinding has kept the problem under control. On the high-speed track, a special inspection car can be spliced into any TGV. A two-weekly inspection of the whole network takes place with inspection cars incorporated into regular trains run at normal speeds.

The TGV has had several derailments but fortunately no deaths. A train, previously involved in an emergency stop, derailed on 14 December 1992 at 270 km/h. The emergency stop had created wheel flats.

Flying ballast injured people waiting at a station when the derailed train went through at speed.

After heavy rain, the last four trailers and rear power car of a TGV derailed at 290 km/h on 21 December 1993. The heavy rain had opened a sinkhole under the track. The sinkhole may have been from WW1 trenches. The train stopped in 2.3 km and remained upright.

A Eurostar train derailed on 5 January 2000 when a transmission assembly failed and a link impacted the track at 290 km/h. The partly derailed train came to a halt 1.5 km down the heavily damaged line. The train remained upright and did not damage the adjacent track.

Canada

The two largest privately owned freight railways, Canadian Pacific Railways (CPR) and CN have had similar experience. CPR started a coal export service through the Rockies to the Pacific coast port in 1970 (**9**). The trains typically consist of over 100 wagons each carrying 95 te of coal and with axle loads of 30 te. Gross wagon weight is now increasing to 130 te. As well as gradients, the route has 133 km of curves with radii less than 312 m.

The initial average wheel life was 280 000 km and rail life 293 million gross tonnes (mgt). Now the wheel life averages 453 000 km and rail life 612 mgt. At the same time, wagon-km costs have fallen by 25 per cent and fuel costs by 15 per cent. These improvements came from taking a long-term and comprehensive view of the problems. The problems to be solved were corrugations, gauge corner shelling and excessive wear. As early as 1982, surface-initiated fatigue defects were believed to account for 12 per cent of the rail replacement costs. This was despite trying to resolve the problem. The main strategy used to finally deal with the issue is *prevention through grinding.* This has been complemented with the use of chrome-alloyed head-hardened rail, cleaner steel, steerable bogies, and improved rail fasteners.

CPR worked with the National Research Council of Canada (NRCC) to carry out analysis and tests for developing the optimum grinding pattern and optimum rail profile. CPR also worked with NRCC to upgrade their lubrication programme.

Monitoring sites and a test site are used to check, measure, and evaluate grinding patterns before applying them to the whole line. Grinding aims to keep within 0.018 mm (0.007 in) of the planned profile.

The number of fractures per year has dropped from over 250 in 1997 and 1998 to just over 100 per year on the 750 km track.

CPR sum up their experience (16) as showing the need to:

- *have vigilant management of the grinding programme;*
- *maintain permanent records of track work and dimensions;*
- *make sure all staff and contractors have all the information.*

USA

A US government agency, the National Transportation Safety Board, was concerned at the increase in derailments from 1741 in 1997 to 2059 in 2000. The question *Are railroads cutting corners?* was posed. The industry says they are addressing the issue and point out that derailments have reduced by 73 per cent since the 1970s. Over half the derailments are at slow speed in yards and sidings.

In 2001, there were a total of 967 fatalities on all US railroads. Half of them were trespassers, and most of the remaining half were at highway/rail crossings. Between 1998 and 2000, one death was the result of track problems. In the previous three years, four deaths were due to track.

Rail/wheel issues have been key to the economics of the predominant heavy-haul railroads. As a result, there have been many seminars and meetings on the subject over the past couple of decades. The general method of handling the situation is by grinding and by lubrication in curves.

An interesting example is from when Burlington Northern (BN) merged with Santa Fe in the mid-1990s. BN had used grinding for several decades. With the merger, the traffic increased and there was less time for grinding. As a result, there was more rail fatigue damage. In 1997, BNSF asked NRCC to help them improve their grinding programme through better implementation of existing rail-grinding resources. As part of this effort, NRCC metallurgically examined rail samples to try to determine the optimum metal-removal rate. BNSF use their test site to confirm options.

BNSF use a risk-based programme to manage maintenance and limit derailments (17). This programme recognizes that not all defects will be found during a given inspection. Combining crack detection probability with crack growth rates and track history allows track work to be optimized.

Australia

Australia has a mixture of passenger and freight traffic together with some heavy-haul mine railways. One of these is BHP who operate a heavy-haul railway. In 1972, the BHP Institute of Railway

Technology opened in Monash University. In 2001, BHP ran a train of 682 wagons hauled by eight locomotives distributed along the train. The train was 7.3 km long and had a gross weight of about 100 000 te. Rail/wheel problems are kept in check by a combination of inspection, grinding, and lubrication. Collaboration between Monash and BHP ensures the sharing of the technology and track experience. Six universities, including Monash, have joined with six industrial companies to share research and information on wheel/rail technology.

Hong Kong

The MTR Corp of Hong Kong is a heavily used metro system with a 42 km network. MTR started working with BHP in 1989 to help with rail management. Since 1992, they have been able to halve the amount of rail to be replaced. Rail shelling has been cut by 85 per cent and flange wear by 67 per cent. These results came from using heat-treated rail, grinding, and lubrication. The lesson that MTR learnt is that *the wheel/rail interface has to be actively managed.*

COMMENTS

Complexity

The interaction of a wheel on the track is clearly complex. Figure 1.4.6, modified from a diagram by Magel and Kalousek **(10)** shows some of the interactions and feedback loops.

Permanent way

Neither scientists nor engineers have complete knowledge on wheels and rails. Engineers have let the public think that railway lines are part of the landscape – the *permanent way*. The expression *permanent way* is clouding the truth seeing that, on average, the rail is replaced every 20 years. The expression appears to come from early railway construction. Temporary rails were used to move earth from cuttings to embankments before the *permanent way* was installed.

The need for knowledge

Gauge corner cracking and rolling contact fatigue exploded as topics for discussion by the public and engineers following the Hatfield crash. A comment in the Railway Gazette International, which describes itself as 'the worlds leading journal for senior railway management', is brutal. They said *the saga of the Hatfield derailment emphasizes the shallowness of knowledge in Britain's railway industry* **(18)**. It seems harsh to generalize on the whole industry.

At a lecture *Avoidance of Failures* delivered in 1968 **(19)**, Dr B.J. Nield, Assistant Director of BR Research Division said that one

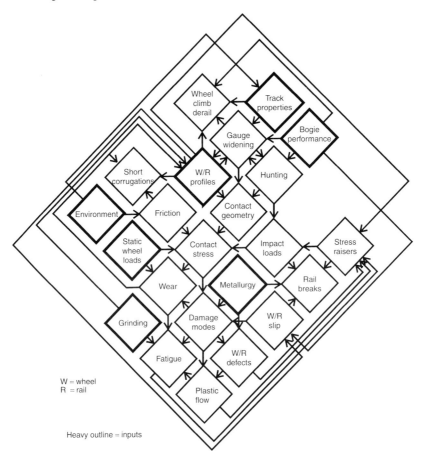

Fig. 1.4.6   Wheel/rail interactions and feedback loops

design problem to watch out for is the *failure to recognize and apply the available relevant information.* His comment is also appropriate for senior management.

There were many conferences around the world on wheel/rail technology in the 1990s. Engineers from the United Kingdom attended some of these. Therefore, railway engineers and academics were aware of the potential for rolling contact fatigue.

After the Hatfield crash, the PM met with Gerald Corbett, the Chief Executive of Railtrack. Following the meeting, the PM's spokesman said that *the problem (at Hatfield) was gauge corner cracking. This was a relatively new phenomenon and the speed at which it had spread through the rail network had surprised the industry* (**20**). The reports by BR and ERRI from around 1990, and the German experience from the 1970s, show that gauge corner cracking had been around for some time and was by no means a new phenomenon. The fundamental research in this area dates from several decades before Hatfield.

At the conclusion of the *Wheels on Rail* symposium in May 2002 **(2)**, Brian Clementson summed up the meeting as a useful sharing of information. He said that progress is being made and there are fewer broken rails. The knowledge and development of people is key. Four important aspects are:

- *data – having it available*;
- *sharing knowledge*;
- *knowing and using the best practices worldwide*;
- *learning.*

As to the future – *engineering is back on the agenda*. Improved communication, understanding the economic consequences and the political expectations all have to be taken into account. His final comment was – *the answer lies with people.*

Andy Doherty, the head of Railtrack systems engineering post Hatfield, pointed out the need for relearning **(2g)** – he had to go out and re-find data. Knowledge tends to reside with people rather than organizations. Once organizations are broken up or have major changes of staff, the knowledge base disappears quickly. The privatization of BR scattered the knowledge to a multitude of companies.

The role of universities, laboratories, and test facilities

In a complex technology, where there are still a lot of technical questions unanswered, there is a need to harness the skills of many players. Experience around the world is showing the advantage of engaging academic researchers to help understand the science and technology. National and industry laboratories fill a gap between the academic work and the practical application. Several countries have found the benefit of using test track to gain experience and knowledge on maintenance procedures before applying them to the whole system.

Relevance of heavy-haul experience for the United Kingdom

Knowledge gained on heavy-haul systems may not have been seen to be relevant to Railtrack. However, the UK rail system is 'curvy' due to the Victorian entrepreneurs building the track along contours wherever possible. This avoided the cost of moving material for cuttings and embankments. When the French and Japanese built their high-speed lines, they built new and straighter track. The United Kingdom used the existing track for the higher-speed InterCity trains. Consequently, the wheel loads are higher in the United Kingdom, and nearer the levels of heavy-haul systems. It should be no surprise that the statistics for United Kingdom are similar to heavy-haul systems.

Maintenance 'holiday'

Delaying maintenance or taking a *maintenance holiday* can save short-term cash but at a risk to the long-term health of the system. Abandoning grinding while not knowing the state of the system was a recipe for disaster in the United Kingdom.

Risk on the rails

The statistics for fatalities due to derailments from broken rails are very low. Good inspection and maintenance can keep the number of breaks to a minimum. And not all rail fractures lead to a derailment. The multiple fracture of the rail at Hatfield was probably caused by repeated use of an already fractured and cracked rail. While the accident is simply stated as being caused by gauge corner cracking, the delays in replacing the defective track are probably a truer statement of the cause of the derailment.

At the time of Hatfield, the Minister, Stephen Byers, said, *safety has to be paramount, not cost.* One can almost hear the chants of approval from the backbenches, and applause from the media and lobby groups. But what does the public really want? The public would probably accept the low level of risk from derailments, if it was sure that the track was well maintained and that all staff were doing their jobs properly. It is impossible to guarantee no broken rails. The public would probably trade a small risk in exchange for fast reliable service.

Regulation – distorting the balance of judgement?

Few would deny the need to have an effective and independent safety organization. The safety organization has to ensure that the rail system meets the standards. What are the appropriate standards? They are stated as reducing risks to as low as reasonably practicable (ALARP). The operator, knowing that their judgement will be challenged, is tempted to move their balance of judgement towards more safety – probably at the expense of other features or cost (21). Then it will be easier to get through a safety assessment. Safety assessments can take precious time and progress can be likened to taking a walk through deep mud. They can generate mountains of paper and quickly become bureaucratic.

So the dynamics of well-intentioned safety staff looking for ALARP can insidiously pressurize the applicant to distort their best judgement. It takes very strong commitment on both sides to keep a rational focus on the true requirements.

The rail industry has had a long-standing approach to using standards. Having standards accepted by all parties is a sound way to manage safety. Safety is best built in from the start of a project. It is far cheaper to design in safety than to try and add it later.

Risk management is based on reducing risk to within a tolerable level or better. The accepted level is usually based on a given cost to be spent for each life saved. Following an accident, there is a tendency for the Regulator to demand the 'extra mile'.

During the rail chaos after Hatfield, Railtrack executives pleaded with the PM to instruct the HSE to reduce their safety demands. They argued that there had only been six fatalities in the last 30 years, including the four at Hatfield. Had they extended the timeframe to 33 years, they would have had to include the 49 fatalities at the Hither Green derailment. A slightly different picture. Nevertheless, it is true that fatalities from recent derailments are very low. But it is unrealistic to expect any politician to overrule a technical decision of a national safety organization.

One danger from the current process is that safety gets relegated to a separate group to generate all the safety documentation, and deal with the interface with the Regulator. This puts an extra interface between the safety regulator and the operations. The emphasis then is on generating the paper and getting approval. The danger is that the quantity of safety-related paperwork detracts from the quality. Management can believe that the safety issue has been delegated and can lose sight of the broader safety perspective – and the need for a comprehensive safety culture.

What else is there to know on wheel/rail interface?

A paper by Professor R.A. Smith, *Rolling Contact Fatigue of Rails: what remains to be done?* summarizes the ongoing issues (**5**). These include getting a better quantification of wear and initial crack development, the impact of fluid entrapment in the cracks, and residual stresses from manufacture, and the laying of continuously welded track. Further work is needed to understand the relationship between laboratory tests and field conditions. Professor Smith emphasizes the need to develop improved inspection methods.

This section has concentrated on the fatigue cracking of rails. What are the other issues at the Wheel/Rail interface? Similar problems to those described for rolling contact fatigue in curves can apply to tangent track, points, and crossings. The track can be damaged at road/rail crossings and by operational accidents. This damage may initiate surface cracks. The Wheel/Rail interface includes the fasteners, sleepers, and track bed. Flooded and frozen ballast can affect rail loads.

There is more to the Wheel/Rail interface than meets the eye. Without a background in railways, it is hard to imagine that the apparently solid rail can be so easily broken by fatigue and end up like the 200 pieces of the Hatfield track.

1.4.13   Lessons learned
Failure to recognize and apply the available relevant information
It is ironic that a BR engineer said these words, albeit referring to design. The words are equally applicable to manufacture, operation, maintenance, and management.

Managing a large engineering project usually involves a multitude of technologies. Managers have to know that relevant information is being used and check that the expertise is available to recognize relevant information.

- Take the time to find and recognize the available information that may be relevant to the task in hand.
- Search out people who might help you in recognizing relevant information that may not be in your field of expertise.
- Apply the relevant information.

Is this information relevant?
There appears to have been the view in some quarters that heavy-haul experience was not relevant to the UK rail system. The warning bells should ring when you hear someone say *that's not relevant for us.* The question that should then be asked is a*re you sure?*

- Do not be quick to say that information is not relevant. Stop and make sure.

Look for the next issue
It is easy to feel elated when a big improvement or solution has been found to a long-standing problem. The danger is that the next limit

may catch you unaware. Improvements to rail wear brought fatigue to the forefront as the main source of failure.

- Once one problem has been solved, look for the next issue so that it can be managed before it becomes a problem.
- Keep asking – what else can go wrong? It may sound to be taking a negative attitude, but it may save you from a failure.
- Be aware of those things that can lead to a change, such as, train service frequency, train mass, speed increases, and changes to maintenance practices for both trains and track.

Knowledge is with people
Knowledge is with people not organizations. The break-up of BR scattered knowledge to many new companies and out of the industry. The knowledge was then not available when it was needed.

- Know what knowledge is necessary for the task in hand, and make sure that people with that knowledge are available.
- Do not break up a group of knowledgeable people until you are sure that you either do not need that knowledge any more, or are sure that you can get replacement skills.
- Learn where the knowledge is available.

Knowledge is worldwide
Even in the early days of railways, the technical knowledge soon evolved in many countries. Today there is a mass of published data. The Internet is awash with raw data. The problem is to be able to find the gems of relevant knowledge without wasting too much time.

- Look to see what relevant knowledge is available from other countries.
- Take time to find good sources.
- Learn how to search for information quickly.

Share knowledge
Knowledge is a two-way street. If you want knowledge, then you should also be prepared to share knowledge. There is always the question of proprietary knowledge to be considered. Many industries have found more benefit from sharing knowledge than keeping it proprietary.

- Take the time to consider what knowledge you can share with others. Think who could benefit.

- There can be a lot of value in attending conferences and technical meetings. Contribute to the exchange and join in the discussion.

Lessons from maintenance holidays

The UK rail problem is a classic example of the disastrous consequences of delaying maintenance. Unfortunately, one can see too many similar examples in other industries and public works. Maintenance holidays are usually a gamble – and not a good idea.

- If you are tempted to take a maintenance holiday, remember the photograph of the shattered Hatfield rail.
- Quantify the risk of taking a maintenance holiday. With this information, it will be easier to argue the case for not taking a maintenance holiday.

Lessons from railway safety regulation

- Manage the safety issue as an integral part of the business. Work with the regulator to establish and agree standards as early as possible.
- Be vigilant and continue to evolve a safety culture. The safety culture needs to be consistent and understood amongst all those involved in the work.
- Work to achieve a consistent understanding of the safety standards and practices throughout the industry, and from top to bottom.
- Communicate, let the public know the true facts. After all it is their risk.

References
(1) **Timoshenko, S.P.** (1953) *History of Strength of Materials*, Dover Publications, New York, ISBN 0 486 61187 6.
(2) Wheels on rails – an update, understanding and managing the wheel/rail interface, *IMechE Symposium*, 23 April 2002.
  (2a) **Kapoor, A.** *Shakedown Theory and Rolling Contact Fatigue*.
  (2b) **Evans, E.** and **Iwnicki, S.** *Vehicle Dynamics and the Wheel/Rail Interface*.
  (2c) **Dobell, M.** *London Underground's Experience at the Wheel/Rail Interface*.
  (2d) **Clark, S.L.** and **Dembosky, M.A.** *Rail Head Checking on the British Railway System*.
  (2e) **Eickhoff, B.** Lifetime Issues.

(2f) **Watson, C., Tremble, G.** and **Schmid, F.** *Managing the Wheel Rail Interface: Operational Issues in Wheel Life Management*.

(2g) **Doherty, A.** Keynote Address.

(2h) **Wasserman, S.** *Monitoring the Wheel and the Rail*.

(3) **Carter, F.W.** (1926) On the action of a locomotive driving wheel, *Proc. R. Soc. London*, **A112**, 151.

(4) **Johnson, K.L.** *Contact Mechanics*, Cambridge University Press, Cambridge, Reprint 1987 ISBN 0 521134 796 3.

(5) **Smith, R.A.** (2001) Rolling contact fatigue of rails: What remains to be done? *Proceedings of the World Congress on Railway Research*, Cologne, November 2001.

(6) **Kalousek, J.** and **Magel, E.** (1997) Achieving a balance, The 'magic' wear rate, *Railway Track and Structures*, March 1997.

(7) **Smallwood, R., Sinclair, J.C.** and **Sawley, K.J.** (1987) An optimisation technique to minimise rail contact stresses, *Wear*, 373–384.

(8) **de Vries, R., Sroba, P.** and **Magel, E.** (2001) Preventative grinding moves into the 21st century on Canadian Pacific railway, *AREMA Conference*, Chicago, September 2001.

(9) **Stanford, J., Sroba, P.** and **Magel, E.** (1999) Grinding helps handle heavy haul traffic, *Railway Track and Structures*, December 1999.

(10) **Magel, E.** and **Kalousek, J.** (2000) The application of contact mechanics to wheel/rail profile design and rail grinding, *5th International Conference on Contact Mechanics & Wear of Wheel/Rail Systems*, Tokyo, July 2000.

(11) **Hiensch, E.J.M., Kapoor, A., Josefson, B.L., Ringsberg, J.W., Nielson, J.C.O.** and **Franklin, F.J.** Two-material rail development to prevent rolling contact fatigue and reduce noise levels in curved track, *Proceedings of the World Congress on Railway Research*, Cologne (2001), updated at the Edinburgh Congress in 2003.

(12) **Wolmar, C.** (2001) *Broken Rails – How Privatisation Wrecked Britain's Railways*, Aurum Press, ISBN 1 85410 857 3.

(13) **Kondo, K., Yarazaka, K.** and **Sato, Y.** Cause, increase, diagnosis, countermeasures and elimination of Shinkansen shelling, *Wear*, **191**(1–2).

(14) **Muster, H.** *et al.* (1996) Rail rolling contact fatigue. The performance of naturally hard and heat-hardened rail, *Wear*, **191**(1–2).

(15) Grassie, S.L. (2001) Preventative grinding control and rolling contact fatigue defects, *Int. Railway J.*

(16) Roney, M.O. and **Meyler, D.K.** (2001) CPR's long-term strategy pays off, *Int. Railway J.*

(17) Palese, J.W. and **Zarembski, A.M.** (2001) BNSF tests risk-based ultrasonic detection, *Railway Track and Structures*, February 2001.

(18) *Railway Gazette Int.*, **157**(6), 2001.

(19) Nield, B.J. (1988) *Avoidance of Failures*, Joint Materials and Metallurgy Trust and IMechE, ISBN 0 44619 653 6.

(20) #10 Newsroom, 9 November 2000, www.number-10.gov.uk.

(21) Sixsmith, E. (2002) Achieving safety on the rail networks, *Delivering Safety in Changing Times Symposium, IMechE*, 17 April 2002, London.

Fig. 1.5.1   60 MW turbine generator

Fig. 1.5.2   The damage to the Uskmouth turbine

A picture is worth a thousand words. This photograph is easy to remember so that its lesson is not forgotten.

1.5.1   The failure

On 18 January 1956, the number 5 turbine at the Uskmouth Power Station suffered a catastrophic accident. Two men were killed and nine injured. The turbine generator was one of six identical 60 MW units in the turbine hall (Fig. 1.5.1).

Damage was considerable (Fig. 1.5.2). The low-pressure rotor had been hurled out of the building and ended up several hundred yards from the station.

1.5.2   Circumstances surrounding the failure

The machines were in the process of being transferred from the suppliers commissioning staff to the utilities operations. The first four units had been in service for some time. Unit 5 had been operating at full load for 60 h. Unit 6 was being shut down for adjustments following commissioning tests.

The exciter for unit 5 was accidentally tripped instead of unit 6. With the load removed, unit 5 rapidly increased in speed. In a few seconds it disintegrated.

1.5.3   What should have happened?

With the tripping of the load from the generator, the solenoid trip in the turbine-governing gear should have operated. It was designed to activate the oil control system to shut all steam valves. While the steam valves are closing, there will be a transient speed increase. It should not exceed seven per cent of the normal speed of 3000 r/min.

1.5.4   The investigation

Immediately following the accident, a *Committee of Investigation* comprising all the involved parties was established. Experts were invited to attend. Everyone was given unrestricted access to all the findings and evidence.

The investigators faced the surprising fact that there had been no escaping steam when the unit disintegrated. This initially suggested that the steam valves had operated and that the failure might have been caused by a metallurgical failure of the low-pressure turbine under the small overspeed transient during the trip.

The governing and tripping mechanisms of the wrecked machine were badly damaged in the accident and could not be tested. However, this equipment on the other units was tested and no design, operating, or construction faults were found.

How fast had the turbine rotated?

The next step was to determine how fast the turbine had rotated. The low-pressure turbine was heavily damaged by being thrown out of the turbine hall. However, the high-pressure turbine was still in its casing and provided valuable evidence.

Yielding in the bore of the high-pressure wheels gave a clue that the rotor had reached a high speed. The next evidence came from the rivets holding the shrouding on the blades. They had been progressively sheared. Tests on similar rivets, together with calculations showed that the high-pressure rotor had reached speeds of around 5000 r/min.

The next question was whether the high-pressure rotor had uncoupled from the low-pressure rotor before it had reached this speed. This might have allowed the high-pressure rotor to reach higher speeds than the low-pressure rotor. Examination of the debris from the low-pressure turbine showed that some wheels had grown by 3/16 in. This showed that the whole machine had reached a speed of around 5000 r/min – about 70 per cent overspeed.

Suspicion returns to the governing and tripping mechanism

The steam valves would have to be open for 13 s after the exciter was tripped for the turbine to reach 5000 r/min. The question was *why had the steam valves stayed open for these 13 s before closing?*

The reason was that the primary governing relay piston had seized in its bore. The shock from the break up of the machine must have jolted the piston so that it eventually moved.

Hard black oxide deposits had built up on the piston and in the bore. Examination of the oil showed the presence of extremely fine particles, about 0.05 microns.

Where had the black oxide come from?

The black oxide came from inside the oil pipes. The oxide could have been produced by contamination of the oil with a small amount of saline water, combined with a long-period shutdown. The saline water had probably entered the system from a small leak in an oil cooler. During the commissioning delays of several months, the oil system had been drained. Small pockets of contaminated oil were then in contact with air in the pipes. This produced conditions that were favourable for producing magnetite. On refilling the oil system, the magnetite was pushed into the relay valve piston. The small movement of the piston in its bore compacted the fine particles onto the moving surfaces until they seized when called upon to make the larger movement required to trip the set.

Had there been any warning signs of problems?
The operation and maintenance records showed that some stickiness of the relay valves had been reported. Maintenance reports said that the pipe scale had been removed from the secondary relay cylinders.

Operating staff commented that the speed-changing response of this turbine had been sluggish, and it had been getting worse in the weeks before the accident.

Conclusion of the investigation
The investigation concluded that the formation of black oxide, due to the unusual conditions to which the unit had been subjected, had contaminated the oil and resulted in the seizure of the governing valve. The Committee of Investigation concluded that the *cause of the disaster and the phenomenon is believed to be unique.*

1.5.5   The technical paper and discussion
A full description of the investigation of the accident is given in the IMechE paper by Sir Arnold Lindley and Sir Stanley Brown presented at a meeting of the Institution on 21 March 1958 (1). Following the presentation, there was an interesting discussion.

Opening the discussion, Dr Forrest commented that *the decision to publish it at all was a very wise one.* He had known that companies adopt a conspiracy of silence about their troubles and failures. *But, surely, it could only be in everybody's benefit in the long run if a frank objective statement of the findings after a failure were published and openly discussed.*

Another speaker pointed out that an accident like this gave the opportunity to learn a lot from examining equipment that had been exposed to extreme conditions. Examining overstressed components could give some indication of the accuracy of design calculations.

The authors commented on the investigation, and the difficulty of getting reliable evidence from witnesses. *There was considerable conflict in the information given by those who were present, and a lapse of a period of a day following the accident was sufficient time to affect a person's memory as to what had happened.*

The authors also reported that since writing the paper, the oil coolers were found to be leak tight. They had to assume that the contamination came during the washing of some components with contaminated liquid. Corrosion is very dependent on the detailed conditions. Small chemistry differences can change the type of corrosion product.

COMMENTS
Rotating equipment
A basic engineering concern is to beware of the energy stored in rotating machinery. The rotor of this 60 MW unit had a lot of energy when it took off. Today's steam turbines have outputs 20 times greater than this 1950s machine.

Design failure?
A turbine should not explode in service. Despite the bland conclusions of the investigation, the design failed to prevent the overspeed. The Uskmouth turbine, in common with established practice, had two systems that should have prevented overspeed. Since the 1950s, there have been great advances in safety engineering. The philosophy of independent protection systems has been well developed. Design practice now ensures less chance of common mode failures – and systems are designed to take account of degradation due to age or use.

   The aircraft and nuclear industries have led the development of safety engineering. Some of their solutions may be considered expensive and over-elaborate for general engineering application. Nevertheless, some of the concepts can be applied to improve the safety of more routine engineering. This is an example of the value of keeping abreast of developments in leading industries to see whether they can be applied in other work.

The importance of cleanliness
While the circumstances of this accident were unique, it is worth asking *are there other situations that can lead to a similar disaster?* Any fine dirt in oil control systems can cause a similar problem, and highlights the need for clean conditions around equipment with fine tolerances. This includes most oil or air control systems, as well as bearings and fine mechanisms.

   Filters are one way to protect against dirt. But oil in dead ends, such as the space in front of a piston, does not circulate through the filter. Dirt just gets pushed backwards and forwards in front of the piston, always at risk of being trapped between the piston and the bore.

The role of corrosion
In this example, corrosion in hidden parts of the oil system created the fine black oxide. Corrosion can be the bane of an engineer's life. Atmospheric conditions as well as many substances can cause corrosion on commonly used materials. The designer has to think of all the circumstances the equipment might be subjected to in storage and

in service. All those involved in manufacture, storage, transport, and operation have to watch for situations where corrosion might occur.

Getting equipment into service

The stage from assembling an item such as a turbine to getting it into routine service can be critical. The erection and commissioning engineers have to be aware of what the designer was aiming to achieve so that any problems or shortfalls can be highlighted for resolution. Once the equipment is handed over to the operation and maintenance staff, the responsibility is theirs. They need to know the idiosyncrasies of the equipment. This requires good communication between all parties.

When equipment is going through the test and early operating phase, it is likely that the designers are off on some other task. Unfortunately, there are many pressures that keep the designers from visiting sites. Yet, this is the time to observe the result of their work. To see for themselves, and hear the views of those who have been testing, operating, and maintaining their equipment. This personal presence can be so much more fruitful than just reading the reports and seeing photographs. Talking to all the people, from fitters to engineers, is always a great learning experience, and is cost effective. The cost is trivial compared with the capital costs and the cost of any problem.

Warning signs

There were warning signs of problems with the governing equipment. These warnings were not given sufficient attention. It is not clear whether the *stickiness* and *sluggish* behaviour was communicated to the suppliers' engineers on site, and in turn to the designers. Engineers tend to hate paperwork – other than drawings. Most projects and equipment runs smoothly and so most paperwork reporting is mundane, and contains little of concern. How do you highlight the warning signs of problems? Part of the answer has to be with those generating the reports. They have to highlight concerns without falling into the trap of crying wolf. Another part of the answer is for the designer to identify what is critical, and the importance of various aspects of the design to safe operation.

Reporting problems

Vibration problems were being experienced on a turbine at another power station. The station was also in the stage of commissioning, with utility and contractor staff working together. The turbine operator on the night shift filed his report before leaving site. He reported vibrations in one of the turbines. He graphically reported the situation by writing in his report that when he stood on the bearing pedestal, the vibration *made his balls jangle*. Before lunchtime next day, everyone on the site knew about the problem. It got prompt attention.

Juxtaposition of control panels
Uskmouth turbine 5 was accidentally tripped instead of number 6 because the two control panels were close together. The location of control panels should not result in confusion on which unit is being controlled. However, the oxide in the governor equipment of the number 5 machine was a latent problem waiting to surface.

1.5.6 Lessons learned
Photographs can help recall a lesson

- Photographs of failures can be a quick and telling way to remind people of a potential risk.

Lessons for design
When equipment fails, there is nearly always some blame to be laid on the designers. Sometimes this can seem harsh. A first reaction is to say *they were not supposed to do that* and hence it is not a design issue. In the ultimate, if equipment fails, then the designers did not meet their goal.

- Always think about how the equipment can fail and make sure that effective protective mechanisms are in place.
- Beware of the effects of dirt and corrosion.
- Beware of the stored energy in rotating machinery.
- Make sure there can be no confusion between the control system and the equipment it is controlling.
- Keep abreast of developments in other industries.
- Learn from other industries.
- Make sure those using, building, and maintaining the equipment you designed know the essential data and understand the limits.
- Go look, see, and feel how good your design is performing in the field.

Good communications are essential
This disaster highlights the need for good communication. Good communication and good working relationships between all parties increases the chance that warning signs are noticed.

- Make sure your reports get your key messages across.
- Always be alert for warning signs of potential trouble.
- Be well informed and keep others well informed.

Be open and discuss failures

Despite the concerns that people may have, history shows great benefits by openly reporting failures and being prepared to discuss them.

- Take part in the discussions of failures.
- Be prepared to share your own experiences.

Reference

(1) **Lindley, A.L.G.** and **Brown, F.H.S.** (1958) Failure of a 60MW steam turbine at Uskmouth power station, *Proc. IMechE*, **172**(7).

Dr Richard Feynman and the Challenger Shuttle Inquiry

A lot has been written about the *Challenger* disaster (**1**). Documents on the disaster, held by NASA, fill a two-storey warehouse. The aim here is to see what lessons can be learnt from the experiences of Richard Feynman during the Shuttle Inquiry.

1.6.1 The Presidential Commission

On January 28, 1986, the Space Shuttle *Challenger* exploded shortly after launch, killing seven astronauts. President Reagan formed a Presidential Commission. A former Secretary of State, William P. Rogers, was appointed the Chairman. The Commission had 12 members: politicians, astronauts, military engineers, and one pure scientist – Dr Richard Feynman. The NASA Acting Administrator, Dr William Graham who had been a student of Dr Feynman, recommended him for a position on the Commission.

1.6.2 Dr Richard Feynman (1918–1988)

Throughout his career, Richard Feynman showed a missionary zeal in his search for an understanding of the basic building blocks of science. He rubbed shoulders with scientific leaders like Einstein, Pauli, von Neumann, and Bethe as he progressed through MIT, Princeton, Los Alamos, Cornell, and Caltech. His work on quantum electrodynamics earned him a Nobel Prize in 1965. One of his achievements was the development of simple diagrams to describe the complex behaviour of subatomic particles.

Dr Feynman had a persistence to get to the truth (**2**). He valued honesty, independence, the willingness to admit ignorance and to learn from others. The techniques of scientific investigation were important to him, the acceptance of uncertainty – the courage to doubt. He liked to search out all the facts and views, then debate and critique the ideas before coming to a conclusion. He enjoyed sharing his passion for knowledge with fellow scientists, friends, and people in all walks of life. His writings show him as a warm and caring individual, with a great sense of fun.

1.6.3   Culture clash

Feynman's description of his days on the Commission in Washington are interesting and amusing (**3**). There was a cultural divide between his basic scientific approach and the legal/political style of the Chairman. Despite Feynman's Nobel Prize, his extensive travelling, and his experience of debate with the brightest of minds, he found the methods of Washington to be a new world. The Chairman was determined to run a politically correct process and aimed to control the proceedings and activities of the Commission. This clashed with Feynman's style of unrestrained and objective searching for evidence, and open debate.

1.6.4   The working methods of the Commission

The Commission had public meetings wherever possible. This was to avoid leaks to the press that might be seen as signs of a cover-up. The Commissioners were given briefing books and set piece technical presentations. They were issued with a dictionary of acronyms – a very thick book. Feynman was unfamiliar with the overhead style of presentation, as well as the propensity to use bullets. He hated being steered through a prepared presentation. He wanted to explore information through a detailed question and answer session.

Feynman quickly got into the details. When questions were outside the presenter's expertise, the standard answer was *We'll get that information to you later*. Feynman was frustrated by the responses, which came packaged in a file. You could track the question being passed down through the hierarchy. The answer was sandwiched between the notes passing it back up the line. The process often resulted in the response not answering the original question, although useful information was often included.

Persistence paid off, and Feynman talked directly to the engineers and technicians with first-hand practical experience, as well as with designers. His enthusiasm and personal commitment appears to have overcome any reluctance on the part of the people he met to be open and frank.

Feynman was shocked to see the Chairman distracted in Commission meetings by secretaries bringing in papers to be signed.

1.6.5   The Space Shuttle and its solid booster rockets

Two solid booster rockets (SBRs) are attached to the launch platform. They support the external fuel tank, which holds the liquid oxygen and liquid hydrogen fuel. The shuttle is mounted on the fuel tank between the two SBRs (Fig. 1.6.1).

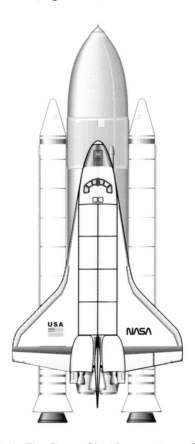

Fig. 1.6.1   The Space Shuttle, courtesy of NASA

The SBRs are 149.2 ft long, 12.17 ft in diameter and weigh 1 300 000 lb at launch. Each has a thrust of 3 300 000 lb. The SBRs provide just over 70 per cent of the thrust at lift off. In principle, the SBRs are similar to the successful Titan rocket. However, the SBRs are larger and more powerful. Other differences are that the SBRs are recoverable, reusable, and operate at one-third higher pressures.

Thiokol, later to be merged with Morton to become Morton–Thiokol (and now ATK Thiokol), was awarded the contract for the SBRs in 1973. NASA accepted the design in 1976. Thiokol's factory is in Utah. The SBRs are transported to the Kennedy Space Center (KSC) in sections, by rail. The diameter of the sections is fixed by the size of the rail tunnel en route. The rocket motor sections, filled with propellant in Utah, are assembled at KSC with field joints.

1.6.6 The SBR field joints

The field joints were based on the Titan rocket design, but upside down. A secondary seal was added to provide an interspace to leak check the primary seal and provide redundancy (Fig. 1.6.2).

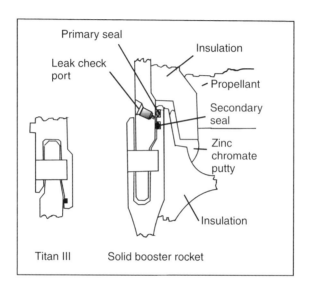

Fig. 1.6.2   Field joints

From the start of the design, Marshall Space Flight Center (MSFC) engineers had trouble with Thiokol on the joints (**4**). There were ongoing reviews, disagreements, dissension, and some changes – but with no great comfort on either side.

In hydro-burst, over-pressure tests in 1977, the rocket structure withstood the test – but the seal failed. A gap opened as the rocket pressurized. This was because the wall is thinner than the joint section and expanded more under pressure. The two sides of the joint rotate under the bending loads (Fig. 1.6.3). At 1004 psig, the gap opened by 0.042 in to 0.060 in. The designers had thought that the pressure would force the joint together. However, the MSFC engineers had expected the joint to open. At the time, the Chief Rocket Engineer at MSFC wrote, *I personally believe that our first choice should be to correct the design in a way that eliminates the possibility of 'O' ring clearance.* There was no redesign, but the seal diameter was increased slightly, tolerances tightened, and shims wedged into the joint to maintain concentricity.

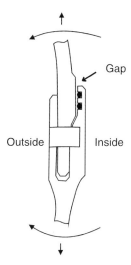

Gap

Outside   Inside

Fig. 1.6.3   Joint rotation under pressure (exaggerated)

During 1978–80, three qualification motors were fired and there were no leaks.

1.6.7   Putty
There is a layer of insulation between the propellant and the steel rocket wall. A continuous thermal barrier across the field joint is provided by zinc chromate putty.

In 1983, NASA had to make a change that, in hindsight, exacerbated the seal problem. The original supplier discontinued making the putty. They were concerned that the asbestos content would create potential liabilities. The replacement putty, from a different manufacturer, also had asbestos filling, but in smaller quantities. The original putty was sticky and adhered well. The replacement putty was more difficult to pack, became harder at lower temperatures and had less thermal protection capability.

1.6.8   Seal test pressure
After seven static firings and nine launches, 'O' ring problems only occurred in four joints. The decision was taken to concentrate on the leak test procedure to prove that the primary seal was properly

seated. The leak test in the interspace between the primary and secondary 'O' rings was initially 50 psig. The putty might seal at this pressure and mask a defective seal. The test pressure was raised, first to 100 psig and then 200 psig. Unfortunately, testing the 'O' rings at the higher pressures could allow some initial gas flow to blow pencil-sized holes through the putty. At launch, this let hot gas reach the 'O' ring. The impact of this was not appreciated until much later. The statistics were:

Test pressure (psig)	Number of flights	Flights with seal anomalies
50	7	1
100	2	0
200	15	8

1.6.9   Anomalies and erosion

In the records, the word *anomalies* was used to describe seal damage. *Anomalies* included hot gas passing the primary seal and erosion of the seal. Rockets were launched successfully, while the seals showed *anomalies* on inspection after recovery. This led to *anomalies* becoming acceptable. A new criterion was established to define acceptable erosion. Estimates were made of gas flow rate and heat input. Experiments were conducted. In these tests, up to 30 per cent of the seal eroded. The conclusion was that an adequate safety margin was present.

1.6.10   Preparation for the launch

Prior to the January 1986 launch, there had been 24 successful launches. Planning for this mission started in 1984. The initial schedule was to launch in July 1985. This date was postponed to accommodate changes in the mission payload.

NASA has a detailed Flight Readiness Review (FRR) process before each flight. In this process, all questions are addressed, and a check is made that the responsible managers in NASA have certified all equipment.

The launch was delayed three times. Finally, the launch was scheduled for the morning of 28 January. The overnight temperature prior to launch was predicted to be below freezing.

1.6.11   Raising concerns about the low temperature

No questions had been raised at the FRR about the field joint seals. When it became known that the launch would take place in unusually cold weather, a manager at MSFC in Huntsville asked the local Morton–Thiokol engineer whether Morton–Thiokol had any concerns. The local engineer contacted his managers in Utah. A conference call was organized between Utah, Huntsville, and Kennedy, connecting staff of Morton–Thiokol, MSFC, and KSC.

In the first conference call, the Morton–Thiokol engineers expressed their concern at the low temperatures. The managers recommended that the launch not take place until the temperature was equal to or greater than 53°F. This was the lowest temperature from previous launches.

NASA did not like creating a new launch criterion on the eve of a launch, especially after the FRR had approved the launch. They questioned Morton–Thiokol's recommendation. Morton–Thiokol managers caucused with their engineers. The debate focused on how strong the evidence was that the seals would fail under these conditions. The engineers stuck to their recommendation not to launch at low temperatures, but the information they presented was not conclusive. It did not show significantly worse risk at lower temperatures. The managers felt they did not have enough firm facts to stop the launch. As a result, they formally approved the launch.

1.6.12   Accident sequence

The rocket sections had been assembled at KSC. The sections were transported by rail on their side. This sometimes produced out-of-roundness due to the weight of propellant. In this case, the aft rocket section was 0.393 in out of round and had to be jacked into a better geometry to fit the mating part. The nominal gap between the mating parts at the factory was 0.008 in.

At the time of launch, the sun was shining on the right SBR. The seal temperature in the sun was estimated at 53°F. Where the sections had to be jacked to fit was in the shade, and the temperature was estimated to be 28 ±5°F. This was also near the lower aft strut joining the SBR and external fuel tank. It was at this point that the leak occurred.

Evidence from previous launches had shown that water could enter the joint. There had been seven inches of rain during the 38 days that *Challenger* had been on the pad. The overnight temperatures would have turned the water to ice.

Video showed smoke puffing out of the lower aft field joint between 0.7 and 3.4 s after launch. The first sign of a flame was at 58.7 s. At 60 s, the flame impinged on the liquid fuel tank, which breached at 64.7 s. The lower aft strut failed at 72 s.

The launch was in gusty winds that exceeded previous flight conditions in the subsonic region. These gusts were estimated to be well within structural design limits, although they would have increased the joint opening.

1.6.13   Dr Feynman at the inquiry

It was largely through his separate discussions with engineers that Dr Feynman homed in on the core of the problem. The problem was trying to seal a widening gap at low temperatures when the 'O' ring was not very resilient.

Feynman became frustrated when the obvious answer was not being brought out clearly in the meetings. One of the meetings was planned as a public session with the media present. Dr Feynman decided to conduct a simple experiment. He would clamp a piece of 'O' ring and cool it in a glass of iced water. Once it was near freezing temperature, he would remove the clamp and show that it took some time for the 'O' ring to return to its round shape.

The media quickly latched onto the unusual experience of seeing an experiment during an inquiry. The story broke that the shuttle crashed because the launch temperature was too low for the seal.

1.6.14   Dr Feynman and Roger Boisjoly

Roger Boisjoly was the seal engineer at Morton–Thiokol. He had been concerned for some years about the seal design. He was one of the engineers who recommended that they wait until the temperature increased before launching. Dr Feynman had several discussions with him (5), both in the formal meetings and in separate discussions. Feynman saw the difficulty that Boisjoly was in during the pre-launch conference call. He was being pressed to give proof that the seal would fail at the low temperatures. Feynman pointed out, and Boisjoly agreed, that this was not possible. Seals had worked on some occasions at low temperatures.

1.6.15   Figures of fantasy

During his search for information, Dr Feynman heard a variety of views on the probability of failure. He explored these for two components, the SBR and the liquid fuel engine.

Feynman heard from the Range Safety Officer that out of 2900 firings of rockets 121 had failed; a probability of roughly one in 25. This figure included early problems. With more mature technology, the figure may be 1 in 50, and with special attention, 1 in 100. The official NASA figure for the SBR was one in 100 000. This appears to be a judgement following a string of successes rather than a reliability analysis. Feynman saw this as fantasy, and that belief in the fantasy led NASA to playing Russian roulette.

On the shuttle's main liquid fuel engines Feynman heard a long list of technical problems. In discussions with engineers from MSFC, he got an estimate of the probability of failure around one in 300. The manufacturer's estimate was one in 10 000, while NASA claimed one in 100 000.

This range of numbers was beyond belief. He thought the greater risk to be nearer the truth. This was a judgement rather than a scientific assessment.

Dr Feynman concluded that NASA officials should *deal in a world of reality in understanding technological weaknesses and imperfections well enough to be actively trying to eliminate them.*

1.6.16   Dr Feynman and the report writing

Dr Feynman thought it correct to write reports of his findings to aid the work of the Commission. He asked the secretariat to distribute his reports to other Commissioners. It infuriated him to find that his reports had not been circulated until he repeated his request. He also did not like wordsmithing taking precedence over a real discussion of ideas.

There was agreement on the main conclusion. The accident was due to the failure of the 'O' ring seal, caused by a faulty joint design that was unacceptably sensitive to a number of factors.

1.6.17   The recommendations

A tentative list of recommendations was tabled. Feynman felt that he was being railroaded – there was not enough debate and thinking for his liking. For example, he felt that recommending new safety boards was inappropriate before discussing the effectiveness of the current boards.

Eventually nine recommendations were agreed upon. The first is that the faulty seal must be changed, by redesign or eliminating the joint.

A 10th recommendation was floated, recommending that NASA continue to receive support from the Administration and the nation. Feynman did not like the way this recommendation had been brought up, nor did he consider it consistent with the other recommendations. Negotiations failed to achieve what he wanted. He formally asked for his signature to be taken off the Report unless the tenth recommend-ation was removed, and his own report included without editing. This action produced the results. The tenth recommendation became the Concluding Thought, and Feynman's report became Appendix F of volume 2. However, Appendix F was not included in the initial release of the Commission's Report (**6**). It was available several months later.

1.6.18   Dr Feynman's afterthoughts

In his book, *What do you care what other people think* (**3**), Dr Feyn-man presents some of his thoughts from after the Commission's Report had been handed to President Reagan. He acknowledged that he must have irritated Chairman Rogers.

COMMENTS

A summary of the failure

The seal was a disaster waiting to happen, with so many factors that could lead to a failure. The diagrams in Fig. 1.6.4 summarize the situation. The first diagram shows the seal before launch. With the combination of 'O' ring size, fit between the SBR sections, and the tolerances, it is likely that the seal was in contact with all four sides a, b, c, and d. It was probably compressed between 27 and 33 per cent.

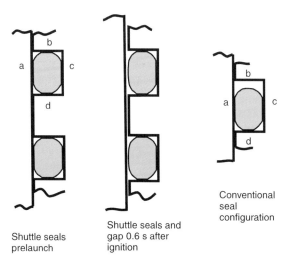

Shuttle seals prelaunch

Shuttle seals and gap 0.6 s after ignition

Conventional seal configuration

Fig. 1.6.4   'O' ring seal

At 600 milliseconds after launch, the gap to be sealed would have fully opened because of the build-up of internal pressure to 770 psig from the rocket firing. The second diagram shows both the gap due to joint rotation and the amount the seal will have relaxed from its initial shape at 28°F. It then depends on the dynamics, the rate of build-up of pressure in the seal space, and Bernoulli effects of the hot gas escaping through the gap to determine whether the seal moves over to seal.

The dynamics of gap opening and seal relaxation

Plotting the gap opening and the seal relaxation data together (Fig. 1.6.5) shows that the primary seal would need temperatures over 75°F to relax enough to seal the gap. The secondary seal would relax enough to seal at temperatures above 55°F. This is oversimplified because it does not take into account the fact that the seal could not start to relax until the joint opens. The overlay shows that the rate-of-gap-opening is faster than the rate-of-seal-relaxation at any of the temperatures plotted. The conclusion is that there will always be a transient gap allowing leakage until the seal relaxes.

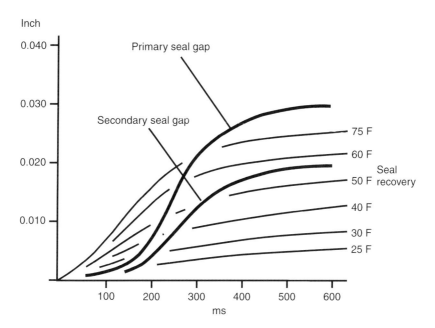

Fig. 1.6.5   The calculated seal gap opening overlaid on the measured seal recovery from 0.040 in compression

Only one level of initial compression is shown in Fig. 1.6.5 and this could vary with tolerances on the groove and seal. Also, it ignores any gas flow or pressure effects. Tests conducted after the accident showed that at pressures above about 200 psi, the pressure forces the seal into the groove rather than seal the gap. The pressure build-up at the seal is dependent on the internal rocket pressure. Holes in the putty can affect the rate of build-up of pressure seen by the seal. These dynamic effects are not easy to predict and will affect whether the seal moves to seal or is pressed further into the groove. Debris from the putty or propellant may block the leak path, at least temporarily.

The operation of the seal was unpredictable. So many variables and dynamics affected the performance. While low temperatures had an adverse effect, other variables could combine and produce a seal failure at any temperature. There was no redundancy in most situations.

Six seconds after ignition, the bolts holding the rockets to the launch pad are released. The twang on lift off, and other dynamic loads from the struts increases the seal gap a further 0.025 in.

In retrospect, it was only a question of how many launches took place before a sequence of coincidences resulted in a failure. It would be hard to produce a design, where the chance of sealing was more random.

Normal application of 'O' ring seals
A normal 'O' ring application is shown in the third diagram in Fig. 1.6.4. It should be compressed 15 to 20 per cent. The seal should take up only 75 per cent of the groove. It should not be in contact with face b. The pressure bearing on the top face of the 'O' ring will force it against face d. This squeezes the 'O' ring, producing a load between faces a and c. This creates a pressure-assisted sealing condition. It is not normal to try to seal a widening gap.

Why have a field joint?
There were four bids in 1973 for the SBR contract: Lockheed, Thiokol, UTC, and Aerojet. Aerojet offered a jointless, monolithic structure, while all the others had segments and field joints. The Aerojet proposal warned that problems with rubber seals in field joints could jeopardize safety. In the overall assessment, NASA found the Aerojet proposal wanting in other aspects, and it came last in the total evaluation of technology, manufacturing, and management.

In hindsight, it is a pity that the jointless concept was not pursued. A good way to deal with a design problem is to eliminate the feature giving the problem.

Reporting and analysis of the disaster
As well as the Rogers Commission, there was a review by the US Congress House Committee on Science and Technology. In general, these two formal reviews agreed. Both saw that no one in NASA

or Thiokol fully understood the operation of the joint. Both said that the NASA organization was implicated. The Rogers Commission emphasized poor communication and inadequate procedures, while the House Committee felt that procedures had been followed and that information on the problem was widely available. The House Committee blamed people making poor technical decisions over a period of years.

The NASA web page (7) has a bibliography of 71 books and articles on the disaster. Searching for *Challenger* Accident on the Google Internet search engine produces 35 600 hits. Many of these are academic studies from both staff and students. They tend to concentrate on three aspects – engineering reviews, ethics, and management issues.

Engineering reviews

The majority of the short reviews covering the engineering aspects state that the cause of the accident was launching at too low a temperature. There is no doubt that this view is influenced by Dr Feynman's experiment and the media attention it attracted. It is ironic that Dr Feynman, who cherished the search for the real facts, contributed to the misconception of the proximate cause – low temperature rather than bad seal design. The full engineering story is well described on some web sites, for example, at the University of Texas site (8).

Ethics

The ethical aspects of the *Challenger* launch procedure have been adopted as course material for many academic studies. The role of Roger Boisjoly is debated. Most reviews concentrate on the events around the pre-launch conference. These events are seen as good case-study material for a debate on whistle-blowing.

A more interesting debate for engineers is the ethics of allowing an ill-conceived design to get to the production stage. Unfortunately, the last-minute conference call makes a more dramatic study.

Management problems

The discussion of the management problems at NASA has continued. Some almost go as far as charging NASA with criminal negligence. A different view is taken in the history of the Marshall Space Flight Center (4). Their view is that the Rogers Commission oversimplified complex events. The Marshall engineers and managers did not get the opportunity to tell their side of the story. They saw the Commission reading history backwards and ignoring positive information about the joint.

In 1986, **Challenger** had at least 8000 components that had been classified critical. NASA management had categorized hazards. They had 277 hazards to resolve prior to launch. With previous successful launches, the SBR field joint seals were not seen as one of the hazards. One of the managers said, *I don't think there is a single launch where some group of sub-system engineers didn't get up and say, 'Don't fly'. You always have arguments.*

The management questions raised by the Inquiry, and the concerns in Feynman's mind, have evolved into detailed management studies – and debate. Psychologists and sociologists have researched group-think, the process of decision taking and the concept of *normal* accidents.

Group-think

Motivating a team to *think positive* combined with production pressure can result in *group-think*. Trying to create a cohesive group with consistent thinking leads to striving for agreement. Agreement then takes precedent over assessment.

Irving Janis, a Yale social psychologist was fascinated by why teams of professional staff can make disastrous mistakes **(9)**. His symptoms of *group-think* are:

- *illusion of invulnerability of the group;*
- *belief in inherent morality of the group;*
- *collective rationalization;*
- *stereotype those outside the group;*
- *self-censorship to avoid appearing to criticize the group;*
- *illusion of unanimity;*
- *pressure on dissenters to conform;*
- *self-appointed mindguards to protect the group.*

Group-think creates a comfortable and seductive environment. Anyone challenging the status quo is seen as negative and difficult.

Normal accidents

Yale sociologist Charles Perrow coined the phrase *normal accidents* **(10)**. Perrow's hypothesis is that accidents are not the result of unprofessional or criminal behaviour but are caused by the system. He believes that accidents are inevitable with complex and tightly coupled systems.

Sociologists' views on the Challenger accident

Diane Vaughan, a Professor of Sociology at Boston College, has carried out the most extensive sociological review of the ***Challenger*** disaster (**11**). Her study challenges the view of the Rogers Commission that wrongdoing and production pressures on middle management led to violating their own rules. Instead, she concludes that the cause of the disaster was *a mistake embedded in the banality of organizational life.* The structure of power, and the power of structure and culture, can make mistakes more likely. She saw that the actions and motives were infinitely more complex than the simplistic analysis in most reports.

Handling masses of data

NASA staff and managers were faced with an enormous volume of information. Their means of creating some semblance of order was to institute a bureaucratic process – the FRR. It handles a mountain of data and anomalies starting at the lowest level. At each level of review, the items are disposed as either *go* or *no go*. Actions are assigned on *no go* items to make them *go*. Items cleared at one level are not raised at the next level. This reduces the amount of detail to be reviewed at the senior levels. The process was claimed to be open with all participants able to have a say. In practice, it required a lot of guts to make a challenge. Challenges to decisions had to be backed by analysis, facts, and figures. Hunches and uncertainty were discouraged. While this process can be intense, all participants were party to a *can do* culture. Change was discouraged. If something worked on a previous flight, it should be expected to work on the next. NASA became *bureaupathological*.

Structured secrecy

The FRR process transformed disorder into order with a simple singular answer – launch or don't launch. The method of distilling the data tended to produce a limited knowledge as one went up the organization. Vaughan saw this as structured secrecy. She saw a similar problem with NASA internal safety reviews. The safety organization had a limited breadth of activities. It looked for serious mistakes and did discover quite a few. They tended to rely on the record of decisions for information and so had difficulty in being an effective independent reviewer. To be thoroughly effective, they would have had to go through the same mountain of facts as the FRR process. They did not have the staff to do that. Therefore, the internal safety review process could not be guaranteed to surface problems like the field joint.

The eve of flight conference call

Vaughan sees the eve of flight conference call as far more complex than most reviewers. There were only *weak signals* in an environment where people demanded hard data before a decision could be changed. The style of the conference call prevented visual clues. Attitude, facial expression, and body language all contribute in face-to-face meetings.

Interestingly, Vaughan uncovered the fact that both NASA and MSFC were expecting the Morton–Thiokol managers to reconfirm the recommendation not to launch until the temperatures increased.

Perrow's view on Vaughan's analysis

Charles Perrow challenges Vaughan's analysis. Perrow sees NASA as a *damaged organization that allowed production pressures to override safety concerns.* The safety culture was *corrupted* and the technical culture was *suppressed* by the use of organizational power. Image management and control games prospered.

Turner's failure of foresight

A *failure of foresight* is a common thread in accidents studied by Barry Turner (12). He saw that disasters often follow a long incubation period in which faults remain latent. A period when a variety of events can signal potential danger. But over time the data are overlooked and misinterpreted. The accumulating impact is not noticed.

An engineering view on the sociologists' findings

Sociologists' research into how engineers reach decisions is a valuable area of study. It can give insight into decision taking. A couple of engineering observations can add to the assessment of the sociologists. The first is the nature of identifying and reviewing an engineering problem.

The failure to follow up on concerns raised at the design review

At the initial design review between Morton–Thiokol and MSFC, the issue was to understand the design and assess it. The fact that Morton–Thiokol thought that the joint would close on pressurizing should have been a warning that they did not know how the joint worked. The MSFC staff had correctly seen that a gap would open. It seems incredible that Morton–Thiokol appears not to have made a structural calculation that would have shown how the joint opened. They only recognized this after the hydro-burst test. These strong signals should have led both parties to improve the design.

When a problem becomes just an item on an agenda

Once the field joint passed the design review, it became just another item in the review process. It became a black box with a title. Subsequent reviews just looked at the input actions on the black box and the apparent reaction. Hence many, maybe hundreds, of engineers and project staff over the years, talked about the joint as part of the general review process without understanding what was in the black box. It is here that a Feynman is invaluable. He would have said, *tell me how it works*. And then pursued the issue until he had a full understanding. In the NASA culture, despite the claim of openness, it would have taken guts to say, *stop the process until I understand what's going on.*

The probability nature of engineering

Engineering is not a precise science, it is probabilistic. But the FRR process does not recognize this. It disposes of points as either *go* or *no go*. There is no flexibility to give a probability as an answer. Then the psychology takes over. While at the point of decision there may be doubts, once the *go* decision is taken, the doubts tend to be relegated to the back of mind. After time, and more successful launches, the doubts diminish further. This process makes it even more difficult to call for a halt.

Engineers versus managers

Several of the sociologists make a clear distinction between engineers and managers. The Rogers Commission also drew the distinction. Engineers were seen as doing technical work, whereas managers looked at cost and schedule. Some reviewers have distorted the differences further and postulated that engineers were honest, while managers were unethical. This distinction overplays the differences, and Vaughan shows this is not true. Engineers also have to take account of cost and schedule as part of their technical work. The managers at NASA were themselves engineers who managed a wider range of technical items within their teams.

Richard Feynman's style

Richard Feynman was the antithesis of *group-think*. His courage to challenge information until it can be authenticated was a valuable asset. A skill that is useful in any team and does not need a Nobel physicist.

He would have been seen as a disruptive element in the NASA style bureaucratic project management system. He would have distracted the FRR process by wanting to fully understand the issues. But that is one way to avoid disasters.

It is ironic that a bureaucratic and controlling style was also present in the workings of the Rogers Commission. Feynman's presence on the Commission avoided the worst effects.

Communication aids

Feynman used his experiment with the 'O' ring clamped in a glass of iced water to make his point. In his writings, Feynman shows his distaste for advertising and PR. He sees them as using selected facts to distort an argument in order to persuade a customer to buy. This was against his scientific philosophy on the use of information. It is ironic that his experiment can be classed as a piece of PR. He would probably be shocked to learn that his experiment has resulted in many people believing that the low temperature caused the accident rather than bad design.

Some commentators have questioned whether Roger Boisjoly failed to communicate his concerns. Should he have used better visual aids? This sounds like harsh criticism but unfortunately he failed to be persuasive. On the other hand, you can be the best communicator in the world and fail if the other party is not prepared to listen.

Getting the facts

Feynman hated set piece technical presentations and preferred free-ranging question and answer sessions. Both approaches have their merits. The set piece can get the point across quickly and in a logical manner but may miss a key secondary issue. It tends to control debate to the logic being presented. The question and answer session can be chaotic. It will probably take more time and be more difficult to draw to a conclusion. But with the right participants, it can explore the issues fully. The best answer probably lies in using a blend of both approaches.

Sorting the facts

Some reviewers claimed that there was poor communication. Others said that there were plenty of red flags. NASA staff pointed out that there are a large number of concerns prior to each launch. They aimed to address all the red flags. After the disaster, it is easy to find the red flag that was not fully addressed. No one seems to have a foolproof method of ensuring that no key red flags are missed, or that a red flag is incorrectly diagnosed.

Final comment

The poor joint design was the proximate cause of failure. The basic engineering requirement that you understand what you are designing and reviewing appears to have been at the root of the problem. It was hardly rocket science.

1.6.19   Lessons learned

Have a Feynman on your team

Feynman recognized he could be a pain. But every team, no matter how small or large, can suffer from some symptoms of *group-think*. The answer is to have a Feynman on your team.

- Encourage yourself or someone on your team to go out and search for all the facts.
- Constructively challenge the status quo.

Pay attention to communication

- Engineering is a team event. Members of the team have to communicate well, especially for tabling potential problems. Develop and encourage communication skills.
- Look for the best communication aids to get your point across. Use all the tools at your command to be persuasive.

Do not drift into trouble

The issue of the field joint, unresolved over many years, led to the disaster. The successful launches led management to believe in their invulnerability. There was a continuous shift towards lower safety factors as modifications were made and tests were successful.

- Beware of becoming complacent.
- Look for all signals, and be especially cautious when they continue to show problems, no matter how small.
- Do not fine tune or rationalize the odds until something breaks.

Success can cloud your judgement

- Beware of overconfidence created by successful tests.

Pay attention to the details

The failure to understand how the field joint worked lay at the heart of the disaster.

- Take time to understand the details, and how they work.
- Do not let a weak detail get past the first design review.

Work in the world of reality

Dr Feynman saw that the exaggerated reliability figures quoted by NASA stemmed from their senior staff being detached from reality.

- Take time to check that you are still working in the world of reality.

Ask how good is your decision making?
Engineering progress is a continuum of decisions.

- Take time to review how effective your decisions have been.
- Study the work of psychologists and sociologists to see what can be learned about the decision-making process.

POSTSCRIPT

Following the Commissions Report, the field joint was redesigned. The redesign, Fig. 1.6.6, uses several of the ideas that had been discussed before the disaster. A capture latch with an additional seal has been added. This limits the opening of the seal gap on pressurization. The putty has been deleted. A thermal barrier across the joint is provided by an overlapping insulation joint. A flap seal is forced closed by the pressure from ignition. This is similar to the arrangement in the earlier Titan rocket. Heaters keep the seal above 75°F, and a weather seal prevents water from entering the joint.

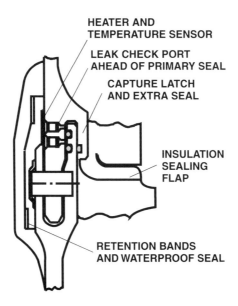

Fig. 1.6.6   New design of field joint

All the key requirements appear to have been met; but it cannot be called an elegant design!

References

(1) Information on the STS-51L/Challenger Accident. On www.hq. nasa.gov/office/pao/History/sts51l.html.

(2) Feynman, R.P. (1999) *The Pleasure of Finding Things Out*, Helix Books, Cambridge, Massachusetts, ISBN 0 7382 0349 1.

(3) Feynman, R.P. (1988) *What Do You Care What Other People Think?*, W.W. Norton & Co, New York, ISBN 0 393 32092 8 pbk.

(4) Dunar, A.J. and **Waring, S.P.** (1999) *Power to Explore: History of MSFC 1960–1990*, NASA US Government Printing Office SP-4313.

(5) Roger Boisjoly on the Challenger Disaster, http://onlineethics. org/moral/boisjoly/RB-intro.html and www.scs-intl.com/online/ frameload.htm?/online/boisjoly.htm.

(6) Information on the PRESIDENTIAL COMMISSION on the Space Shuttle Challenger Accident, June 1986. Available on http://history.nasa.gov/rogersrep/genindex.htm.

(7) NASA Challenger Accident and Aftermath bibliography, www. hq.nasa.gov/office/pao/History/Shuttlebib/ch7.html.

(8) Haisler, M.A. and **Throop, R.** (1997) The Challenger Accident, An Analysis of the Mechanical and Administrative Causes of the Accident and the Redesign Process that Followed. On www.me.utexas.edu/~uer/challenger/challtoc.html.

(9) Griffin, E. (1997) *A First Look at Communication Theory*, McGraw-Hill, ISBN 0 07248 392 X, Chapter 18 Groupthink of Irving Janis, www.afirstlook.com/archive/groupthink.cfm? source=archter.

(10) Perrow, C. (1999) *Normal Accidents: Living with High Risk Technologies*, Princeton University, New Jersey, ISBN 0 691 00412 9.

(11) Vaughan, D. (1996) *The Challenger Launch Decision; Risky Technology, Culture, and Deviance at NASA*, University of Chicago Press, Chicago, ISBN 0 226 85175 3.

(12) Turner, B.A. (1997) *Man-made Disasters*, Butterworth-Heinemann, London, ASIN 0 75062 087 0.

This section summarizes a report (**1**) from the Aerospace Safety Advisory Panel (ASAP) – an independent group established by Congress to report its findings to both NASA management and Congress. The report was written after the ***Challenger*** disaster but before ***Columbia*** crashed. This would normally have been in Part 2, but its direct relevance to the two disasters makes it logical to position it here between the two Shuttle reviews. The report says:

1.7.1 Preface

The extent to which the problems described here repeatedly occur should eliminate any complacency on the part of any manager or designer that such shortcomings apply only to others.

As a program matures, it is advantageous to pause and reflect on the lessons learned... and to record these while they are fresh in mind so that other programmes can benefit.

1.7.2 Technical and administrative management

Large programs have to be divided into manageable units. This can result in fragmentation and compartmentalization of information. Strive for a structure that makes the whole greater than the sum of the parts. Continuously emphasize the prime philosophy – *technical excellence.*

Identify all items at the start of the programme. This prevents problems with minor items from adversely impacting the progress later.

Uncertainty creates problems. Firm program goals must be established and maintained. But the schedule must be realistic and accepted by all parties.

It is important to achieve cross-fertilization and broaden the experience of both engineers and managers. To be effective, design reviews need input from test staff and operators.

Relationships, communications, responsiveness, and team spirit become more impaired as resources, schedules, and performance requirements tighten up. This leads to *buck-passing* and an attitude that *it's your problem, not mine.* An elaborate review structure can lead to the feeling: *not to worry, the other guys will catch it down the road.*

Risk documentation can evolve into listing all possible concerns so that retrospective examination can find no basis for criticizing thoroughness. It is better to document the evaluation of risks assessed at each management level.

Management should keep the organization for reliability, safety, and quality organizations within the decision loops.

1.7.3   The funding trap

Funding constraints, coupled with enthusiasm to deliver, lead to the *funding trap*. Enthusiasm produces a success-oriented culture. This can result in unrealistic budgets and schedules. These place an extraordinary burden on programme managers to minimize testing and rely on less-than-optimal system designs. The result is an increased acceptance of risks.

Doing things right the first time takes time and money. Solving problems later costs even more. But the programme should not be based on everything working right the first time. Time and funds must be available to allow for changes. Development projects often need to have alternative designs explored. Sometimes it would be prudent to continue with a back up design.

In summary, a major program should not be attempted without the appropriate funding and support.

1.7.4   Aggregate risk

Risk identification and assessment are a common thread in a programme. With tasks divided into separate groups, it is not easy to measure the aggregate risk. There are many methods for identifying risk, but it can only be quantified as a result of tests. Judgement alone is not an adequate assessment tool. Testing philosophy must be constantly reviewed to make sure it is pertinent, adequate, and necessary.

Risk assessment is a key part of the work of the engineering organization. However, the formal risk assessment for the whole programme should be independent and not part of an engineering design function.

Configuration management and interface management reduce risk. Managing the overall risk requires continuous effort to ensure good communication and rigour in meeting specifications.

1.7.5   Achieving adequate safety levels

Safety reviews tend to concentrate on the engineering design and quality control aspects of safety. These are important but so are the manufacturing practices, organizational structure, facilities, and human attitudes. Management has to balance schedule, cost, design, development, and testing with safety.

The programme has, on the one hand, to produce the output, and on the other, to identify uncertainties and assess the risk. These require different mental attitudes. The desire to complete a design makes it difficult to be objective about what may be wrong with it.

1.7.6 Some of the small issues that can have a large impact

Operating and maintenance staff are expert at finding ways to work around problems. Their skill then becomes a normal part of the work. But the most effective answer is to remove the problem. The design team needs to be given feedback on operation and maintenance problems so that these can be resolved.

Good instrumentation benefits from feedback from operations and test staff. A programme needs an overall instrumentation philosophy and standards.

Welding, to a great extent, is an art. Designers must provide good access for both welding and inspection.

Contamination – particles, corrosion, and so on *are little things that speak softly and cause very large problems*. There needs to be a good understanding of all the conditions that the equipment will experience, in manufacture, testing, storage, transport, and use. All along the line, the staff needs to know what conditions the equipment can tolerate.

It is best not to use Swiss-watch tolerances for locomotive-sized hardware.

Changes must be controlled. A change library and retrieval system is essential. All staff, including contractors, must have ready access to change information.

1.7.7 Software/computers

NASA's experience, in the 1980s, was that half of the software contracts had cost and schedule overruns. Also, 45 per cent of contracted software could not be used and only two per cent was useable as delivered. Management should pay careful attention to the integration of hardware and software design and how it fits into the whole programme.

Individual subroutine modules and packages should be able to be isolated, corrected, and improved without changing the whole system.

Computers cannot solve all problems. They are limited by the data that are put into them, and are incapable of manipulating these data except in the way that they are programmed to.

1.7.8   Summary

It has been customary in NASA to publish lessons learned on major programs. It is often done at the conclusion of a program. Unfortunately, the time pressure of starting the next task loses many of the good thoughts.

Design excellence and manufacturing quality are the result of engineers and managers using the experience bank built up over the years. Lessons learned are key to that bank of knowledge. Some lessons we would prefer to forget – but that is not wise.

The President of the Flight Safety Foundation said *Contrary to what many of us wish were otherwise, there is a correlation between economics and achieved safety levels. It cannot be quantitatively defined in most cases, but, qualitatively, the relationship is abundantly demonstrated. We live in an economic world, and it is up to us to be clever enough to wrest the highest level of safety from the many economic restraints that hamper its full achievement.*

COMMENTS

Space programmes have engineering and management issues similar to more mundane and down-to-earth projects.

Roth commented that candid revelations might allow some critics to draw incorrect inferences on the efficacy of the organization. However, the value of passing on the lessons learned is worth the risk.

Roth's report dates from 1986. This was a tough time for NASA, following the **Challenger** accident. The Harvard Business Review (**2**) says that NASA was a *lumbering risk-averse behemoth* but has now been transformed into a *lean, fast, model government agency.* The credit for the change goes to Daniel Goldin who stepped down in 2001 after nine years as NASA's Administrator. Goldin found that there was *this incredible fear of failure* of the huge, expensive, and high-profile programs.

Goldin's answer was to *tolerate some acceptable level of failure. Succeeding 10 times out of 10 is not acceptable because it means you're not pushing hard enough, and you're setting mediocre goals. We'll make failure acceptable by breaking programs into smaller pieces and increasing the number and diversity of programs.* The concept was to ensure that failure of any one subprogramme did not result in failure of the whole programme.

Goldin then reinforced this concept by addressing the blame game. *It happens all the time: When there's a failure, search for the people who failed and hold them up to public ridicule and destroy their desire to take risks. I established a policy that says; if there's no malice of forethought and no gross management incompetence, if a major failure occurred, I would take responsibility. That freed people up to dream and take risks.*

References
(1) Roth, G.L. (1986) *LESSONS LEARNED, An Experience Data Base for Space Design, Test and Flight Operations*, November 1986. Available on http://history.nasa.gov/asap/lessonslearned.pdf.
(2) Conversation with Daniel Goldin, HBR, May 2002.

On 1 February 2003, the orbiter, Space Shuttle *Columbia,* was re-entering the atmosphere at 8:44 a.m., 16 minutes from touchdown, when it broke up. All seven crew members died. The debris fell in a swath 550 miles long, covering over 2000 square miles of Texas and Louisiana. This was the 113th orbiter flight. *Columbia* was the first orbiter to fly. This was its 28th mission, launched on 16 January as a multidisciplinary science mission.

1.8.1 The investigation board

Sean O'Keefe, the Administrator of NASA, was waiting at the Kennedy Space Center for *Columbia* to land. Recognizing that there had been a disaster, he activated the Contingency Action Plan. The plan was established after the *Challenger* disaster. Within one and a quarter hours, O'Keefe had initiated the Mishap Interagency Investigations Board. He named retired Admiral Harold W. Gehman as Chairman. The Board chose the name Columbia Accident Investigation Board (CAIB) and co-opted a total of 13 Board members.

Of the 13 members, five were from the military and four from academia, five were pilots and four had experience reviewing major accidents. One current member of NASA, Dr G. Scott Hubbard, Director of the Ames Research Center, and one former astronaut, Dr Sally Ride, Professor of Physics at the University of California, were appointed Board members. Dr Ride had also served on the Rogers Commission. The members held an array of degrees in subjects as diverse as aeronautics, astronautics, industrial engineering, systems management, physics, psychology, human relations, political science, management, and law.

A staff of 120, together with 400 NASA engineers, supported the CAIB. They examined over 30 000 documents, conducted 200 formal interviews, and heard from dozens of experts as well as the public. The first, and main volume, of the Board's report (1) was issued in August 2003, seven months after the disaster. Five other volumes provided backing data.

1.8.2   The physical cause of the disaster

CAIB summarized the physical cause of the disaster as *a breach in the Thermal Protection System on the leading edge of the left wing, caused by a piece of insulating foam which separated from the left bipod ramp section of the External Tank at 81.7 s after launch, and struck the wing in the vicinity of the lower half of Reinforced Carbon-Carbon panel number 8. During re-entry the breach... allowed superheated air to penetrate... and progressively melt the aluminium structure... resulting in break-up of the Orbiter.*

1.8.3   The debris

An intense search for debris started immediately. More than 25 000 people took part. Finally, 84 000 pieces of orbiter debris were collected, weighing about 38 tons. This represented 38 per cent of *Columbia* by weight. Fortunately, the debris fell in a sparsely populated area and did little damage and no one was hurt. A 600 lb piece of the main engine dug a 6 ft wide hole in Fort Polk golf course. The largest piece weighed 800 lb. It hit the ground at an estimated 1400 miles/h. Another piece landed between two natural gas tanks that were only a few feet apart. An important find was the Modular Auxiliary Data System, which recorded output from 800 sensors. This provided key data for deducing the final cause and sequence of the accident. Aluminium with tiles attached was found from the right wing, but no aluminium was found from the left wing.

1.8.4   The bipod and its foam insulation

A 'V'-shaped bipod strut (Fig. 1.8.1) attached the nose of the orbiter to the external tank. The orbiter connected with the point of the 'V', and the ends of the two arms were fastened to the structure of the external tank. The tank had three main components, the liquid oxygen tank at the top and the liquid hydrogen tank at the bottom, with an intertank separating the two fuel tanks and providing the structural strength. The bipod strut was fastened to the external tank at the joint between the hydrogen tank and the intertank. The liquid hydrogen was at – 423°F (−217°C). The external tank was coated with a two-layer thermal protection system – one layer to dissipate heat and the other, a low-density closed cell foam, to provide insulation.

Most of the external tank insulation was machine sprayed. The complex geometry around the bipod connection was first coated with a super-lightweight ablator to dissipate heat and then hand sprayed with polyurethane foam. Around the bipod connection, the foam

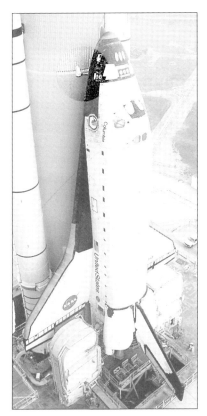

Fig. 1.8.1   The highlighted area in the picture on the left shows the foam ramp at the end of the bipod strut. The picture on the right, looking down on the nose, shows the bipod strut attaching *Columbia* to the fuel tank. Photographs courtesy of NASA

was manually shaved into the ramp shape. The polyurethane foam is 90 per cent air.

Over the years there have been design changes. The shape of the ramp facing the airflow has changed from 45° to between 22° and 30°. The foam chemistry has changed because of Environmental Protection Agency regulations and material availability. These modifications have reduced chlorofluorocarbon emissions.

Design requirements for the foam
The insulation has to reduce the amount of heat reaching the liquid fuels. The outside temperature must be high enough to prevent ice forming because of atmospheric moisture or rain while the shuttle is on the ramp. The foam has to withstand the loads from launch

and the aerodynamic loads from going from standstill to reaching high Mach numbers. The foam and its bond to other materials have to withstand stresses due to movements from temperature changes during launch. During ascent, the outside surface of the foam reaches 600°F (315°C). And the original specification called for no foam being shed during launch.

CAIB noted that the design originated in the 1970s when design tasks were typically handled one at a time. For example, the structural design was completed first and then followed by the thermal design and finally the aerodynamic design. This led to a complex geometry around the bipod which had to be insulated. Today's integrated design teams are more likely to arrive at a design that better balances the various requirements, without any feature being too cumbersome.

At the design stage, tests were conducted. Material characteristics, strength, and insulating properties were measured. Wind tunnel tests were made on the geometry rather than on actual samples made by the same process as used for production.

In-service experience with foam

Foam shedding occurred on the first shuttle fight – by *Columbia* in 1981. As a result, 300 thermal tiles had to be replaced prior to the next flight. Most of the following missions had insulating foam shed during ascent. This was seen in over 80 per cent of the missions for which photographs were available. The size of the foam shed ranged from popcorn- to briefcase-sized chunks. Initially, the foam shedding caused concern and actions were placed to investigate.

The quality of the foam consistently improved on most areas of the external tank. There was better control of the machine spraying. Removing hidden faults from the area over the stringers of the intertank, and bipod connection, was more elusive.

Foam failure mechanisms

NASA thought that either cryopumping or cryoingestion might be the cause of shedding foam. Cryopumping is where atmospheric air enters cracks in the foam and liquefies when it gets near the surface of the cryogenic tanks. As the temperature increases on launch, the trapped liquid flash evaporates and builds pressure in the foam until it breaks off. Cryoingestion is similar and is due to liquid nitrogen, temporarily used in the intertank while loading fuel, seeping into the lower levels of the foam. As a result of these hypotheses, NASA poked small vent holes through the foam to allow trapped gas to escape.

One of the CAIB members, Douglas Osheroff, a Nobel Prize winner, set out to understand the possible failure mechanisms. He recognized the difficult nature of the problem with the complex interaction between many components. But the basic question was: *How does the foam fail structurally?* He experimented with foam samples glued to brass plates. In the centre of the plate, a small tube was used to apply pressure with a tyre pump to the underside of the foam. Instead of the explosive break-up and shedding of the foam as expected by NASA, a tear-shaped crack developed. This released the pressure to the surface, without any foam being shed. Dr Osheroff concluded that hydrostatic pressure would not generally lead to foam shedding.

CAIB had three bipod foam ramps dissected and found that they all had interior defects. It concluded that the defects were not due to poor workmanship but due to the difficult geometry.

CAIB concluded that *the precise reason why the left bipod foam ramp was lost may never be known.* A combination of conditions including higher than typical wind shear during the climb (although within specification), liquid oxygen sloshing in the external tank leading to oscillating loads, and possible inherent defects contributed to the failure. It added that, despite commonly held perceptions, numerous tests showed that moisture absorption and ice formation in the foam appeared negligible.

1.8.5 Shuttle damage
Design requirements for the wing leading edge
The shuttle wing is covered with thermal tiles to protect the aluminium wing structure from the ~3000°F (1650°C) boundary layer air temperature on re-entering the atmosphere. Air above this layer is around 10 000°F (~5500°C). The wing leading edge is constructed of 22 Reinforced Carbon–Carbon (RCC) panels sealed against each other to prevent hot air entering the wing cavity. The RCC panels are designed to withstand 1.4 times the loads expected in operation. These loads include a very small kinetic impact load. The specification states that the leading edge would not need to withstand impact from debris or ice *since these objects would not pose a threat during the launch phase.*

Impact tests to prove a point
CAIB initiated tests. By now, they were convinced that the foam had caused the accident, but NASA still refused to believe this explanation. A piece of foam was fired from a nitrogen gun at a section

of wing leading edge. Photographs from the launch showed that the detached foam was between 21–27 in long and 12–18 in wide with a volume of 1200 in³. Once it broke away at Mach 2.5 (~1600 miles/h), it quickly slowed down. *Columbia's* wing hit it with a large speed differential, estimated at 545 miles/h.

The tests used a piece of foam $19 \times 11\frac{1}{2} \times 5\frac{1}{2}$ in, weighing 1.67 lb. A test on panel 6 resulted in its cracking and moving slightly on its fastenings, but it did not leave a leak path sufficient to cause serious damage. The best estimate of where the foam hit the wing was on RCC panel 8, just below the leading edge. When foam was shot at panel 8, it made a hole approximately 16 in by 17 in (Fig. 1.8.2). The exact size of the hole in *Columbia* may never be known. Analysis suggests that the hole was around 10 in.

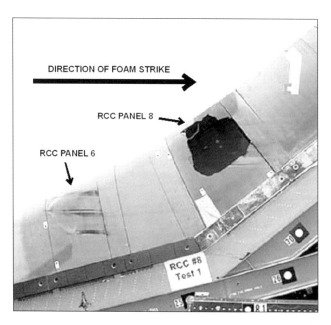

DIRECTION OF FOAM STRIKE

RCC PANEL 8

RCC PANEL 6

RCC #8 Test 1

Fig. 1.8.2   Test impact damage on RCC panels 6 and 8. Photograph courtesy of CAIB

Damage to shuttles on previous flights

CAIB noted that damage caused by debris had occurred on every shuttle flight. This required maintenance after each flight. On average, there were 143 divots on tiles after a flight, with 31 of these having a dimension greater than 1 in. On a mission in June 1992, *Columbia* hit a piece of foam from the left bipod ramp, measuring 26 in by 10 in. It left a $9 \times 4\frac{1}{2} \times \frac{1}{2}$ in deep dent in a tile. In October 2002, a piece of foam from the left bipod hit the SBR attachment. The foam shed 33 s into launch when the shuttle was at 0.75 Mach. When the SBR was recovered, it was seen that the impact had made a 4 in wide × 3 in deep hole in the insulation.

1.8.6   Statistics
NASA Headquarters Safety Office did a review after the October 2002 foam damage. They estimated a 99 per cent probability of foam not being shed from the same area in future flights. CAIB considered this prediction a *sleight of hand.*

CAIB restudied the photographs of launches and found two cases of left bipod foam shedding that NASA had not noticed. In 72 missions, for which photographs were available, there were seven cases of left bipod foam shedding. So there was nearly a one in ten chance of foam being shed from the left bipod ramp.

The left bipod ramp is the source of the largest foam shedding debris. The right bipod ramp seemed to be less liable to shed foam. This may be due to the different airflow pattern near the adjacent liquid oxygen feedline.

1.8.7   Mission Management's role in the disaster
The technical cause of the disaster was the damage to RCC panel 8 caused when it was hit by the insulating foam – but what role did the engineers and managers play?

1.8.8   Attitude to foam shedding prior to this mission
There was concern about foam shedding after the first shuttle flight. With each successful flight, both engineers and mission management became less concerned. Foam shedding was accepted as inevitable. It was judged to be either an *acceptable risk* or *unlikely to jeopardize safety.*

The Intercenter Photographic Working Group (IPWG) monitored each take-off. Both still and video pictures of each launch were taken from several ground stations. Among other tasks, it looked for foam shedding. After the October 2002 mission, the IPWG reported the largest foam shedding seen to date from the left bipod ramp. It recommended that it be classed as an *in-flight anomaly.* But the Program Requirements Control Board decided against this and down-graded it to an action to investigate. Two further missions, including the fatal **Columbia** mission, flew without this action being completed.

The issue was not raised at the Flight Readiness Review for this mission.

1.8.9   The photographic record

When *Columbia* took off, there were no immediate or obvious signs of any problem. A quick scan of the photographs did not reveal any foam shedding. But a more detailed review on Day 2 of the mission showed foam had shed at around 66 000 ft – one large piece and at least two smaller pieces. Unfortunately, one of the cameras filming that stage of launch was out of focus due to a maintenance error. From the shower of foam debris coming from under the wing, it was clear that it had hit the lower surface of the wing. The images only showed the upper wing surface. The head of the IPWG immediately alerted senior programme managers. In addition, he made an informal request to see whether the Department of Defense could use its capability to provide images of *Columbia* in flight to look for damage. The shuttle crew had no means of inspecting that area of the wing.

It had become routine for astronauts to film the external tank as it separated from the shuttle. This gave a record of any missing foam. On this flight, some of the film was downloaded to Mission Control but none of the bipod ramp. No request was made by Mission Management to see whether film of the ramp area was available.

In its research, CAIB asked the US Space Command whether they had any evidence of damage to *Columbia.* They searched their records and found that for several days an object estimated as about 100 in^2, was orbiting close to *Columbia.* It is possible that this was part of the broken RCC panel 8. On impact, during launch, the broken section might have been pushed into the wing. When the object was seen, the shuttle was flying backwards. Use of the manoeuvring rockets might have dislodged the broken section and allowed it to float out of the wing cavity.

1.8.10   The engineer's assessment of the damage

The knowledge that a large piece of foam had been shed, and that *Columbia* had hit it, was widely circulated among the engineers. Several groups self-initiated work to assess the likely impact. A quick estimate was made of the debris size, weight, speed of impact, and the likely point of impact. Later analysis by CAIB showed this early assessment to be remarkably accurate.

On Day 2, Mission Management formed an engineering Debris Assessment Team (DAT). The Team examined possible areas that might be damaged – RCC panels, tiles, and the seal around the undercarriage door. The main focus of attention was on the tiles. Tiles had

had the most damage from foam on previous flights. A computer programme – Crater – had been developed in the 1970s to assess impact damage.

1.8.11   Crater – a tool outside its range

Crater predicted that tile damage would be deeper than the tile thickness. But Crater was known as a program that had given conservative answers in the past. And, when dented, the tile was denser and thought to be better able to withstand impact. The conclusion was that there might be heavy denting of the tiles, but this was unlikely to lead to excessive heating of the wing structure.

Crater had been validated by test. But what was ignored was the fact that the size of the foam striking *Columbia* was 400 times larger than the test samples used to validate Crater.

1.8.12   Presentation of engineering analysis to Mission Management

CAIB noted that there had been, over time, a drift to using Power-Point briefing slides rather than comprehensive peer-reviewed technical papers. Furthermore, the slides for the *Columbia* investigation were potentially misleading. The Board quoted the presentation on the Crater analysis. The slide heading was **Review of Test Data Indicates Conservatism for Tile Penetration.** The text used qualitative words like *significant* rather than quantifying the point. The fact that the debris size was well outside the test data on which Crater was validated was hidden on the bottom of the slide.

1.8.13   Mission Management's view and review of engineering input

The reaction of Mission Management to the reports of the foam shedding was 'We've seen it before' and 'It's a maintenance issue'. It was not considered a safety-of-flight issue. Discussions with engineering managers reinforced that view. The engineering managers hypothesized from past experience. This appeared to indicate that the shuttle would withstand foam impact. Their immediate comments were given more prominence than the fact that their staff were still looking at the incident in more detail and trying to analyse the consequences. In effect, the incident was disposed of as a maintenance issue before any analysis had been carried out.

CAIB saw that Mission Management did not effectively manage the engineering support work. It did not provide direction, nor did it take the time to understand the engineering analysis, or question the results. It did not ask whether there were any minority views that differed from the broad summary.

1.8.14   Requests for photographs

CAIB investigation noted three specific requests for imagery to try and see whether any damage had occurred. Imagery would come from specialized Department of Defense capability. Several unofficial requests were made to the Department of Defense who had started to investigate how it could help.

On Flight Day 7, Mission Management formally asked the Department of Defense to try to obtain images. But 90 minutes later cancelled its request. This was because quick polling of some managers showed that no one said it was a firm requirement – just a wish. The decision may have been influenced by the belief that they were only expecting minor damage, like tile denting, which would be difficult to see. Manoeuvring to get the images would detract from *Columbia's* mission and extend the flight. CAIB concluded that Mission Management was making critical decisions about imaging capability on the basis of little knowledge.

The engineers quietly accepted Mission Management's decision. NASA Head Quarters Safety and Mission Assurance Office had enquired about obtaining images but did not pursue the issue. In CAIB's view, they remained passive.

1.8.15   Mission Management meetings

NASA procedures call for the Mission Management to have a daily meeting during a mission. On this 16-day mission, the Mission Management only met on five days.

The Mission Management Team was chaired by Linda Ham. William Langewiesche, writing in the Atlantic Monthly (2), describes her as embodying NASA's arrogance and insularity. She had power in the hierarchy. As a result, her decisions carried, without question or dissent. In an email, she commented that the rationale for handling the loss of foam decision for the previous flight was *lousy,* but this did not trigger deeper examination.

The response to the requests for images in orbit was to ask who had been making unofficial requests to the Department of Defense, rather than finding out why staff wanted the images.

1.8.16   Message to the crew

Mission Control sent an email to the crew of *Columbia* on the eighth day of the mission telling them that there had been a foam strike. They were told in case of questions from the press on landing – otherwise it would *not be worth mentioning. Experts have reviewed the high-speed photography and there is no concern for RCC or tile damage. We have seen this phenomenon on several flights and there is absolutely no concern for entry.*

1.8.17   Management view post-disaster

In the days immediately after the disaster, the media were asking about the foam shedding. NASA officials said the foam was too light to have caused any damage. Administrator O'Keefe, no doubt ill advised by junior staff, dismissed the media focus on the insulation as the work of *foamologists.*

1.8.18   CAIB's summary of management decisions

*Management decisions made during **Columbia's** final flight reflect missed opportunities, blocked or ineffective communications channels, flawed analysis, and ineffective leadership. . . Management displayed no interest in understanding a problem and its implications. Managers failed to avail themselves of a wide range of expertise and opinion. . . Management techniques unknowingly imposed barriers that kept at bay both engineering concerns and dissenting views, and ultimately helped create "blind spots" that prevented them from seeing the danger the foam strike posed.*

But management was trapped in NASA's culture. CAIB recognized that blaming and replacing the local management was no cure. The roots lay deeper in the organization.

1.8.19   Organizational flaws
The background history

The space age started with the first large-scale rockets built by Germany in WWII. Rocket development took off with the Cold War. By the 1950s, the United States and Soviet Union were racing to develop intercontinental ballistic missiles (ICBM). The Soviet Union surprised the world in 1957 by launching the first satellite – *Sputnik.* The United States decided to focus its space activities by creating NASA, in 1958. NASA inherited the research-oriented National Advisory Committee for Aeronautics (NACA) with its staff and facilities.

NASA's first project was to launch a manned spacecraft. This was achieved in 1961 with Alan Shephard on board. A month earlier, the Russians had put Yuri Gagarin into orbit. President Kennedy then announced that America would *place a man on the moon before the end of the decade.* Apollo 11 landed Neil Armstrong and Edwin

Aldrin on the moon in July 1969. The Apollo Program made five more lunar landings.

Following the successful Apollo Program, NASA embarked on plans for space stations and space transport. These plans were inconsistent with the political reality of the time. By now, the United States was embroiled in Vietnam, and space-station plans were thrown out. NASA redrafted its case to President Nixon in 1970. It argued for a reusable space shuttle that would economically launch scientific, commercial, and military payloads. Nixon authorized the Shuttle Program on the basis of the political benefits of the jobs it would create rather than on NASA's aim to *revolutionize transportation into near Space, by routinizing it.*

The commitments NASA made to get the Shuttle Program authorized resulted in a difficult specification with conflicting requirements – to be achieved on an inadequate budget. The shuttle project slipped behind schedule and was over budget. President Carter's decision to continue was based on his priority to launch intelligence satellites to verify the SALT II arms control treaty.

The first orbiter – ***Columbia –*** was launched in April 1981. When the fourth mission landed on 4 July 1982, President Reagan declared the orbiters to be fully operational and able to provide *economical and routine access to Space.*

The Shuttle Program now faced competition from the European Space Agency, which was using Ariane rockets to launch commercial satellites without the need for manned space flight. During the 1980s, the Shuttle Program gave the outward appearance of working well – but the shuttle was expensive, difficult to operate, and needed more maintenance than was expected.

Then came the ***Challenger*** disaster and the Rogers Commission. One outcome of that was the decision that the shuttle would no longer launch commercial satellites or military payloads. The future of the Shuttle Program became linked to carrying heavy payloads and astronauts to the International Space Station (ISS).

1.8.20   Budget and staff cuts

In the midst of the race to the moon, NASA's budget in 1965 was over $24 billion (in 2002 dollars). By 1993, the budget was $17 billion, and $15 billion in 2002. But this budget covered all NASA activities, including the United States' share of building the ISS. The budget for the shuttle has reduced by 40 per cent during the past decade.

In most years, Congress marginally cut the NASA budget requested by the President. In addition, Congress has a habit of earmarking funds out of approved budgets for topics it chooses. These are frequently outside the requested budget, which puts more pressure on existing projects like the shuttle.

In 1999, NASA Administrator Daniel Goldin wrote to the White House saying that the United States faced a Space Launch Crisis. An additional budget for safety upgrades was agreed on, but by 2003, it had been cut by 34 per cent.

The ISS was also under budget pressure. In 2001, Sean O'Keefe, then Deputy Director of the White House Office of Management and Budget, presented the Administration's plans for bringing ISS costs under control. The X-38 crew return vehicle was cancelled. This decision resulted in the ISS crew being reduced from seven to three. Three is the most that can be returned to earth by the Russian Soyuz capsule.

The Bush White House and Congress effectively put the ISS, the Shuttle Program, and NASA on probation while they monitored the performance in completing Node 2 of the ISS, which was scheduled for 19 February 2004.

Contractors and NASA staff working on the shuttle have decreased from 32 425 in 1991 to 17 462 in 2002. The reduction came from a hiring freeze, buy-outs, and early retirement. This has left the Program with fewer and less experienced staff and no new engineers.

Throughout the reviews of NASA, there have been two key questions in the background – *How long will the Shuttles remain in service?* and *What will replace the Shuttles?* Neither question was clearly answered.

1.8.21   Management of NASA

The senior position in NASA is the Administrator, who is a political appointment. There have been ten Administrators. The first was Dr Keith Glennan, an engineer. James E. Webb followed him and led NASA through the Apollo Program. Webb had experience as a pilot, lawyer, industry executive, Director of the Bureau of Budget, and Under Secretary of State. He was a Washington insider – a key attribute for getting the necessary support and financing. While he oversaw the Moon-landing Program, he believed that NASA should not be a one-shot organization. He saw NASA's role as leading a long-term exploration and science programme in space. This would be a catalyst for scientific and engineering development. President Kennedy, when he appointed Webb, made it clear that he saw the Administrator's role as a policy job.

Dr Thomas Paine, an engineer from GE, succeeded Webb and was followed by Dr Fletcher and Dr Frosch, both scientists. Dr Fletcher served two terms – 1971–1977 and 1986–1989. Administrators James Beggs and Richard Truly came from the Navy. Truly was a Navy aviator and also an astronaut.

The longest serving Administrator, from 1992–2001, was Daniel Goldin, an aerospace engineer and executive from TRW. Goldin described himself as an agent-for-change in the post-*Challenger* era. His tenure has been described as one of continual turmoil – which produced both improvements and controversy. He was a strong believer in Dr Edwards Deming's philosophy (3). In particular, he saw that there were too many checks and balances that were often counter-productive. He required all staff to take responsibility for their work. He changed the role of NASA HQ to setting strategy and transferred control to the units carrying out the work. Goldin pursued outsourcing. He propounded throughout NASA the catch phrase – *faster, better, cheaper*. Goldin pioneered the US work on the ISS, and during his tenure, the Hubble Telescope was repaired and missions were sent into outer space.

In 2002, Sean O'Keefe, a career public administrator and trouble-shooter was appointed NASA Administrator. O'Keefe received a clear mandate from Congress to cut the massive overhead.

1.8.22   Schedule pressure

The question 'Was this mission under schedule pressure?' can be answered by both 'yes' and 'no'. As a scientific mission, there was no particular time pressure. But shuttle launches had to fit into the overall plan for the Kennedy Space Center facilities. NASA was under pressure to meet the schedule for the ISS and especially Node 2 delivery. NASA HQ had circulated a computer screen saver showing the countdown to the 19 February 2004 deadline. Mission managers already knew that this deadline would be difficult to meet.

1.8.23   Previous investigations, reviews, and reports

CAIB reviewed more than 50 previous reports on NASA and the Shuttle Program. Many of these had accepted the promoted view that the shuttle was proven technology. Others clearly identified the shuttle as still being a development project. No effort appears to have been made to reconcile these conflicting views.

The Augustine Committee found NASA overcommitted. They estimated that to meet NASA's plans the budget would have to increase, and reach $40 billion in 2000. The actual budget in 2000 was $13.6 billion.

The Government Office of Technology Assessment in 1989 said that *Shuttle reliability is uncertain, but has been estimated to be in the range 97 per cent to 99 per cent. If the Shuttle is 98 per cent, there would be a 50:50 chance of losing an Orbiter within 34 flights.* There were about 80 flights between when that prediction was made and the loss of **Columbia**.

The Augustine Committee said in 1990 *although it is a subject that meets with reluctance to open discussion, and has therefore been too often relegated to silence, the statistical evidence indicates that we are likely to lose another Space Shuttle in the next several years. NASA has not been sufficiently responsive to valid criticism and the need for change.*

The Chairman of the Space and Aeronautics Subcommittee of the House Science Committee, Dona Rohrabacher, said *I have witnessed time and again NASA over-promising, over-marketing and underestimating costs. It's a pattern, NASA goes for the grandiose, ignoring doable, more affordable alternatives.*

1.8.24 Safety organization

Following the Apollo fire on the launch pad in 1967, Congress pressed NASA to establish the Aerospace Safety Advisory Panel, as a senior advisory committee. Congress also required a Safety and Reliability Office at NASA HQ, and at each Center. The Rogers Commission saw that the intent of Congress had not been met. There was no independent safety oversight. They recommended an Office of Safety, Reliability, and QA reporting directly to the Administrator. But once again, the arrangements did not meet the intent. The HQ Office did not have direct authority. HQ was saying what had to be done – but not how. This was left to the Centers where safety funding and responsibility became linked with general management. In summary, the safety system had no clout.

1.8.25 Safety culture

There is no doubt that everyone from the Administrator down was committed to safety. But commitment and actions to meet that commitment are two different things. The heavily promoted motto within NASA during the 1990s – *faster, better, cheaper* became a strong signal that the Administrator wanted efficiency. However, couple this with personnel cuts in all areas, including safety, and then safety becomes a weak signal. This is unlikely to have been the aim of the Administrator – but it became the unintentional consequence.

1.8.26   Can-do culture
During the Apollo Moon Program, the mission and goal were simple – easily understood and readily accepted by everyone. There was a collegial team spirit. The task, though difficult was simple, and the organization was simple. Everyone knew what was expected and what he or she had to do. The success produced a *can-do* culture and a belief in both the engineering capability and the strength of the operating systems. This culture is very resilient and has withstood all the outside pressures and budget cuts. CAIB noted that the can-do culture, when taken too far, creates a reluctance to say that something cannot be done. Emotion overrides sound engineering and management judgement. The can-do culture produced a climate of invincibility, in contrast to reality. It fed a belief, sold to NASA Administrators, politicians, and the public, that anything was possible.

1.8.27   Engineering practises
In both shuttle disasters, the engineers were aware of the problems. They saw what was happening but did not understand the fundamentals. They did not delve deep enough to find out why. They lived with the problems until it was too late. Changes were made without being fully proven by test. If a flight with changes went well, then the changes were assumed to be sound and proven. No non-destructive tests were called for to check the quality of the difficult job of laying the foam on the bipod ramp.

   With budget pressures, some basic engineering principles like keeping good records, monitoring trends, and having drawings up-to-date were allowed to slip. Drawings had errors. Many changes made in the last two decades had not been incorporated in the drawings. CAIB found that it took up to four weeks to get drawings – which were still paper based, not digitized. The engineers worked with an information system with over 5000 critical items and 3200 waivers. NASA has a Lessons Learned System, but input is strictly voluntary.

1.8.28   *Challenger* and *Columbia* similar disasters?
CAIB saw both disasters fitting Barry Turner's description as *failures-of-foresight*. Both failures were with simple and common engineering features. In both cases, there was no redundancy, and local management failed to prevent the disasters. In both cases, the problems were rooted in NASA's history. Changes had been made after the

Rogers Commission report. Some improvements followed. In some cases, they were short-lived, and then there was a reversion to the previous practices.

Ironically, both disaster investigations had Nobel Prize winners conducting simple experiments – an 'O' ring in a glass of iced water and foam stuck on small plates. These simple experiments clarified the issues.

1.8.29   Insights from organizational theory

CAIB found useful ideas from both High Reliability Theory and the theory of Normal Accidents (more details of these are given in Part 2.). High Reliability Theory reinforced the Board's view on the necessity for a strong safety culture. The Theory is based on the principle that serious accidents can be prevented by a willingness to learn from mistakes, from technology, and from others. A key to High Reliability is the ability to operate in both centralized and decentralized modes as the situation dictates. Realistic training empowers staff to know when to centralize, and when to decentralize, for operations and problem solving.

Normal Accident theory argues that complex interactions make organizations more accident-prone. More redundancy can mean more complexity and more risk rather than less. Therefore, blind faith should not be placed on redundancy. Despite best intentions, staff can act in counterproductive ways when placed in difficult circumstances.

1.8.30   Insights from experience in other high-tech, high-risk industries

CAIB looked at experience from the US Navy Reactor Safety Programs, the Navy's SUBSAFE Program, and the Air Force work with the Aerospace Corp. This experience shows that success depends on:

- keeping engineering and safety independent of operations and relatively immune to budget pressures;
- keeping safety requirements and responsibilities clear, consistent, and uniform;
- having concise and timely communication of problems, aided by formal written reports with peer review;
- ensuring good communication. Using both formal and informal avenues to spread information;
- facing facts objectively with attention to detail;
- insisting on airing minority opinions;

- learning from mistakes – one's own and other people's – with recurring, relentless, and innovative training. (The Naval Reactor Program has, since 1996, educated more than 5000 of its personnel on the lesson learnt from the *Challenger* disaster);
- making sure that knowledge is retained in the organization, by careful rotation of staff, record keeping, training, and informational meetings;
- carrying out quantitative assessments of risk and evaluating worst-case scenarios;
- managing change and dealing with obsolescence.

While NASA has paid some attention to similar items, there has been no consistent and systematic approach to these issues.

1.8.31   Discussions with Dr Diane Vaughan

CAIB had a discussion with Diane Vaughan to hear her views, based on her nine-year study of the *Challenger* disaster. She summarized that disaster as a failure of the organizational system – an incremental descent into poor judgement: not group-think, not incompetent engineers, not unethical or incompetent managers. It was a normalization of deviance.

Dr Vaughan said we tend to look backwards from accidents rather than looking from the source and seeing how they unfold. NASA engineers were working against a backdrop where problems were expected – having a problem was normal. Their job was to resolve problems for the Mission Managers, to keep orbiters flying.

Decisions and budget limits by Congress and the White House had affected how all people made decisions. NASA had converted from an R&D organization getting a man to the moon into a space-transport business. During the man-to-the-moon phase, the main job was cutting-edge engineering. There was a lot of hands-on work and a deference to engineers and their expertise. On the other hand, routine shuttle work became bureaucratically accountable with rules, forms, and the tyranny of hard data. Intuition and hunches, which are integral to good engineering, carried no weight. The safety system had been weakened. Few people had hands-on experience of what was going on and of the risks. Desk work dominated. The priority shifted to Mission Management. The engineers became support for Mission Management and were lower in the hierarchy, and hierarchy dominated. People were disempowered from speaking up. Moreover, the organizational structure interfered with good communication.

Dr Vaughan pointed out that accident investigations typically look for the technical problem and the technical culprit. The organizational system failures go untouched. It is more difficult to identify flaws in the organizational system. They are harder to pin down and more challenging to correct.

Cultures are hard to change, but leaders must try. The leaders must be in touch with all the risks. NASA has a different perception of risk at the top than at the working level. We think we understand cultures, but they act invisibly on us, and so we cannot really identify what their effects are. What is the effect of rules on culture? Rules are necessary in complex organizations. But does blindly following the rules limit the spotting of unexpected risks? Posting signs for safety, signs saying safety is an attitude, and encouraging staff to speak up might convince some that there is a strong safety culture. When you explore how the organization thinks and makes decisions, you may not find the strong safety culture that you expect or want.

The greater the complexity of systems, the greater the possibility of failure. The same is true of organizations. When looking for solutions the aim should be to spot the little mistakes before they turn into catastrophes. Engineers' concerns have to be surfaced and dealt with – without swamping the system with a multitude of issues. Safety people have to be given more clout. Learning and training is key – NASA as an organization did not learn from previous mistakes. Trend analysis helps in the recognition of patterns. Try to simplify complex systems so that people can respond quickly and short cut the hierarchy. Try studying a social fault tree analysis for failures. Make a diagram for decision-making systems as a tool to benchmark how the system is operating.

Dr Vaughan said that the problems that existed at the time of **Challenger** had not been fixed. And in both cases, they did not have hard supporting data – either on the effects of low temperature or the risk from foam strikes. In both cases, signals were missed.

1.8.32   CAIB's summary of organizational issues

While having had some impressive successes, the NASA organization is flawed. *The organizational causes... are rooted in the Space Shuttle Program's history and culture, including the original compromises that were required to gain approval for the Shuttle, subsequent years of resource constraints, fluctuating priorities, schedule pressures, mischaracterization of the Shuttle as operational rather than developmental, and lack of an agreed national vision for human space flight. Cultural traits and organizational practices detrimental to safety were allowed to develop, including: reliance on past success as a substitute for sound engineering practices; organizational barriers that prevented effective communication of critical safety information and stifled professional differences of opinion; lack of integrated management across program elements; and the evolution of an informal chain of command and decision-making processes that operated outside the organization's rules.*

NASA is a politicized and vulnerable agency, dependent on key political players who accepted NASA's ambitious proposals and then imposed strict budget limits.

The past decisions of national leaders – the White House, Congress and NASA Headquarters – set the Columbia accident in motion by creating resource and schedule strains that compromised the principles of a high-risk organization.

1.8.33   Other facts and issues
Could the damage have been repaired or could the crew have been rescued?
CAIB initiated a study based on the premise that the damage was known while **Columbia** was still in orbit. It concluded that repair in orbit might have been possible – but would have been difficult and risky. The crew could have been rescued by launching **Atlantis** immediately and transferring the crew in orbit. This would have been an untried operation. It would also have been risky because the foam shedding problem had not been resolved, and foam could hit **Atlantis** during launch.

Other issues
In its broad investigation, CAIB found other potential problems. It did not just home in on the foam but also conducted a fault tree analysis with over 3000 elements. Some issues were exposed. Explosive bolts are used for separating components in launch. The device for capturing the pieces was found to be below specified strength, although unlikely to fail. Kapton wiring was showing arcing damage to the insulation. Infrastructure facilities were in a poor state of maintenance. Paint had peeled from the launch tower. Rain had leached the exposed zinc primer and splashed it onto shuttle tiles. The heat from re-entry causes the zinc to create pinholes in the tiles. Computer hardware and software were obsolete.

COMMENTS
The words of Yogi Berra say it all – *It's déjà vu all over again.*

The investigation
The CAIB report is worth reading in full as an example of how to conduct a thorough investigation, and how to communicate the

findings in a clear manner. It recognized the complexities and was determined to table all the issues, together with pointers on how to correct the problems. If one could criticize the report, it is that it has not responded to the underlying and exceedingly tough issue that surfaced, which is, how are the responsibilities for a long-term national programme to be shared among the White House, Congress, and NASA Administration and how are the weaknesses to be corrected? An associated question is how do you rationalize the short-term political decision-making and annual national budget process with the need for a long-term commitment and funding for such a technological programme? Without resolving these questions, any future programme will still be at risk. *The Economist* on 30 August, reporting on the CAIB findings, concluded that *NASA is lost in Space.* But those to whom NASA reports are also lost, by having no clear idea of what they expect from NASA.

Issues surrounding the investigation

NASA initiated the CAIB. The media soon questioned whether a Presidential Commission should cover the ***Columbia*** disaster, as had been the case for ***Challenger.*** Some in Congress called Admiral Gehman a captive investigator. The exchange of letters and draft terms of reference show that O'Keefe initially had CAIB taking direction from himself and working to NASA policies and practices. Gehman negotiated out of these restraints and defined CAIB as an independent board. CAIB has been hard-hitting on all weaknesses. Would a Presidential Commission have been as hard on the White House and Congress? NASA created an internal review using the management staff immediately associated with the disaster and decreed that they should interface with CAIB. CAIB saw this as a conflict and persuaded the NASA administration to drop the idea. Gehman told a reporter that *bureaucracies will do anything to defend themselves.*

The report and summary

The report categorized the failures in three areas – the physical cause, Mission Management, and broad organizational flaws. The stark simplicity of the findings is frightening. The main findings are as follows:

- the initial design requirements were flawed. The wing was not designed for impact. It was assumed that the foam would stay intact;

- from the first shuttle flight, experience had showed that foam comes off and impacts the wing. Yet, shuttles were continued to be flown outside their own specification;
- the organization duly recorded foam impacts. It bureaucratically put a label on it, which did not require immediate attention. With time and cost pressures, only items for immediate attention got any action;
- the organization of NASA was incomprehensible. It was supposed to be a matrix of engineering and project roles. But time and cost were controlled by Mission Control, that is, by project management. This gave it power in the system and led to an informal caste system working outside the formal rules. Over the decades, this changed NASA from an engineering organization to one managed by project staff incapable of dealing with the engineering challenges;
- safety came under a project-management money squeeze;
- NASA lost its clear vision. Congress and the White House had no clear mission for NASA, but this did not stop them from continuous meddling.

The frightening part is trying to understand how so many competent people could create this scenario and live with it for so long until something broke.

In identifying the change from engineering to project management, CAIB unearthed a root cause of the problem. The *Challenger* investigations saw the problem as the difference between engineers and managers. CAIB saw that the real issue was the conflict between engineers and project staff and the relative power each had.

The report comments that complex systems almost always fail in complex ways and it would be wrong to reduce the answer to a simple explanation. Unfortunately, tabling all the issues could allow some people to focus inappropriately on just one aspect.

Unheeded advice
Advice from the Aerospace Safety Advisory Panel and from working-level engineers went unheeded in NASA. The summary in the previous chapter showed that ASAP recognized many of the problems. Warnings from the Augustine Committee and the Government Office of Technology Assessment went unheeded in both NASA and Congress. The Rogers Commission's recommendations to the President were not fully implemented. What does one have to do to get action? One constraint is that those recommending

changes may not have the power to get them implemented. Funding might be limited. Most reviews only have funding until they report. Clout might be limited. One answer is to explore power – first of all the power of good communication and thus to find out where the power is in the hierarchy to get action. The power of networking support should also be recognized. There is no easy answer, especially if the recipients are not motivated to take the advice. It will be interesting to see whether CAIB gets better results than previous investigations.

The Senate's Commerce, Science, and Transportation Committee is holding meetings on the Space Program. Admiral Gehman reminded the Committee that Congress and the White House shared responsibility for the constraints that were the cultural causes of the accident.

Responsibility

Langewiesche reflected in his article **(2)** that no one in NASA has stood up and accepted personal responsibility for contributing to the disaster. O'Keefe has challenged this criticism and says that he has taken responsibility. He did say in a TV interview after the disaster that there was national regret. No one in the White House or Congress appears to have accepted any responsibility.

Mission

President Kennedy articulated a very clear message – call it a vision, goal, or mission – *To place a man on the Moon by the end of the decade.* In April 2002 **(3)**, Administrator O'Keefe presented the new NASA Vision for the future:

- *to improve life here;*
- *to extend life there;*
- *to find life beyond.*

At the same time, he said that NASA's mission is:

- *to understand and protect our planet;*
- *to explore the Universe and search for life;*
- *to inspire the next generation of explorers... as only NASA can.*

On the same occasion, he re-introduced the idea of sending a teacher into space.

Which of these approaches gives the clearest motivation for the staff of NASA?

Questioning the role of man in space

In 1961, Congress asked Simon Ramo, the R in TRW, for his view on manned space flight. Dr Ramo had led the development of the ICBM (**4**). He was hesitant in supporting manned space activities. He saw it as being expensive and risky. Many scientific explorations could be achieved using robotics rather than astronauts. On the other hand, he recognized the emotional boost from having the first man on the moon – particularly in beating the USSR in the Cold War era. Then he had the conflict where his staff and colleagues would love to work on a new space challenge. But he kept coming back to the rational argument that if one rocket in ten were to fail, then the Program would merely cost 10 per cent more. But if astronauts were killed, then the Space Program would grind to a halt. He felt that the Program would suffer from unintended risk-taking, common with government projects. Government contractors outdo each other in competitive bidding to promise better performance, lower costs, and faster schedules. Then costs overrun, schedules slip, and risks are taken all around to try and salvage the Program. People get committed and corners get cut.

Writing shortly after the ***Challenger*** disaster, Dr Ramo pointed out that the Rogers Commission had focussed on the 'O' ring and did not examine the nation's decision-making process when the Shuttle Program was initiated (**3**).

In Congressional hearings in 2003, several scientists repeated the view that few of the space science experiments really needed astronauts and could be conducted remotely. While there may be spin-off benefits, an attempt should not be made to justify the Program on the basis of scientific advances.

Despite these views, there was an underlying opinion, shared by CAIB, that manned exploration of space should continue. This is driven by man's impulse to explore.

The design philosophy

The process of getting the Shuttle Program launched resulted in a specification that was all things to all people. This led to a complex design. A harsh but objective assessment showed that many of the specification points were not met. Starting with a simpler concept might have produced better results.

Comparisons between the shuttle and Soyuz will raise emotions. Some might describe it as comparing the sophisticated with the crude. A rough estimate of the cost of launching a shuttle flight is $500 million and that of a Soyuz flight is $50 million. The shuttle can carry seven astronauts and Soyuz three. The shuttle can carry large and heavy loads into space – maybe two or three times the weight that Soyuz rockets can launch. Soyuz was used to construct Mir; both are being used for the ISS. And both approaches have led to astronauts being killed.

One conclusion is that the first step is to define and agree on the Space Mission before committing a new transport vehicle. The questions *'What is it needed for?'* and *'What is it expected to do?'* have to be answered before starting on the engineering concept. Then allow enough time to develop a sound and preferably simple concept before beginning detailed design.

The dangers of pride
The old adage *pride comes before the fall* is alive and well. It is easier to see it in others than to recognize it in oneself. NASA is in a league of its own, with a long and distinguished history unmatched by anyone else. In that situation, it is reasonable for NASA to take pride in its achievements. It is easy to adopt a frame of mind that says there is no competition – we are so far out front that we cannot learn from anyone, they have to learn from us. But CAIB showed that there were lessons to be learned from the Navy Reactor Program and others. And that exploring high-reliability organizations and normal accident theory, as studied by sociologists and psychologists, could be of benefit.

The future
On 14 January 2004, President Bush announced a new vision for NASA. The vision was described by three goals:

- *first, America will complete its work on the ISS by 2010;*
- *second, the United States will develop a new manned exploration vehicle – the Crew Exploration Vehicle (CEV); and*
- *third, America will return to the Moon as early as 2015 and no later than 2020 and use it as a stepping stone for more ambitious missions.*

To accomplish the first goal, NASA will return the shuttle to flight consistent with safety concerns and the recommendations of the CAIB. The shuttle will be retired by the end of this decade (2010). On the second goal, the CEV will be tested by 2008 and have its first manned mission no later than 2014. The CEV will service the ISS after the shuttle retires. The third goal will have extended human presence on the moon for exploration and developing new technologies. NASA will increase the use of robotic exploration.

Funding for the new programme will total $12 billion over the first five years. $11 billion of this comes from reallocation of current programmes. President Bush proposes to increase NASA's budget by five per cent per year for the first three years and one per cent or less per year in the following two years.

At a press conference following the President's announcement, Administrator O'Keefe said that NASA got a mandate where exploration and discovery were the central objectives. He said that the new tasks *will require a different way of doing business.* In an answer to a question on the CEV, O'Keefe said, *We've got to avoid getting fond of a design. We ought to get fond of the exploration requirement and agenda... One approach... we are very much attracted to... is to engage in a spiral development program, which by increments, demonstrates capacities that are necessary.*

Does this new vision, programme, goals, and funding meet the hopes and expectations of CAIB? The funding and schedule for the CEV has been fixed before the specification and concept. Is this putting the cart before the horse? The CEV will be the largest and most prominent engineering project of the decade. It will be interesting to see how NASA learns from its history. NASA will need strong and clear leadership – capable of managing the political interface, getting sufficient budget, reallocating internal budget and staff, and building a new culture with safety as a priority. A real challenge!

1.8.34   Lessons learned
Lessons from experience with projects
These lessons come from one of the largest projects of the age. But the lessons are relevant for most projects – large and small.

- No commitment should be made to achieve a programme without adequate funding. Hoping that a budget will come later when it is required, is usually wishful thinking.
- Compromises made to obtain approval and initial budget can come back to haunt one.
- Only schedules that are consistent with available resources should be adopted. Although deadlines are an important management tool, they must be regularly evaluated to ensure that

any additional risk incurred to meet the schedule is recognized, understood, and acceptable.

- Beware of the risks of mischaracterizing a product as operational when it is still in the development phase.
- Ensure that those managing projects are capable of understanding the engineering issues.

Lessons on organization

- Changes in organizational structure should be made only after careful consideration of their possible unintended consequences.
- Changes in organizations usually redistribute power. The power structure in the organizations should be checked to see whether it is in line with the needs to get the work done – and has not just created local fiefdoms.
- Unintended centres of power can create their own informal rules. Being informal, they are more difficult for the executive to spot and manage.
- Changes that make the organization more complicated may create new ways that it can fail.
- Clarity, consistency, and continuing commitment are key to getting results in complex organizations.
- Watch for subtle changes, occurring slowly over time, that change the character of the operation. Don't let this evolution allow the organization's skills to degenerate.
- Leaders have to understand and manage the culture in their organization. Reorganization probably will not change the culture.

The importance of good communications

- Time should be taken to make sure communications are thorough and complete. Time should be made to prepare, present, question, debate, and listen.
- People who might appear marginal or powerless in an organization may have useful information or opinions that they do not readily express.
- When there appears to be unanimous consent and when the decisions are said to be obvious, it is worth playing *devil's advocate*. Quietly ask – *prove it to me*. It would have helped if someone had asked *prove there has been no damage from the foam strike*.

The importance of retaining knowledge and memory

- Both engineers and project staff must keep drawings and documentation correct, up-to-date, and easily and quickly accessible.
- Properly recording knowledge needs both discipline and time.
- Institutional memory is lost as personnel and records are moved and replaced.
- Consider how well the knowledge base is being retained when staff are posted or retire.

Lessons in managing risk

- A safety culture is essential in high-risk ventures.
- A safety culture needs relentless and innovative training that highlights and debates lessons that can be learned from past mistakes and near misses.
- Strategies must increase the clarity, strength, and presence of signals that challenge assumptions about risk.
- One must be aware of the cumulative effect of decisions that in themselves appear correct, routine, insignificant, and unremarkable, yet in retrospect have a stunning cumulative effect.
- Separate the Technical Engineering Authority and Safety Authority from the project functions to ensure they do not become subservient to project pressures and budget.
- Always have intellectual curiosity and scepticism when reviewing safety.
- Honestly study worst-case scenarios looking for weaknesses.

And finally three general lessons:

- Accidents are the end result of a chain of events and decisions that occurred over a long time frame.
- Do not just know what is happening – find out why.
- Get the initial concept right – for the engineering specification and concept, as well as the organization. Take time to get it right at this early stage – it will pay handsome dividends.

References

(1) Columbia Accident Investigation Board Report, Vol. 1, August 2003, S/N 033-000-01260-8, US Government Printing Office, Washington DC, Vols. II to VI, October 2003. Accessible at www.caib.us/ and at www.nasa.gov/columbia/home/index.html

(2) **Langewiesche, W.** (2003) *Columbia's Last Flight, The Atlantic Monthly*, Boston, November 2003, p. 58.

(3) Edwards Deming, W. (1982) In *"Out of Crisis"*, ASIN 0 91137 901 0 he tabled 14 points for improving management and commerce.

(4) Ramo, S. (1988) *The Business of Science: Winning and Losing in the High-Tech Age*, Hill and Wang, New York, ISBN 0 80903 255 4.

.

Roll-on/Roll-off (ro–ro) vessels and particularly the vehicle/passenger ferries are one of the most successful types of ship today, especially for short sea crossings. The safety concern is that they have an undivided deck running from stem to stern. They can rapidly lose stability and roll over if water enters the vehicle deck.

1.9.1 History of ro–ro ships

Road vehicles have been transported across sheltered waters on barges since the earliest times. But the first ship with a continuous deck, accessed by doors, came with the development of the railways. She was the rail ferry across the Firth of Forth in 1851. The ferry was the invention of Thomas Grainger. Thomas Bouch, later the designer of the ill-fated Tay Bridge, designed the loading docks – no easy task considering the wide tidal range he had to take into account.

By the 1890s, ro–ro train ferries were in operation between Sweden and Denmark, and across Lake Michigan. In 1895, Sir W.G. Armstrong Whitworth & Company Ltd. of Newcastle-upon-Tyne won the order for an icebreaker train ferry to cross Lake Baikal in Siberia. This ferry provided the connecting link for the Trans-Siberian Railroad. The ferry was shipped in over 7000 pieces and reassembled on Lake Baikal under Armstrong Whitworth technicians. The ferry had three parallel tracks, was double hulled, and had numerous watertight subdivisions.

Surprisingly, it was not until 1936 that the Southern Railway and SNCF established the first ro–ro train ferry across the English Channel. The *Golden Arrow* train took this route.

The big development in ro–ro ferries came in the 1950s and was triggered by the successful experience with tank-landing craft during WWII. The Port of Dover provides a typical example. Dover had handled 10 000 cars per year, loaded by crane. The first drive-on service started in 1953. In the first year, traffic increased over ten times and by 1994 had risen to 4.5 million vehicles/year.

There are around 4500 ro–ro ships in service around the world, representing nearly eight per cent of the world's fleet.

1.9.2 Accidents

The first major loss was the **Princess Victoria**, a British Railways (then government owned) ferry en route from Scotland to Northern Ireland, in 1953. She was retreating from a violent storm when heavy seas buckled the stern door, and 133 lives were lost.

The cargo ro–ro **Hero** was lost in 1977, partly as a result of water entering through a leaking stern door. In 1994, over 900 drowned when the **Estonia** had her bow visor torn off in heavy seas, and she quickly capsized when water entered the vehicle deck (**1**). The **Estonia** was travelling from Tallinn to Stockholm at night.

Accidents have happened in sheltered waters, and in good weather. The **Straitsman** sank when the stern door was opened too soon. In 1987, the **Herald of Free Enterprise** rolled over outside Zeebrugge harbour in less than two minutes, and 193 passengers and crew drowned. In 2000, the **Express Samina** sank quickly off the Greek Islands when she ran onto rocks at night, and 80 died. In 2002, the **Stena Gothica** took on water on the vehicle deck when she collided with the jetty at Immingham after the Master underestimated the tidal current.

Ro–ro ships that have sunk rapidly as a result of collisions include the **Jolly Azzurro** (1978), **Collo** (1980), **Tollan** (1980), **Sloman Ranger** (1980), **Ems** (1981), **European Gateway** (1983), and **Mount Louis** (1984) (**2**). In 2002, the cargo ro–ro **Tricolor** sank in the English Channel after a collision in fog. All the crew escaped, but nearly 3000 luxury cars sank to the bottom when she capsized.

The **Seaspeed Dora** capsized in 1977 when movement of the cargo caused a list sufficient to allow water to enter through an open bunkering door. Other ships lost because of a shift in cargo are the **Espresso Sardegna** (1973), **Zenobia** (1980), and **Meckhanik Tarasov** (1982, in very bad weather) (**2**).

In 1988, the **Scandinavian Star** suffered a major fire in the passenger accommodation section en route from Norway to Denmark and 165 died.

1.9.3 Herald of Free Enterprise

There are many aspects to the sinking of the **Herald of Free Enterprise**. She was operating in a highly competitive, cut-throat environment. The owners, Townsend Thoresen, had been taken over by P&O a few months before the accident.

The **Herald of Free Enterprise** had two vehicle decks. The facilities at Zeebrugge could only easily handle loading from the lower deck. So the practice was to fill the ballast tanks to lower the bows so that the ramp could reach the upper deck. Once the upper deck was loaded, the ballast tanks should have been trimmed to balance the ship before sailing, and the loading should have been recorded. But the ballast tank pumps took too long to trim the vessel, and so she routinely left port without being trimmed.

The assistant bosun handled the bow-door operations. On this occasion, after unloading, he went to his cabin for a brief break. He fell asleep and did not hear the warning signal for departure. The bow-doors could not be seen from the bridge. An officer thought he had seen the assistant bosun go towards the doors. The bridge assumed the doors were closed and sailed. The bow was 2–3 ft lower than the correct balance. As the ferry accelerated out of the harbour, the bow wave entered the vehicle deck at a rate of 200 tons/min. She immediately listed 30 degrees, and in 90 s, rolled over onto her side on a sand bar.

Justice Sheen ran the inquiry into the sinking, under the Merchant Shipping Act (3). It emerged that there had been plenty of warning signs of risky practices. On several occasions, ferries had sailed before the bow-doors had been closed. Crew had suggested that a switch be attached to the doors with a light on the bridge showing when the doors were closed. Management had rejected this proposal. The number of officers on board had been reduced from four to three. With shift rotation, they did not regularly work with each other or with the same crew.

Justice Sheen concluded that the company, right up to the Board of Directors *did not appreciate their responsibility for the safe management of their ships.* From top to bottom, there was a *disease of sloppiness.* Shore management took little notice of what was said by ship's Masters.

The company directors were charged with manslaughter and reckless conduct. This was the first time that directors had been charged with manslaughter. The case did not succeed, largely on technicalities as a result of the maritime law not being up to date with the technical evolution of ro–ro vessels (4).

1.9.4 Basic safety principles
Safety of engineering structures and components follows a hierarchy:

1. the basic concept precludes major accidents; or
2. features are added to prevent a major accident; or
3. safety equipment is provided to mitigate the effects of an accident; or

4. control or evacuation procedures are introduced to limit the risks from an accident;

5. and, if all else fails – put up a warning sign.

The aim should be to work as high up this list as possible.

1.9.5   How do ro–ro ships meet these safety steps?

The basic concept of a ro–ro vessel, having a through deck, breaches step 1. A fundamental tenet of Naval Architects is to subdivide ships into watertight compartments. This is a long-standing philosophy, despite the loss of the *Titanic*, but cannot be easily applied to ro–ro ships.

Water on deck

The next step is to limit the water on the deck and improve stability. Making sure that doors are closed and strong enough are obvious steps – especially in hindsight. But water can also enter as a result of a collision. Then the design task is to limit the amount, and movement, of the water. The amount on board can be reduced by pumping it back overboard. Strathclyde University proposed a scheme where the water drains from the side of the deck into central tanks in the hull and is pumped out from there.

The stability can be improved by fitting transverse bulkheads. But these have to be moveable to allow drive-on/off facilities. They also add another constraint to getting the maximum vehicle loading. Longitudinal bulkheads can also improve stability. Studies at Portsmouth University (5) show that one central bulkhead can reduce the lack of stability due to water on board by a factor of four. Another innovative idea they came up with to stop water slopping about on the deck is to use a perforated deck. Water drains into sub-floor tanks that can then be pumped out with powerful pumps (Fig. 1.9.1).

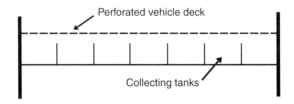

Fig. 1.9.1   Concept of perforated deck to drain flood water (6)

It is interesting that bulkheads were fitted to ro–ro vessels commandeered for the Falklands War, but were removed when they returned to commercial use.

Stability

Predicting and controlling stability have been exercising the designers of ro–ro vessels for some time now. The key factors are the sea state – height of waves, freeboard, and metacentric height (Fig. 1.9.2). The centre of buoyancy B moves to B′ as the ship heels. The metacentre M is where the vertical line through the buoyancy meets the centre line. The distance between the centre of gravity G and metacentre M is the metacentric height GM. Rolling stability is provided by the moment arm GZ.

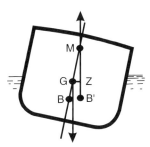

Fig. 1.9.2   Diagrammatic definition of metacentric height GM

A recent assessment of many ro–ro vessels operating in the 1970s showed that they could capsize quickly in heavy seas if any water gathered on the vehicle deck.

Evacuation

The aim is to enable complete evacuation before the vessel lists or capsizes. The target for ro–ro ferries is to remain stable for a minimum of between 45 minutes and an hour. The catamaran ferry *St. Malo* highlighted the problem of predicting evacuation times. In trials under controlled conditions, it took just eight minutes to evacuate 357 passengers. Later the same vessel ran aground and it took 1 hour 17 minutes to evacuate. Predicting evacuation capability has now become a speciality of academic groups such as the one at Strathclyde (**6**).

Design features like handrails, adequate space, and off-loading facilities can aid evacuation. So can the ability of the crew to handle the emergency and control the crowd.

Lifeboats

A long-standing marine safety measure is to provide enough lifeboats. Photographs of the ***Herald of Free Enterprise*** and ***Tricolor*** show lifeboats still in their stowed position – unable to be launched from a rapidly listing ship. Modifying the position and type of lifeboat can improve the situation. Stern launching goes some way to curing this problem.

Another issue when passengers are in the water is hypothermia and the need to pick up passengers as quickly as possible. One concept is to have fast rescue launches on board. But the crew have to be experienced in the use of these craft otherwise they may create more risk than they remove.

Several safety ideas have been tabled to aid quick evacuation. Ideas similar to aircraft evacuation chutes that can quickly get passengers with life jackets into the water and clear of the vessel.

Summary

In total, there are plenty of innovative ideas to help improve safety. These cover the range of safety principles from limiting the risk of capsizing to improving the chance of survival.

1.9.6   Who calls the tune?

There are a number of parties involved with any shipping business. They are the:

- owner;
- designer;
- builder;
- ship registry;
- classification society; and
- insurer.

These players are all linked together by money. The owner generally needs financing. The financier will not be satisfied unless the vessel is registered, classified, and insured. The designer and builder have to satisfy the other players.

Ship registry

In the United Kingdom, the Merchant Shipping Act of 1786 required all ships to register with the customs in their home port. Maritime nations all have their own ship registries. But owners can now choose where to have their ships registered and which flag to fly. The old traditional shipping nations kept high fees, high taxes, and strict rules. Flags of convenience became popular in the second half of the twentieth century. They might be described as 'flags of freedom' – freedom from high registration fees, freedom from high taxes, and freedom from strict rules – particularly on labour relations and wages. About half the world's fleet uses flags of convenience. The standards of flags-of-convenience nations vary. Some are not obvious. Liberia is the second largest registry but is managed by a private US company with the President's office on Park Avenue, New York. Registries for the Marshall Islands and Panama are also managed from the United States. Registries rely on classification societies to place a vessel in a classification and carry out surveys.

Classification societies
Marine insurance originated in early commercial history – even in Babylonian times. But underwriters needed some basis for knowing the state of the ships they were insuring. Edward Lloyd came up with a register in the mid-eighteenth century. Hulls were classified as A, E, I, O or U according to how good they were. Machinery was classed 1, 2 or 3. Hence, *A1 at Lloyd's* means the best. By the mid-nineteenth century, several classification societies had formed as neutral third parties to set rules for construction and conduct surveys. There is an International Association of Classification Societies (IACS) with ten members and one associate.

In recent decades, there has been criticism of classification societies as being too prescriptive, not being current with the advances in ship technology, and not being rigorous enough with surveys. IACS is addressing these issues but also points out that classification societies have no powers of enforcement (**7**). Responsible ship owners should themselves be employing qualified superintendents to ensure quality is maintained.

The shipping industry
In effect, classification societies provide the industry with a degree of self-regulation but without anyone to enforce the rules. The shipping industry operates in the broader environment of transport, trade, and society at large. As a result, it is subject to the various laws, rules, and regulations set by politicians and regulators.

1.9.7   Regulations and regulators
Governments
Laws applying to the seas have evolved over centuries and in a haphazard manner as political pressures dictated. In the United Kingdom, there was concern in the early nineteenth century about the number of shipwrecks. The Merchant Shipping Act of 1854 introduced statutory control over ship safety. The Act gave the Board of Trade the right

to inspect ship construction and required every steam ship to have a certified engineer on board. After lobbying by Samuel Plimsoll MP, the Unseaworthy Ships Bill of 1876 made the Plimsoll line a legal requirement. At the same time, it was defined as a crime to send a ship to sea in an unseaworthy condition.

Following the sinking of the *Titanic*, the UK government called a Safety of Lives at Sea Convention (SOLAS) in 1914. The Convention established an International Ice Patrol and set minimum standards for radio-communication and life-saving equipment for passenger ships. WWI interrupted the ratification process, and another convention was held in 1929 that concentrated on fire-resistant bulkheads. By 1933, the required number of countries had ratified the 1929 Convention for it to come into force. The United States feared outside control and did not ratify until 1936. Ratification only occurred because of public pressure on Congress after the loss of 124 lives in the fire aboard the *Morro Castle* off the New Jersey coast.

After the *Herald of Free Enterprise* accident, the UK government enacted the Merchant Shipping Act of 1988. This omnibus Act updated previous acts going back to 1894 but more particularly strengthened the responsibility of owners and Masters for safety and seaworthiness. The aim was to make it clear that everyone involved with ship management, both on land and at sea, could face prosecution. It also established a Marine Accident Investigation Branch.

IMCO/IMO

The United Nations Conference of 1948 formally established an Intergovernmental Maritime Consultative Organization (IMCO) based in London. The IMCO took over the SOLAS Conventions, and its first major task was the 1960 SOLAS Convention that came into force in 1965. The 1960 SOLAS Convention continued the maritime tradition, used by the classification societies, to use prescriptive and deterministic rules.

In 1982, the IMCO was renamed the International Maritime Organization (IMO). The earlier title reflected the modus operandi of the organization – to persuade the member states, by consultation, to take co-ordinated actions. IMCO/IMO works through committees to generate views and recommendations. These then have to be tabled at conventions where, through debate, the aim is to get them accepted and ratified by members. IMO has no regulatory powers and is dependent on the governments of member states to pass national laws to implement and enforce the decisions. The whole process is lengthy. Initially, it took nearly ten years to get IMCO decisions into effect.

Members were hesitant to move forward. Now the time frame from convention to ratification is around two to five years.

Following the **Herald of Free Enterprise** accident, a series of requirements to improve the stability of passenger ships were included in the SOLAS 90 standard. For the first time, the pressure to take action reduced the time from adoption to entering into force to 18 months.

The **Estonia** accident led the Secretary General of IMO, William A. O'Neil, to call for a review of ro–ro safety. He established a Panel of Experts. Its proposals were considered at the 1995 SOLAS Conference. Several recommendations were adopted, in particular, the one stating that existing ro–ro ferries must meet the SOLAS 90 damage stability standard. Previously most IMO requirements had not been retroactive. It was agreed that the SOLAS 90 standards should be implemented over a period up to October 2005. But the recommendation that the standard be enhanced to include the ability to handle a given amount of water on the vehicle deck met with controversy. The Conference, after a lot of back-room negotiation, accepted a resolution by eight north-west European countries that they could adopt the higher standard within the framework of IMO. It became known as the Stockholm Agreement. The signatories were Norway, Sweden, Denmark, Finland, Germany, the United Kingdom, Ireland, and the Netherlands.

The Stockholm Agreement

The Stockholm Agreement standard is SOLAS 90 plus up to 50 cm of water on the vehicle deck – SOLAS 90 + 50. While the standard is deterministic, the resolution has allowed an alternative 'equivalent' standard. This allowed the demonstration of adequacy by a combination of numerical analysis and model testing.

The Agreement, for the first time, explicitly required water on board and wave action to be taken into account in assessing ship safety. In addition, it introduced performance-based standards.

The signatories decided to reduce the time for complete compliance to 2002.

The European Union/Commission

Within the European Union (EU), the European Commission has a long-range Transport Policy for all forms of transport. As part of this policy, it has a white paper aimed at raising the safety standard of passenger ships by 2010.

The EU has observer status at IMO. The IMO, like other UN agencies, has membership from signatory states. The EU is not comfortable with this situation and would prefer direct membership rather than trying to ensure that all EU nations work together within IMO.

The EU has concentrated on ensuring that all IMO decisions are implemented and enforced by EU member states. However, they feel that some IMO standards have been diluted during the negotiation stage in order to achieve ratification.

The policy of the EU is to have consistency among member states. Hence, the Stockholm Agreement, only applying to the Irish Sea, North Sea, and Baltic, is anathema to Commission staff. The loss of the *Express Samina* in Greek waters in 2000 highlighted their concern. As a result, Directive 2003/25 was tabled on 14 April 2003. This extends the Stockholm Agreement standards to the whole of the EU by 2010, for international trips. Ships not meeting the standard will be phased out after 30 years of service or by the deadline of 2015.

Requirements for voyages in domestic waters have been separated into four categories depending on the worst sea states. Ships operating domestic voyages in category A and B will also have to meet the Stockholm Agreement standards by 2010.

The Commission requires EU member states to bring their laws into force to comply with this directive by November 2004.

1.9.8   Technical developments

Ro–ro ships are complex to design. Predicting the stability of both intact and damaged ships with water on the deck further complicates the design work. The sinking of the *Herald of Free Enterprise* and *Estonia* highlighted the stability issue and led to major R & D programmes, particularly in Europe. The UK Department of Transport and EU helped fund these programmes. Hydrodynamic tank testing facilities and specialist academic groups spearheaded the research. The work has followed two broad lines, model testing and analysis. Both routes were used to make comparisons with the prescriptive criteria adopted by SOLAS 90 and the Stockholm Agreement.

Early work showed that a simple separation of safe/unsafe conditions could be established by plotting metacentric height against the significant wave height. The safe/unsafe boundary was not a single line but a band of uncertainty.

Modes of capsize
Capsizing can result from several paths: loss of static stability, loss of dynamic stability, and broaching (**8**).

- Loss of static stability occurs when the righting arm GZ is critically reduced. This can occur when a following wave of critical length and steepness overtakes the ship.
- Loss of dynamic stability takes several forms:
  - dynamic rolling with coupling of roll, surge, sway, and yaw. The typical motion is rolling heavily to leeward at the crest of the wave and windward in the trough;
  - parametric excitation can occur with coupling of rolling motion and wave action. It can occur when the wavelength is of the same order as the ship's length in both head and following seas;
  - resonant excitation occurs when the natural dynamics of the ship are excited by wave action;
  - impact excitation when steep and breaking waves can cause severe rolling;
  - bifurcation when one of the above modes switches to another.
- Broaching occurs when the wave conditions result in the inability to keep course and leads to an unintended change in direction. This can allow breaking waves from a beam sea to swamp the ship.

Analysis and model testing

Predicting the risk of capsizing is no simple task. It is highly dynamic, non-linear, and probabilistic. Despite these difficulties, there have been some significant advances. Calculations have been able to predict and analyse the results from model testing with reasonable agreement. As a result, the combination of analysis and model testing has shown that the prescriptive requirements of SOLAS 90 give an acceptable protection against capsizing. In the more complicated case of the Stockholm Agreement with water on deck, the prescriptive requirement of 50 cm is unrealistic and over stringent. However, the *equivalence* route allows for a more realistic assessment.

The *equivalence* route has allowed most existing ro–ro passenger ships to be assessed for meeting the sea conditions on the specific route that each vessel operates. With effective co-operation within the industry, the Stockholm Agreement has been implemented with fewer problems than many feared. Ship owners and designers have found innovative ways to upgrade where necessary.

Design approach

The experience with analysis and model testing of existing ships, and the growth of new knowledge and understanding, have been the catalyst for changing the approach to ship design. Professor Vassalos, the Director of the Ship Stability Research Centre at Strathclyde University, is promoting a first-principles approach by adopting a risk-based design, with safety as a prime objective (9). This is a fundamental change from the traditional approach of first completing the ship design and then adding on safety features to meet prescriptive code requirements. Professor Vassalos aims to have safety assessed along with resistance/speed, sea keeping, stability, and load/unloading features. Dealing with safety *up-front* in the design process allows for greater freedom of choice and for safety to be *built-in* before the design is frozen. Adding safety features later is much more costly.

An example of meeting the Stockholm Agreement standards

The Danish ferry company DFDS has upgraded its fleet of ro−ro passenger ferries plying the Baltic and North Sea by using the *equivalence* approach to calculations and model testing. The company has described the modifications to nine ships built between 1975 and 2002 (10). Ships were upgraded for specific routes. The significant wave height varied between 3.1 and 4 m depending on the routes. Four of the ships have had a transverse flood control door added and another vessel has had two doors added. Multiple watertight compartment tanks or sponsons have been added to the sides of five ships. These generally run the length of the ship and are welded to the sides. The company has increased the metacentric height by between $1\frac{1}{4}$ and 4 m. This reduces the roll period and stabilizer efficiency that in turn reduces passenger comfort. Changes to tanks and buoyancy boxes within the hull have been made to three ships. Stern appendix, ducktail, and changes to the bulbous bow have been made on four ships to improve hydraulic performance and offset some of the penalties of the other changes.

1.9.9 Actions by some other countries outside the Stockholm Agreement

Australia

Australia is not a signatory to the Stockholm Agreement but has adopted the standards form that Agreement for its passenger ro−ro ships.

Canada

Canada has ro−ro passenger ferries on both the Atlantic and Pacific coasts and on the Great Lakes. As a result, Transport Canada, a Federal Agency, has closely followed international developments as well as conducting a major experimental program. From this work, Canada did not consider SOLAS 90 + 50 a rational standard and did not sign the Stockholm Agreement. Tank tests showed that flapped freeing ports are an effective way of getting the bulk of floodwater back overboard.

In terms of analytical approaches, Transport Canada reviewed the international work rather than developing a new approach. The Static Equivalent Method (SEM) developed at Strathclyde University offered the best potential. This method is based on the observation from model testing and calculation that a vessel capsizes close to the angle where the righting moment is at a maximum. Before this point, accumulation of water on deck slowly increases the rolling motion. After this point – the point of no return – additional water exponentially increases the roll until the vessel capsizes. This observation allowed the critical volume of water on board to be calculated from the hydrostatic properties of the hull, without having to go through complex dynamic non-linear analysis. Meanwhile, the likely accumulation of water can be estimated from the significant wave height and compared with the critical volume. This approach can also estimate the critical sea conditions for a particular ship and hence define operating limits. Transport Canada recognized that more work was needed to understand damage stability and that more needed to be done to gather reliable data and statistics on accidents.

1.9.10 Maximum wave

The significant wave height is defined as the average of the vertical distance between the crest and trough of the third highest observed wave height over a given time. The report on the sinking of the *Estonia* showed how difficult it is to know exact wave conditions. Various Baltic Agencies and Institutes had estimated the significant wave height at the accident site at between 4.1 and 5.4 m. But waves can exceed this height. The probability of a wave being 50 per cent higher was quoted as 1:100. Masters of ships that came to the area to assist estimated the seas to be 4–6 m. One helicopter pilot reported a single very high wave of 12 m measured by radar.

Douglas Faulkner has been promoting an extensive study to understand the real probability of rogue waves for shipping in general (11). A MaxWave programme has been initiated in ten projects in six countries. Initial results show that a maximum wave can reach 2.9 times the significant wave height. Furthermore, the shape of the wave affects the loads it places on the ship, as does the velocity of the crest of a breaking wave. This evolving knowledge is showing that current ships are underdesigned and cannot handle maximum waves.

1.9.11   Statistics

How safe are passengers on ro–ro ferries? The statistics for voyages from the United Kingdom were studied by John Spouge (12) in 1988. He looked at the records from 1950, covering 28 million ferry crossings. Assuming a typical passenger takes a return ferry trip once a year for vacation, then the risk of death was 5×10^{-7}. UK road accidents in 1988 resulted in a fatality rate of 1×10^{-4}. Hence, a ro–ro journey was about three orders of magnitude safer than the annual risk of driving on the roads.

Another view is shown in Fig. 1.9.3 (13). The frequency of accidents is plotted against the number of fatalities. This shows that ro–ro passenger ferries are safer than non-ro–ro ferries.

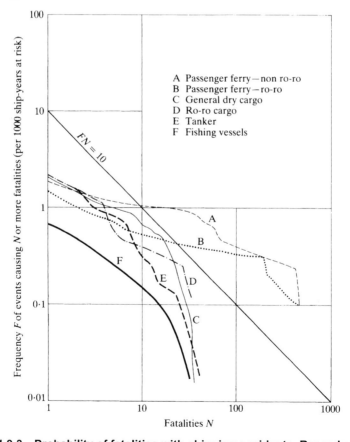

Fig. 1.9.3   **Probability of fatalities with shipping accidents. Reproduced courtesy of Lloyd's Register**

It is generally assumed that the public expects a reduction in the probability of an accident with an increase in the number of fatalities. Hence the reference line FN = 10. This shows the need to reduce the risk of large accidents for all ferries by about three times. The statistics cover the world and include some horrendous accidents with overloaded ferries in third world countries. This highlights the need to continue to reduce the risk of major accidents.

The future

The continuing popularity of ro–ro ferries appears assured. They are now used for cargo-vehicle service as well as for passenger/vehicle ferry service. There are commercial advantages in larger sizes and higher speeds on the busy routes. The harbour facilities often limit the length, so the trend is for more decks and higher superstructures.

Current ro–ro passenger ferries have around 3 km of vehicle lanes on board. Newer ferries are raising this to around 4 km with four or more decks. Speeds have increased to around 23 knots. In the smaller size ro–ro passenger ferries, speeds up to 40 knots are coming into service use. They have a variety of hull designs – mono-hulls, multi-hulls and hydrofoils.

Cargo ro–ro vessels serve the trans-ocean market and are also growing in size. They carry items that cannot easily be fitted into containers. An example is the *Tamesis* built by Daewoo in South Korea, for Wilhelmsen. She can carry a range of vehicles from locomotives to cars, to trailer trucks, and construction equipment. *Tamesis* entered service in 2000 and has a double-skinned hull and nine car decks capable of carrying around 5500 cars. The service speed is 20 knots. Wilhelmsen ordered more ro–ro cargo ships from Mitsubishi in April 2003. These will have the ability to carry up to 6400 cars.

COMMENTS

A lot has changed in the last couple of decades. The aftermath of the sinking of the *Herald of Free Enterprise* and *Estonia* could have been limited to strengthening the responsibility of ferry owners, providing indication of the closure of sea doors on the bridge, and strengthening of door-locking mechanisms. But naval architects were expressing the view that the ro–ro vessels existing at the time had an inadequate probability of surviving more than 30 minutes in many sea states that they were designed to operate in. And their stability was inadequate in both the undamaged and damaged state.

There has been a sea change in the attitude towards safety, by all players. The evolution of the technology from prescriptive to probabilistic to risk performance has been remarkably rapid. As has the move towards a first-principles approach to design. Following the accidents, few would have predicted that such a strong role in these changes would come from academia.

Is this a more scientific approach?

Design and analysis has moved to a more scientific approach. There has been a vastly improved understanding of the stability of ro–ro vessels. But this has come from the application of non-linear mathematics and model testing rather than from science. The improved understanding comes from an empirical base.

What has changed?

Over and above the general changes in approach, there have been many changes included in regulations. Some of these are:

- indication of the state of vehicle deck doors has to be available on the bridge – plus an audible alarm;
- CCTV surveillance of loading doors, key areas where water can enter, and vehicle decks has to be available on the bridge and in the engine room;
- confirmation that all doors are properly closed before getting under way;
- stability calculations have to be made and the handling characteristics predicted, and be available on board. This information is aimed at helping the Master make better decisions on speed and heading under extreme conditions. Fortunately, there are few emergencies and experiences of extreme weather. But this means that Masters get few chances of knowing the handling characteristics in extreme conditions;
- provision should be made for operating books, with operating limits identified, to be available on board;
- regular inspections and stability checks to take place;
- improved drainage of water from decks;
- improved strength, securing, and locking of vehicle deck doors – standard issued by IACS;
- data to be recorded for each voyage – in effect a ship's 'black box';
- a unique identification number and transponder to be provided for each vessel similar to aircraft – to aid in recognition in the event of an accident;

- automatic draught gauges to be provided;
- vehicles to be weighed;
- improved life-saving equipment;
- fast rescue boats to be carried on board;
- improved access ways for emergency evacuation – handrails, vertical divisions to be capable of being walked on in the case of extreme heel;
- improved emergency lighting;
- a Safety Management Manual to be produced and a copy kept on board;
- safety procedures to be regularly checked and audited;
- crew to be trained for emergencies;
- crowd management, crisis management training and education in human behaviour to be given to crew;
- adequate language skills to be demonstrated to ensure good communication between all the crew;
- improved communication systems with search and rescue organizations operating within range of the operating route;
- improved fire protection and two-way radios for fire patrols;
- improved smoke detection, alarm, and sprinkler systems;
- identification of all passengers and provision of boarding cards so that the passenger list is accurate;
- passengers to bc banned from the vehicle decks when the ship is under way;
- improved stowage and securing of cargo – including within trailers.

Speed of regulatory decisions

At first sight, the speed of decision taking, particularly through the IMO process, is painfully slow. On the other hand, the sheer number and importance of the changes made is a credit to the IMO. Safety now appears to be a key priority for the whole industry. It is worth being patient to get measurable improvements to safety.

It is ironic that it has taken a century to return to the double hull and watertight compartments used on the Lake Baikal ferry.

The role of the individual

While the changes have no doubt been led by a few key individuals, the broad changes have been achieved by the co-ordinated effort of many. Good communication of the technical developments and their application has come through the work of Institutions such as RINA and SNAME as well as through IMO Conferences.

Are ro–ro passenger ferries safe enough?

The statistics show that journeys on these vessels are safe and within what is considered an acceptable transport risk. The shock of accidents with large loss of life has generated public pressure for improvement. The changes either implemented or in the pipeline should make a significant improvement – particularly in the EU. There are some caveats before comfortably accepting that everything is fine. They are:

- stability knowledge is from 1/40 scale models and analysis. It is not easy to demonstrate how precise the predictions are for ships in the absence of actual capsizing experience;
- the probability of collisions and the resultant damage are still large variables. The fact that two ships ran into the wreck of the *Tricolor* despite its being well posted does not give great confidence in the average level of seamanship in congested waters. With higher speeds, different bow shapes, and hull designs, the type and energy of collisions is a moving target;
- the developments and improvements are mainly taking place in the developed world. How quickly can these be applied in the third world where there are other pressing priorities as well as an inadequate safety culture?
- there is an almost mythical tradition for those who go down to the sea in ships. The hazards of the sea are accepted – accepted as acts of God. Disasters are followed by services and memorials. A respectful rendering of 'for those in peril on the sea' and then its on with life as usual. The answer to storms at sea in the 107th psalm is to pray. In this climate, there is little recognition that naval architects and marine engineers can and should know more about extreme conditions, and design for them. The MaxWave exercise has the promise of challenging this outdated acceptance of rogue weather accidents;
- the future growth of ro–ro ships looks assured, as well as being rapid. Can the regulatory process keep up with the pace of these changes?
- is there a danger of confusion with the various levels of regulation? Is over-regulation a possibility with competing jurisdictions each trying to take the lead?
- fortunately ro–ro ships have not experienced a major fire on the vehicle deck. The potential for this is high with so many vehicles each with a fuel tank, as well as flammable loads, all

in a confined space. The margarine and flour fire in the Mont Blanc Tunnel is the type of situation that needs to be avoided;

- while significant improvements have been made to safety, the basic design of ro–ro vessels still allows a sudden inrush of water in an accident.

1.9.12   Lessons learned

- The application of mathematics and sound engineering practices is a powerful means of gaining a sound understanding of a problem.
- In a growing technology relying on rules based on past practise is inadequate for preventing the next accident in a growing technology. Particularly, if the background for the past rules is not based on a broad understanding of the problem.
- Complicated problems inevitably need dynamic analysis. A thorough dynamic analysis may be both expensive and time consuming. So look for simple static equivalent assessments that can be effective over at least part of the range.
- Lots of small changes and work by individuals can eventually change attitudes within an industry. But it might take time. Have patience and persistence.

References
(1) Final Report of the capsizing on 28 September 1984 in the Baltic Sea of the ro-ro passenger vessel MV "Estonia", Joint Accident Investigation Commission, 1997.
(2) Focus on IMO, January 1997.
(3) **Justice Sheen,** *MV Herald of Free Enterprise*, Report of Court No. 8074, Department of Transport, London.
(4) **Crainer, S.** (1993) *Zeebrugge: Learning from Disaster, Lessons in Corporate Responsibility*, ISBN 0 95199 950 8.
(5) **Ross, C.F.T., Stothard, S.** and **Slaney, A.** (2000) Dynamic stability characteristics of model ro-ro ferries, *Marine Tech.*, **37**(1).
(6) **Jasionowski, A., Dodsworth, A., Allan, T., Matthewson, B., Paloyannidis, P.** and **Vassalos, D.** (1999) Time based survival criterion for passenger ro-ro vessels, *Trans. RINA*.
(7) **Hidaka, M.** (2001) The role of class in meeting the safety challenge, *Seatrade Safe Shipping Conference*, April 2001.
(8) The Specialist Committee on Stability, Final Report and Recommendations of 22nd ITTC.
(9) **Vassalos, D.** (2000) *Contemporary Ideas on Ship Stability*, Elsevier, ISBN 0 08043 652 8.

(10) Hansen, K.E. and **Dalgaard, J.** (2002) *Conversion of ro-ro Passenger Ships as a Consequence of SOLAS95*, Danish SNAME, October.

(11) Faulkner, D. (2002) *Shipping Safety, A Matter of Concern*, Ingenia Aug./Sept.

(12) Spouge, J. (1996) Safety assessment of passenger ro-ro vessels, *Int. Sem. RINA*.

(13) Blockley, D. (Ed.) (1992) *Engineering Safety*, McGraw-Hill.

Whenever man has stood at the edge of a river, there has been a desire to get to the other side, a desire to be able to cross quickly and easily, to walk across rather than swim. Bridges have been around for over three millennia. The Romans built roads to Rome from all their far-flung provinces. This created the need for many bridges. The most famous of their arch construction is the aqueduct, Pont-du-Gard in France, built around 19 BC. In the last two centuries, the era of travel by road and rail has brought an explosion of bridge building. Bridge design has evolved from beam through arch to suspension, cantilever, and cable stayed.

Bridges can give enormous economic advantage to a community by making travel easier and by developing trade. We use them all the time, we rely on them. Bridges are symbolic and add to the image of the locale. London, Paris, Venice, New York, San Francisco, and Sydney would be poorer economically and aesthetically without their bridges. Bridges blend art and engineering. A beautiful bridge can lift the spirits as well as transport you quickly over water or a chasm. The great sixteenth-century Italian builder Palladio caught the essence when he said:

> *Bridges should benefit the spirit of the community by exhibiting convenience, strength and delight.*

1.10.1   Bridge failures

Unfortunately, bridges can, and do, fail. There are spectacular bridge failures and more mundane problems. Fortunately, bridge failures have generally not resulted in many deaths, which is why there is little press coverage for the mundane failures.

1.10.2   Status of bridges in the United States

The status of bridges in the United States is probably typical of that in most developed countries. A 1991 survey showed that 823 bridges had failed since 1950 (1). Flow hydraulics, or scour of the foundations, caused 60 per cent of the failures. Scour was caused by tides, flood water, normal river flow, and debris.

A National Bridge Inventory monitors 600 000 bridges on public roads (2). This is one-third of all bridges, many of the rest are small and on private roads and rail tracks. Most of the inventory is concrete followed by steel, iron, and masonry. About half of the bridges are between 25 and 50 years old, with the remainder equally split, older and newer. Over 12 000 bridges are more than 100 years old. The 2003 Inventory lists 25.8 per cent structurally deficient or fundamentally obsolete bridges, which do not meeting current safety standards (3).

Most bridges are inspected every two years. The large number of bridges built in the 1960s as part of the Eisenhower Interstate Highway Building Program are now nearing the end of their design life. The funds needed to improve all bridges are over $200 billion.

The US Department of Transport, Federal Highways Administration considers that the National Bridge Inspection Program is one of the Government's most effective safety programs. They believe that this has resulted in virtually no catastrophic bridge failures in recent years. Most failures that do occur are, in their view, a result of natural disasters – earthquakes and extreme flooding.

1.10.3   The strange case of the bridge at Ynysygwas

In December 1985, the bridge at Ynysygwas fell down under its own weight, without warning. The 18.3 m span crossed the River Afan in South Wales. It was built in 1952 using a new segmental post-tensioned beam construction. This form of construction gave longer and more slender spans.

The failure was due to corrosion of the longitudinal prestressing cables. It had been inspected annually, with particular attention paid to possible scouring. A special inspection had taken place five years before it failed. This was to allow the bridge to be crossed by a heavy transport vehicle carrying 30 ton beams. None of the inspections had shown any sign of problems – no corrosion, staining, cracking, or deflection. The corrosion that caused the failure had come from chlorine ingress into the grouting mortar – possibly from de-icing salt as the bridge was on a fairly steep hill. The area that corroded and failed was buried and not visible for inspection. Bridge engineers in the United States have noted that concrete bridges can deteriorate rapidly towards the end of their life.

There are about 155 000 bridges in the United Kingdom, and 74 264 of these are owned by local authorities. The estimated replacement value is £15 606 million. Local authority bridges are 43 per cent masonry and 38 per cent concrete.

1.10.4   A selection of landmark bridge failures

A study in 1977 by Paul Sibly and Alistair Walker reviewed some landmark historic bridge failures (**4**). These failures have punctuated the evolution of longer and longer bridges. In each case, overconfidence in stretching the design concept led to the failure. The bridges they studied are listed in Table 1.

Table 1   Landmark bridge failures (4, 5)

Bridge (location)	Type	Probable Cause of Failure	Year	Interval years
Dee (Cheshire, United Kingdom)	Trussed girder	Torsional instability	1847	–
Tay (Scotland)	Truss	Unstable in wind	1849	32
Quebec (Canada)	Cantilever	Compressive buckling	1907	28
Tacoma Narrows (United States)	Suspension	Aerodynamic instability	1940	33
Milford Haven (Wales)	Box girder	Plate buckling	1970	30
?	Cable stayed	Instability	c2000	c30

Sibly and Walker noticed that the interval between these major failures seemed to occur at around 30 years, and Petroski (**5**) speculated on the failure that might occur around 2000.

The Dee Bridge collapse

On 24 May 1847, the driver of the 6:15 p.m. Shrewsbury & Chester Railway train en route from Chester to Ruabon felt unusual vibrations, and a sinking feeling, as he crossed the Dee Bridge. He successfully crossed two of the three spans and opened the throttle. He just got off the bridge, leaving the tender behind when the coupling broke. The carriages, together with one bridge girder, fell into the Dee. Five people died, and 16 were injured.

There was a separate bridge for each track, and the engine driver acted with courage and quick thinking. He switched to the intact track and drove back across the bridge to stop traffic and get help from Chester.

The bridge belonged to a different railway – the Chester & Holyhead Railway, then under construction. The Chief Engineer and the designer of the Dee Bridge was Robert Stephenson. He believed that a derailment had damaged the bridge and caused the failure, but this theory was soon disproved.

The bridges for each track had three spans at an angle across the river. Each span was 98 ft long and was formed from three cast iron beams bolted together. A wrought iron tension rod system created a compound trussed girder (Fig. 1.10.1).

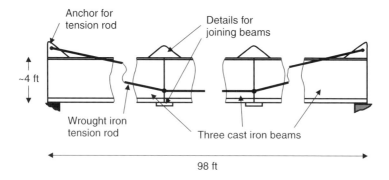

Fig. 1.10.1   Diagram of Dee Bridge compound trussed girder

This rather strange construction had been successfully used for more than 60 bridges in the previous decade – but mainly with shorter spans of 30 to 50 ft. Cast iron was notoriously weak in tension and not a good material for bridge beams. No doubt, the tension rod was aimed at reducing the tensile loads in the cast iron. Today, it is easy to question this concept. The loads are indeterminate.

The investigations

The Dee Bridge was the first iron railway bridge to fail, and it raised a lot of concerns. There was a local inquiry as well as a Royal Commission. Both investigations were widely reported and caused a temporary jolt to the 'Railway Mania'. The investigations showed that the strength of the bridge beams was insufficient to carry the loads.

The bridge opened to traffic in 1846. Stephenson was concerned that hot coals dropped from the engines might set fire to the timber deck. In May 1847, he added a layer of stones over the timber. He did not recalculate the strength of the bridge. It broke when the first train crossed after the stones were added.

The Chief Engineer of the Shrewsbury & Chester Railway claimed that the beams were too weak and challenged Stephenson's design. Stephenson appeared before both investigations. The local inquiry considered manslaughter charges but settled for a verdict of accidental death for the victims.

The Royal Commission broadened the debate, questioned the use of iron in railway structures, and considered imposing design rules. The events were well recorded by the Chester Chronicle reporter. He described Stephenson as being supported by *a cloud of eminent engineers.*

Brunel told the investigation *I am opposed to the laying down of rules and conditions to be observed in the construction of bridges, lest the progress of improvements tomorrow might be embarrassed or shackled by recording or registering as law the prejudices or errors of today.* The combined effort of the bridge engineering lobby avoided specific rules being imposed, but the debate brought out many sound ideas as well as ending the use of cast iron for girders.

Structural analysis

Stephenson had correctly sized the beams to a formula developed in 1831 by Hodgkinson, on the basis of tests of 10 ft long beams. Static loading calculations show that the nominal safety factor for the Dee Bridge was around $1\frac{1}{2}$, much lower than the then traditional safety factors which were in the range $2-7$ (**5**). The steam locomotive would introduce additional dynamic hammer blows from the pistons. The train ran between two beams. The loads from the deck, rails, stones, and train were transmitted to the beam at the edge of the lower flange (Fig. 1.10.2). This introduced additional torsional loading. At these lengths, at the extreme of manufacturing experience, the beams were not very straight or true. The torsional loads would be more significant for the longer span, and larger slenderness ratio of the Dee Bridge than on previous bridges. The design was inadequate.

COMMENTS

The Dee Bridge failure and subsequent investigations must have been a severe blow to the reputation and pride of Robert Stephenson, but his work continued unabated. He had more bridges to build on the Chester & Holyhead Railway, including the bridges at Conway and Menai. He rejected cast iron and used a rectangular tube of riveted wrought iron, with the train running inside the tube. Stephenson involved both Hawthorn and Hodgkinson to conduct experiments and analysis on model sections of the tube. The bridges were proof tested on completion before being opened to commercial traffic. This proved that they were conservatively designed.

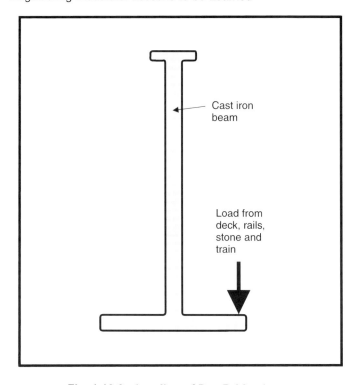

Fig. 1.10.2   Loading of Dee Bridge beam

Maybe his recommendations, repeated at the front of this book, were said with the Dee Bridge experience in mind.

While a tube is an effective structural member, the method used by Stephenson was not an efficient use of materials. Brunel used tubes much more effectively and economically in his elegant Saltash Bridge opened in 1859.

The Tay Bridge disaster

In the late nineteenth century, two wide Firths – the Forth and the Tay – interrupted the rail journey north from Edinburgh along the East Coast of Scotland. Bridges were planned for both, to avoid the tedious ferry crossings. The Tay Bridge was first built in 1878. It was destroyed in a gale on the night of Sunday, 28 December 1879. It failed as the mail train was crossing the high point of the bridge, over the navigational channel. The train, together with 13 spans of the bridge and their piers fell into the Tay with the loss of 75 lives. There were no survivors. The public was shocked, and the high reputation of British engineering was shattered.

Background

Thomas Bouch was chosen to build both bridges because of his reputation to build quickly and economically. The construction contract for the Tay Bridge was let in 1871 to Bouch's design. His original design was for a single track on multiple lattice girder spans of 200 ft supported on brick piers.

It soon became clear that the riverbed was not as solid as shown by the limited preliminary borings. Bouch redesigned the piers to lighten them by using cast iron columns. At the same time, he extended the spans to 245 ft for all but two spans at 227 ft. The bridge was the longest in the world at around two miles. It had 85 piers. The central high-girders gave a clearance of 88 ft above high water in the navigational channel. The rails ran on top of the girders for most of the crossing, but ran between the girders for the 13 high-girders. Three years into the construction, the contractor was changed, while Bouch remained in charge.

The Board of Trade (BOT) inspected the bridge on completion and witnessed the load test of six locomotives coupled together crossing the bridge at 40 miles/h. Queen Victoria crossed the bridge in the royal train. The bridge was seen as a masterpiece of engineering capability, and Queen Victoria knighted Bouch in recognition of his achievement.

The investigations

A Court of Inquiry was formed under a Scottish lawyer, Trayner. The composition of the Court of Inquiry came from simple Victorian logic. The train and bridge were a wreck in the water, so the Commissioner of Wrecks, Henry Rothery, a mathematician and barrister, was appointed Chairman. The accident was on a railway, so Col. Yolland, an engineer and the Chief Inspector of Railways, was on the panel. The accident was to a civil engineering structure, so the President of the Institution of Civil Engineers was brought on board. The Court of Inquiry was thorough and surfaced most of the key issues.

The Inquiry noted that the BOT Inspector added two qualifiers to his approval when the bridge was tested. He recommended that trains be limited to 25 miles/h over the bridge and that there should be an opportunity to observe the effects of high winds when a train was on the bridge. It appears that no attention was paid to either qualifier.

Divers found the train within the girders at the bottom of the Tay. This confirmed that the bridge had collapsed with the train on it rather than it having broken before the train arrived.

The Inquiry completed its work in six months. They found the failure was caused by insufficient cross bracing to withstand the gale and that the supervision after completion was unsatisfactory. The Chairman went further than his engineering colleagues and blamed Bouch for *faulty design, faulty construction and faulty maintenance.*

Aftermath of the inquiry
Bouch was made the scapegoat and had a swift and merciless ending to his career. He was immediately removed from his post as engineer of the Forth Bridge.

The Inquiry initiated the approval of steel for bridges and an end to cast iron columns. They recommended that BOT conduct regular inspections during and following bridge construction. They questioned whether the Government should supervise design and construction of key projects, but made no specific recommendation.

A Royal Commission on *Wind Pressure on Railway Structures* was set up in 1881 and recommended a maximum wind loading of $56 \, lbf/ft^2$. They also proposed that the newly developed anemometers be used to record wind speeds on structures.

A second bridge was started in 1881 and finished in 1885. It was a broader and more stable structure designed for double track. It used new pier foundations. Some of the girders from the old bridge were used on the new bridge. This bridge is still in operation, and passengers can look down on the remains of the piers of the failed bridge – and contemplate the earlier disaster!

There was one survivor from the disaster – the engine. It was recovered from the Tay and saw active service until scrapped in 1919.

Failure theories
While there is general agreement that the gale caused the failure, there is still debate on the details. Pugsley (**6**) in 1966 pointed out that when the Tay Bridge was designed, analysis only took account of loads, and not deflections. He noted that the high-girders had expansion joints on some piers and that these were not tied down onto the bearings. As a train went over a previous girder, the deflection would tend to lift the next girder span and unload the bearing. Transverse loads from high winds could move the girders sideways on the expansion joint bearings. Pugsley also noted that the bridge maintenance engineer replaced broken and missing bolts without telling Bouch. In his view, these were clear signs of potential trouble that went unreported. Lessons that Pugsley took from studying the Tay Disaster were that *structural accidents often cannot be attributed to*

one cause and that *interaction of modes is not easily covered in Codes of Practice.*

A study by Martin and Macleod, reported in the Proceedings of the Institution of Civil Engineers in 1995 (**7**), covers modern 3D computer analysis of the accident. Their results show that the presence of the train in the gale increased the overturning moment on the piers. The bolts at the base of the columns were meant only for location. They came under tension in the gale. They took the first layer of masonry with them. This was seen in photographs taken following the disaster. The diagonal bracing then failed, and the piers collapsed, taking the girders with them. Their analysis used the recently measured strength of the cross bracing that was lower than the original design figures. They showed that the piers would fail in the 10 to 11 Beaufort wind strength that was estimated by experienced Navy Officers on station in the Firth at the time.

The Open University used the Tay Disaster as a case study in their forensic engineering studies (**8**). Bill Dow, a Physics Lecturer at Dundee, noted that two girders had fallen into the Tay during construction. One had been scrapped, but the other had been straightened and used in the region where the accident started. He speculates that the girder may have been weakened or have tended to return to its bent shape. Marks showed that the girder had moved on its bearings. Some of the recovered wooden decking showed wheel marks, indicating that at least one carriage had derailed. This may have taken place before the collapse and led to the impact of the train with the bridge structure.

Peter Lewis, a Senior Materials Lecturer at the Open University, believes that the failure mechanism is more complex – that it is not just a static problem but also a dynamic one. He believes that vibrations caused by trains travelling fast over the bridge were progressively weakening it. Painters and maintenance workers had reported these vibrations to the Inquiry, whereas the BOT Inspector had not reported any vibration when the bridge was proof tested. Broken cast lugs are seen on the pier bases in photographs, which suggests that they had been progressively failing rather than breaking in the disaster. He believes that fatigue cracks developed, and notes that the Inquiry did not study the fracture surfaces.

Wind on structures

Smeaton, nearly a century before the Tay Disaster, estimated wind loads and reported them in a paper to the Royal Society. His estimates ran from $6\,lbf/ft^2$ for high winds to $12\,lbf/ft^2$ for storms. He also suggested $50\,\text{lbf/ft}^2$ for hurricanes. The Tay Inquiry estimated the wind load at the time of the disaster at $40\,\text{lbf/ft}^2$.

Martin and Macleod debated how much blame should be laid on Bouch. While building the Tay Bridge, he was designing the Forth Bridge. With its longer spans, he questioned the probable wind loading, and consulted the eminent engineers of the day. They were not sure, and approached the Astronomer Royal, who recommended $10\,\text{lbf/ft}^2$ on average, but with a possible local gust peak of $40\,\text{lbf/ft}^2$. Bouch also consulted the Chief Inspector of Railways, who said that there was no need to make special allowances for wind load! While Bouch told the Inquiry that he had made no special provision for wind load on the Tay Bridge, he had in fact used the $10\,\text{lbf/ft}^2$ figure. The new computer analysis shows that the current British Standard CP3 wind loading for the conditions at the time of the disaster would be around $46\,\text{lbf/ft}^2$.

When the Tay Bridge was being designed, engineers in France and America were using wind pressure loadings of between 30 and $50\,\text{lbf/ft}^2$. This information was not generally published.

COMMENTS

The assessments of the disaster concentrated on the technical aspects. The design was inadequate for the wind loading in the gale. Many of the other factors listed may have contributed to the failure. Witnesses reported that for both the disaster and the previous train, sparks had been observed along the bridge. This phenomenon has never been fully explained, but may have come from the wheels trying to take the side force from the wind against the rails.

There are other aspects than the technical ones. Bouch had a strong reputation before the disaster. But on the Tay Bridge, he was managing a project that was well over budget and schedule. There was poor quality in design, manufacture, construction, and maintenance. It is not known how many of these issues were known to Bouch.

The effort Bouch made to find the views of contemporary engineers in the United Kingdom on wind loads shows that he was being responsible. The doubts expressed by the engineers he consulted should have suggested that experiments were necessary. He probably thought that they would take too much time and money.

The mood at the time seems to have been to find a scapegoat rather than dismiss the accident as a *natural disaster* caused by extreme weather.

The embarrassments on the bridge at Québec City

The St Lawrence River has been, and still is, a major route for the development and economy of Canada. The cities of Montréal and Québec grew with the trade on the river. Now, with the Seaway, ocean-going ships can go as far as Chicago. In the days before the canals and Seaway, Montréal built a commercial lead over Québec City. The development of the railways exacerbated the competition. The St Lawrence is closed by ice for four months of the year. This gave the incentive to build rail links to the ports at Halifax, Boston, and Portland that were open year round.

Montréal got in first with a rail link to Portland. The link needed a bridge across a shallow part of the St Lawrence at Montréal. The Victoria Bridge was started in 1854. It was 2.7 km long on multiple piers and had to handle the difficult ice conditions. The bridge was the joint effort of Robert Stephenson and Alexander Ross. They used a tubular wrought iron structure.

The Québec City business community wanted to compete with Montréal and started discussing a bridge as early as 1850. At Québec City, the river is tidal and deep. Although there are several potential sites where the river narrows, the span required to cross the navigation channel would be longer than any bridge built at that time.

Various schemes were tabled over the next four decades (**9**). International bridge developments – and failures – influenced the proposals. At one time, there was a general view that suspension bridges were not feasible for rail use. Then there was the development of steel by Bessemer in 1856 and the successful opening of the cantilever Forth Bridge in 1890. Site surveys for the Quebec Bridge started in 1888.

Dramatic political events were taking place while the technical studies were proceeding. The American Civil War took place just to the south from 1861 to 1865. In Canada, the transition from a colony to self-government came with the British North American Act of 1867. The first Canadian PM was elected, and the country came under a balance of Federal and Provincial Government.

The Quebec Bridge Company was incorporated by Act of the Federal Parliament in 1887. In 1891, the Federal Ministry of Railways and Canals showed an interest in the bridge. The same year, the Premier of Québec asked Gustave Eiffel for his views. Eiffel submitted a report recommending a cantilevered bridge. However, the Premier was subsequently dismissed for a separate administrative scandal. This brought the bridge planning to a halt. The next step, in 1897, was when the Mayor of Québec City got a promise of financial assistance from his friend Sir Wilfred Laurier, the recently elected PM in Ottawa.

A slow start to the project

The Quebec Bridge Company appointed a locally resident engineer, Edward Hoare, as their Chief Engineer. Hoare had experience of railways and small bridges. The Board also sought the services of a Consultant Engineer to oversee its massive project.

In the summer of 1897, John Deans, the Chief Engineer of the Phoenix Bridge Company, from Pennsylvania met with Hoare to see what they could offer. The meeting took place at the time of the Annual Meeting of the American Society of Civil Engineers (ASCE), being held in Québec City. The potential bridge was a popular topic at the ASCE Convention. Also attending the Convention was the prominent American bridge engineer, Theodore Cooper. He expressed interest in the project.

In March 1899, six tenders for the bridge superstructure and two for the substructure were received. The Board of the Quebec Bridge Company retained Cooper as Consultant and asked him to review the tenders. On Cooper's recommendation, the contract for the superstructure was awarded to Phoenix, for a cantilever bridge with a main span of 1600 ft. The contract for the piers and substructure was awarded to Davis.

Cooper then recommended that the piers be moved slightly further apart. This would make the foundations easier and reduce possible ice flow problems. The main span increased to 1800 ft, making it the longest span in the world, 90 ft longer than the Forth Bridge. Cooper recommended using higher allowable stress levels for the longer span. Finally, the bridge construction started with a grand opening ceremony in 1900 when PM Laurier laid a corner stone.

The Quebec Bridge Company was in financial troubles. Their bankers were concerned with the financial feasibility of the bridge. The promised Federal Government financial help had not materialized. Work proceeded slowly on the foundations. Three years went by before the Federal Government finally came up with their guarantee of bridge bonds. At last, the contract was signed with the Phoenix Bridge Company. The project went overnight from a lethargic pace to being a crash programme. Phoenix design engineer

Peter Szlapka was in charge of producing the detail design and shop drawings.

With the Federal Government's financial involvement, the Department of Railways and Canals took an interest and started to review the designs. Their bridge engineer criticized the proposed higher allowable stresses. Cooper was outraged, stormed to Ottawa and beat the Department of Railways and Canals into rubber stamping his decisions.

Site work on the superstructure began in earnest in 1904. Cooper worked from his New York office and seldom visited site. He used a young engineer, Norman McClure, as his assistant to visit the Phoenix plant and the site.

Prelude to a disaster

Detail design was being rushed. Unfortunately, no revised weight calculations were made. It was not until 1906 that the Phoenix shop inspector noticed that the shipping weights were substantially heavier than original estimates. Cooper and the Phoenix designers did some hurried calculations and saw that the weight was between 4000 and 5000 tons more than the initial estimate of 31 400 tons. Cooper reluctantly accepted the resulting higher stresses.

By 1907, the south cantilever structure was well advanced. But as early as July 1906, McClure had become concerned that there were distortions in some members. When work restarted after the winter, further signs of trouble started to emerge. By June 1907, there were difficulties in riveting the massive bottom chords. They were up to $\frac{1}{4}$ in out of alignment and were jacked into position with two 75-ton jacks. The case was made that the members were distorted before leaving the plant. Cooper dispatched McClure to investigate. McClure convinced himself that the distortions were caused under stress rather than from manufacturing. A strike on site occurred in the midst of these investigations.

In August, a rib in the lower chord was out of line by $\frac{3}{4}$ in. A week latter it was $2\frac{1}{4}$ in out of line. The Phoenix site foreman, Benjamin Yenser, suggested that McClure and a young site graduate engineer Arthur Birks go to see Cooper in New York for advice. Following a phone call to Hoare, McClure alone was sent to New York.

The events of 29 August 1907

Yenser was concerned at adding load to the bridge and asked Hoare for instructions. Hoare told him to continue work, fearing that workers would leave the site if construction stopped.

McClure met Cooper in New York, unaware that construction was proceeding. Cooper decided that no more load should be added until the problem was investigated. He telegraphed his instructions to the Phoenix head office. McClure promised to send a wire to the site but forgot in his rush to catch the train to Phoenix.

Cooper's wire to the Phoenix office was ignored by a secretary in Deans's absence. Later that afternoon when McClure arrived at the Phoenix office, he met with Deans and Szlapka soon after 5:00 p.m. They decided to act the following day. At 5:30 p.m., the 19 000-ton structure collapsed into the St Lawrence when the lower chord buckled (Figs. 1.10.3 and 1.10.4). Seventy-five construction workers died. Most of the construction workers were from a small local Mohawk Indian Community that was devastated. There were a few lucky survivors – but only 11. One rode a girder 150 ft down to the river.

Fig. 1.10.3   The Québec Bridge on 14 August. Photograph by courtesy of the Hagley Museum, University of Delaware. The South shore is to the left

The enquiry

The day after the collapse, PM Laurier named three prominent Canadian civil engineers to a Royal Commission of Enquiry. Two of the engineers were on site the same day. The Commission interviewed witnesses and all participants. They tested a $\frac{1}{3}$-scale model of the failed chord and lattice bars and rivets. The Federal Department of Railways and Canals asked Charles Schneider to review the design. He recalculated the stresses and found that many members exceeded the specification by up to 24 per cent. The Commission adopted

Fig. 1.10.4   Québec Bridge following collapse on 29 August 1907. The view from the South shore showing the debris up to the main pier. The debris past the main pier is in the river. Courtesy National Archives Ottawa

Schneider's analysis. They presented their report on 20 February 1908 – 25 weeks after the disaster.

The Enquiry was thorough. Their report listed the main deficiencies with the majority of the blame on Cooper and Szlapka. They noted that those working on the project did not appreciate its scale and that tests should have been made on representative samples or models to confirm the designers' judgement. The Commission's work and report was well received and still sets the standard for similar enquiries.

The players

Theodore Cooper was born in rural New York State and graduated in civil engineering. During the Civil War, he served in the Navy and then turned to bridge design and construction. He had a key inspection role on the St Louis Bridge – completed in 1874 and a landmark for its pioneering use of steel. While inspecting the bridge, he fell 90 ft into the Mississippi. His experience in the building of the St Louis Bridge was presented in an ASCE paper.

He ran a bridge-building shop before establishing his own engineering consultancy in New York in 1879. He developed standards for steel railway bridges that were used nationally. In 1889, he published an award-winning paper on *The use of steel for bridges*. In his paper, he said that to avoid failure, there should be a careful calculation of stresses, material testing, and inspection. However, he concluded with a telling comment that *A bridge's stability was more reliant on the engineer's instincts than merely upon a theory of stresses.*

Despite his prominent career, he had never had complete responsibility for a major structure himself. The Québec Bridge provided the opportunity to cap his career. It is not clear why Cooper took the Québec Bridge consultancy with such poor remuneration. He even agreed to it being negotiated down because of the difficult financial position of the Quebec Bridge Company. Despite his earlier hands-on experience, he chose to manage the largest bridge in the world from his office in New York – 600 miles from the site. At the time of the bridge collapse, he was 68 years old and in ill health, although he lived to be 80.

Cooper was outspoken and criticized the Forth Bridge design of Baker as *the clumsiest structure ever designed by man.*

Norman McClure was Cooper's assistant – his eyes and ears on the project both at Phoenix and site. A civil engineering graduate from Princeton, McClure had worked for a couple of railways before joining Cooper a year after graduating. He appeared to recognize most of the problems on the project as they had developed. Either his inexperience or the reputation of Cooper may have been the reason he failed to get the problems resolved.

Peter Szlapka was born in Poland and graduated at Hanover, Germany, before immigrating to the United States. He worked his way from draughtsman to design engineer at Phoenix where he designed several notable bridges. While appearing to have a good working relationship with Cooper, albeit at a distance, some differences emerged at the Enquiry. Szlapka reported that at one stage Cooper had said *there is nobody competent to criticise us.* Szlapka reacted angrily to the findings of the Royal Commission. He is said to have distrusted experiments and preferred analysis.

John Deans spent his entire career at Phoenix and had 30 years of bridge experience at the time of the collapse. He escaped from any serious repercussion to his career, despite being Szlapka's manager.

After the collapse, he went on to be responsible for the construction of the Manhattan Bridge in New York.

Edward Hoare started his engineering career in London before emigrating to Canada in 1869. The next 30 years were spent on railway construction. His bridge experience was limited. The largest he had worked on was 300 ft long. The Enquiry considered that his specification was inadequate and his supervision ineffective.

Benjamin Yenser had worked all his life building bridges. From his experience, he knew that something was wrong but could not get the managers and engineers to pay attention. He died in the collapse.

Arthur Birks was an honours graduate from Princeton and a postgraduate from MIT. He joined Phoenix and worked first in the drafting office and then site erection on a number of major bridges. He was appointed the Site Resident Engineer for the Québec Bridge at the age of 25. Like Yenser, he had concerns about the bridge, but deferred to the experience of Cooper and Szlapka. He lost his life in the collapse.

A second attempt

The Federal Government took over the bridge project in 1908. A committee of three prominent engineers managed the undertaking. They studied a number of designs, conducted extensive tests, and received proposals from around the world. The use of three engineers to jointly manage the project had its moments and disagreements, which are well described by Middleton (9). The construction contract was won by two Canadian companies, Dominion Bridge and Canadian Bridge, joining forces.

The final weight of the first bridge design was 38 500 tons, higher than the original estimate of 31 400 tons and even higher than the re-estimated weight. The second design weighed 66 480 tons. Work started on the second design in 1909. By the end of July 1916, both cantilever arms were complete. The centre span of 640 ft, weighing 500 tons, was brought to the site by barge. Thousands of spectators came to see the span lifted. Suddenly, a casting broke at one of the lifting points, and the whole span fell into the river. It is still there. In the accident, 13 men died.

A new section was built, and in September 1917, it was successfully lifted into position. The first train crossed the bridge on 17 October 1917. On 22 August 1919, the Prince of Wales opened the bridge. In May 1987, ASCE and the Canadian Society for Civil

Engineering designated the bridge an historic landmark. The bridge is still in use.

COMMENTS

The first collapse shocked the engineering community. How could such an eminent engineer as Cooper and experienced bridge contractor as Phoenix be involved in such a catastrophe? Surely, there must be some previously unknown engineering phenomena coming into play in large structures. This was the environment in which the engineers on the Enquiry started their examination. They quickly got to the nub of the problem – the buckling of the chord.

The technical explanation is straightforward – the human errors are more complex. How had Cooper got himself into that position and shown such errors of judgement? It appears his ego took over. Even the most experienced engineers can make mistakes!

The solution is to ensure that there is adequate independent checking on all aspects. The transition from slow to fast pace probably led to the weight re-estimate being missed. It left no time for the prudent testing of models and components.

There were no clearly defined areas of responsibility for the various players. The management was weak or non-existent and did not appreciate what they were getting into.

There were plenty of warning signs of problems. Even the structure itself provided plenty of signals well before its final collapse. It raises the question *How can so many knowledgeable people miss all the warnings?*

An illustration from the magazine *Scientific American*, October 1907, dramatically showed the problem (Fig. 1.10.5).

An engineer's reaction to the illustration is probably an expletive! No thinking time is needed to recognize the potential for buckling – the problem is obvious. What a pity that someone on the project did not produce this illustration before the chord buckled. The Forth Bridge section, which carried a similar load, is shown for comparison.

The aftermath

A couple of features of engineering life in Canada stem in part from the Québec Bridge failure. Professional Engineers have to be registered in the Province in which they work. Other engineers can do engineering work, but they cannot call themselves Professional Engineers. Certain types of engineering work such as steel structures used

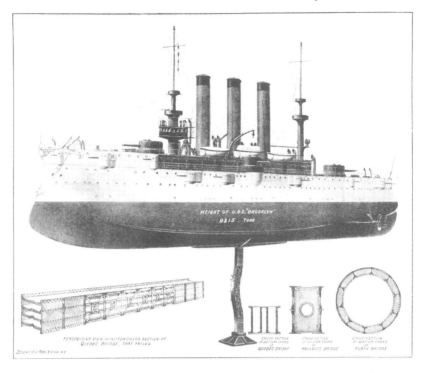

Fig. 1.10.5   From the collection of W.D. Middleton. Reproduced with permission of *Scientific American*, October 1970

by the public and bridges require a Professional Engineers' stamp on the drawings and calculations.

There is an iron ring ceremony where a newly qualified engineer commits to uphold the standards of the profession, to always act in an ethical manner, and to protect the safety of the public in engineering activities. One story is that the iron rings are made from the remains of the Québec Bridge. This is a myth, as the bridge was made of steel, not iron.

Galloping Gertie – the Tacoma Narrows Bridge
Background

There is only one point on Puget Sound where there is a narrow stretch of water between the mainland of Washington State and the Olympic Peninsula. The need for a bridge was obvious to the local community, but the traffic volumes were small. Funding was hard to obtain in the aftermath of the Depression. In response to the local communities, the Washington State Legislature created the Toll Bridge Authority in 1937 to explore the prospect of building a toll bridge. The City of Tacoma sponsored a bridge study and Clarke Eldridge, a bridge engineer with the Washington State Department of Highways, was given the task of designing the bridge.

In 1938, the Public Works Department agreed to finance 45 per cent of the costs and in return retained independent engineering consultants. The State retained the Latvian-born Leon Moisseiff – a world-renowned suspension bridge builder. The 1920s and 30s were exciting decades for American suspension bridge builders. Great bridges were built, such as the George Washington, Bronx-Whitestone, and Golden Gate Bridges. Moisseiff's career had grown in this environment of success.

The design

Moisseiff made significant changes to Eldridge's design. He replaced the 25 ft deep truss girder with 8 ft deep plate girders to give a lighter and cheaper bridge. The bridge design ended up as Eldridge's sub-structure and Moisseiff's superstructure. It was a narrow bridge, only 39 ft wide, with a centre span of 2800 ft and a total structural length of 5935 ft. When it was opened in July 1940, it was the third-longest bridge in the world.

Galloping Gertie

The bridge exhibited unnerving oscillations during construction. This led Burt Farquharson, Professor of Civil Engineering at the University of Washington, to take a close interest. In the first few months of operation, the roadway undulated in various vertical modes under moderate winds as low as 5 mile/h. The maximum amplitude was about 5 ft, with two nodes between the towers and at a frequency of about 12 cycles/minute. This earned the bridge the nickname *Galloping Gertie*. It became a local attraction, and the Toll Bridge Authority thrived on the unanticipated extra traffic. Farquharson recommended extra tie down cables that were installed.

The failure

On Thursday, 7 November 1940, the wind was coming straight through Tacoma Narrows, hitting the bridge at right angles. Initially, the centre span was undulating up to 5 ft. The wind speed increased to around 40 mile/h, the highest the bridge had experienced. At 10:00 a.m., traffic was halted. Soon after the bridge was closed, the nature of the oscillations changed. One of the cable collars loosened. The bridge then took on a torsional mode in two parts, with a node at the centre of the span. The motion grew to 28 ft each side of the horizontal and with the road-deck tilting 45 degrees in each direction at 12 to 14 cycles/minute.

In half an hour, the bridge started to break up. It took a further half hour before the bridge finally plunged into the Narrows. It is still there. Professor Farquharson witnessed the failure.

A cine film recorded the whole episode. With WWII under way, there was little international attention paid to the failure. Post-war, the film became required study for engineering students worldwide. It probably holds the record for the most-watched failure among engineers. Discussion on the analysis of the failure continues.

Leon Moisseiff's first comment following the failure was *I'm completely at a loss to explain the collapse.*

The debate on the cause of failure

The Federal Works Agency quickly appointed a committee to investigate The members were Othmar Ammann and G B Woodruff, both prominent suspension bridge designers, and Theodore von Kármán, the mathematician prodigy and aeronautical engineer. Within a couple of weeks, von Kármán wrote to the journal *Engineering News-Record* with his back-of-envelope calculations showing that aerodynamic instability was the cause. His simple calculations gave a failure wind-speed about 10 mile/h faster than experienced at the collapse. He believed that vortex shedding excited the structure. Vortex shedding behind a bluff body was named after von Kármán after he analysed it mathematically in 1912. Vortex shedding is responsible for the *singing* of telegraph wires and *galloping* of power lines. The committee concluded that the failure was due to:

- *the extreme flexibility of the structure – three times more flexible than previous long-suspension bridges like the Golden Gate Bridge;*
- *its small capacity to absorb dynamic forces;*
- *the lack of realization that aerodynamic forces can be disastrous;*
- *and that the slipping of the cable initiated the torsional oscillations.*

They recommended further experimental and analytical studies.

Von Kármán was outspoken in criticizing the bridge engineers for not being sufficiently aware of aerodynamics. He also criticized them for concentrating on static forces and pressures instead of dealing with the dynamics.

Vortex shedding or flutter?

The predicted vortex-shedding frequency at the wind speed when the bridge failed would be around 60 cycles/minute – or roughly five times the observed oscillating frequency. Hence, vortex shedding is unlikely to be the cause of the final failure, although it could explain the earlier galloping at lower wind speeds.

If not vortex shedding, then what was the explanation? In England, from the 1920s, Pugsley at RAE (Royal Aircraft Establishment) Farnborough and Collar at the NPL had worked on various aeroelastic phenomena and explored the interaction of aerodynamics, structural stiffness, and inertia forces on aircraft. One combination of these forces can produce flutter.

The aerodynamic forces of lift and drag apply at a point roughly one-third the chord, ahead of the centre for structural bending and torsional rotation. The aerodynamic forces will produce a torsional moment that will increase the angle of attack and further increase the aerodynamic forces. When the wing starts to move up, it will reduce the angle of attack between the wing and the relative airflow, and this will reduce the aerodynamic forces. As the torsional forces reduce, the inertia can continue the rotation until the wing tilts in the other direction. Given the right conditions, an oscillation can start with the wing moving up and down and twisting at the same frequency, somewhat like the slats of a Venetian blind vibrating in the wind through an open window. The analysis of flutter can be quite complex, with nonlinearities such as stalling of the aerodynamic forces making analysis even more complex.

Farquharson carried out wind tunnel tests following the collapse. His tests showed various modes. At low wind speeds, there were several different self-limiting oscillating modes. At higher wind speeds, a torsional mode occurred at a frequency that scaled to the frequency he observed at failure.

In the 1970s, Scanlan and Tomko repeated the wind-tunnel tests and demonstrated conclusively that the failure mode was torsional flutter. Various non-linear mode couplings took place at the higher amplitudes and appear to result in limiting the movement. This explains how the bridge was able to continue moving for half an hour before it collapsed.

Investigations are continuing with a somewhat different emphasis. Joe McKenna, Professor of Mathematics at the University of Connecticut, is continuing with complex non-linear analysis. He is trying to develop mathematical tools to make precise predictions.

In the decades since the Tacoma Bridge failure, all long-bridge designers have relied on wind-tunnel testing. It is not easy to get the correct dimensional scaling of aerodynamics, structural stiffness, and inertia all at the same time. Furthermore, wind tunnels run at steady speeds, whereas the wind can gust, buffet, and create turbulence. Fortunately, Professor Alan Davenport has shown that wind turbulence tends to decrease possible excitation, resulting in predictions from steady wind-tunnel tests being conservative.

Was it resonance?

Many papers and even textbooks say that the Tacoma Narrows Bridge failed because of resonance between the wind effects and the structure. Billah and Scanlan have reacted strongly to this misinformation. **(10)** Resonance requires the frequency of the driving force to be close to a natural frequency of the structure. This was clearly not the case in the motion of the bridge at failure.

When the structure was fluttering, it would still have been shedding some vortices. These would have been like the vortices left in the wake of a dinghy if you throw the rudder from side to side. The wake of vortices is the result of your moving the rudder. They are not the driving forces to move the dinghy.

It was not resonance that led to the failure. It was a non-linear mode coupling of wind forces with the flexibility and inertia of the bridge structure.

Was it a new phenomenon?

The simple answer is no. In a 1949 review, Farquharson noted that in the previous seven decades, there had been ten suspension bridges in Europe and North America that had failed because of wind action. While the specific details in each case may have been different, this history should have alerted bridge designers to consider risks from wind effects. Even in the years before the Tacoma failure, there were other suspension bridges in the United States showing movement in the wind, and were being stiffened.

The second and third bridge

In 1950, a new suspension bridge was opened at the same site using the piers from the first bridge. The design has a 60 ft wide deck and therefore is stiffer. The trusses are 25 ft deep – ironically the same depth as the first design by Eldridge. The design was tested in the wind tunnel before construction. An additional bridge, Tacoma 3, is planned for completion in 2007 to take the extra traffic. In May 2003, wind tunnel tests were completed in the NRCC tunnel in Ottawa on the two suspension bridges, Tacoma 2 and 3 side by side.

A record?

Locals claim that the wreckage of the first Tacoma Narrows Bridge at the bottom of Puget Sound is the largest man-made structure lost at sea. They point out that the structure is heavier than the *Titanic!*

COMMENTS

There were plenty of warning signs from the history and recent experience of suspension bridges. Von Kármán's and Pugsley's comments on the lack of use of knowledge from other disciplines is telling, and still relevant in many aspects of engineering. Joe McKenna makes an interesting observation related to his ongoing mathematical studies – *Engineers are by nature forward-looking. When they eliminate a problem, they don't worry too much about the cause.*

Billah and Scanlan complain about what they consider errors by others in considering bridge flutter similar to aircraft wing flutter. There are great differences in the aerodynamics between the smooth airflow on aerofoils and the separated flow around blunt bridge structures, particularly the ⊢—⊣ shape at Tacoma. However, the basic principle of analysing the interaction of aerodynamics, structural and inertia forces is still applicable. Therefore, the view that the bridge designers could have learned from the people working on aircraft flutter is accurate and relevant.

Petroski points out that the design climate can have a profound effect, and that the boundary between success and failure can sometimes be subtle and escape even the greatest of designers (5). The design climate came from a string of successes and a preoccupation with aesthetics. It also started with pressure for a cheap design. Billington, in his review of the Tacoma Disaster in 1977, concluded that *individual personalities play a significant role in the history of structures* and that *history, for structural engineers, is of importance equal to science* (11). His comments started a lot of discussion.

The Milford Haven Bridge collapse
Steel box welded girder bridges

The ability to reliably weld steel structures led to a new design for bridges. Rivets were out and welding in. Welding gave the ability to make box shapes capable of taking high bending and torsional loads. The reconstruction of Europe following WWII saw the widespread use of this technique. These bridges were of modest spans.

The Milford Haven and Yarra River bridges

The Milford Haven Bridge over the Cleddau River in South Wales used a deep trapezoidal section steel box as the main girder. In 1970, during construction, a pre-assembled section was being moved out from a pier in cantilever fashion when it buckled at the pier.

Welded box sections were being used in another bridge that collapsed by buckling during construction in the same year. This was the West Gate Bridge over the Yarra River in Melbourne. The West Gate Bridge failure was different from the Milford Haven Bridge and more complex. The West Gate Bridge had two asymmetrical box sections side by side, one left hand and one right hand. The two were to be joined together later in construction to form the deck of the cable-stayed bridge. The two sections deflected differently under their own weight and were difficult to bring together for joining. Ballast weights were tried to bring the sections into alignment but was not successful. Holding bolts were then relaxed as a further means of achieving alignment. This resulted in a flange buckling, and the bridge collapsed. Unfortunately, there were site offices and workmen's huts under the span that collapsed, and 35 people were killed.

The investigations

The Milford Haven Committee of Inquiry concluded that the design was insufficient against buckling under the loads from the erection process, particularly at the piers. They also noted that the then current British Standard and Codes of Practice were inadequate for that design of bridge.

The Committee prepared interim design and workmanship rules and made 27 recommendations, most of which have been implemented. These included investigating whether it would be cost effective to increase design expenditure. It appeared that a 10 per cent increase in design costs spent on checking could be worthwhile. Independent checking of design and erection methods was recommended. The committee also believed that steps should be taken to check the adequacy of the resources available to the various parties to a project as well as to ensure that key staff are suitably qualified.

An Australian Royal Commission investigated the West Gate failure. They uncovered a sorry tale of the use of unproven construction practices. They were critical of the poor relationship and communication between the parties in Australia and London and saw this as a key fact that led to the failure.

COMMENTS

There is no benefit in having an economic design if it collapses during construction. At least, these failures have resulted in the development of more effective Codes and Standards as well as technical progress in box girder designs and erection methods.

The technical analysis of the buckling of thin-walled structures was pioneered in the aircraft industry where buckling was seen as one of the failure boundaries to beware of. Maybe the same recognition was not present when bridge builders started using thin-walled structures.

The Millennium Bridge failure

Henry Petroski extrapolated the Sibly and Walker pattern of failure (Table 1.10.1) and speculated in 1994 that instability of a cable-stayed bridge might be the next landmark bridge failure occurring around 2000. There were several bridge failures around 2000, but the failure that can be considered a landmark is the Millennium Footbridge in London. Petroski's speculation about a cable-stayed bridge failing probably ensured that sufficient attention was paid that no such failure has occurred! The Millennium Bridge is a shallow suspension bridge not cable stayed. However, his probable cause – an instability – is not far off the mark with the Millennium Bridge. While it did not collapse, it did oscillate sideways sufficiently to be closed to the public and earn the name *the wobbly bridge*. In 2001, R. Scott suggests that the Millennium Bridge failure is a good candidate to add to the cyclic pattern of bridge failures **(12)**.

The background to the Millennium Bridge

In 1996, a competition was sponsored by the *Financial Times*, the Borough of Southwark, and the Royal Institute of British Architects to design a footbridge across the Thames. The footbridge would link the activities on the South bank, particularly the Tate Modern Art Gallery with the St Paul's environs on the North bank. Over 200 designs were submitted. The collaboration between the architect Sir Norman Foster, the sculptor Sir Anthony Caro, and the engineers at Arup won the competition.

The bridge is a unique structure, described as a *blade of light* due to its low profile and bright aluminium deck. It has three spans of 81, 144, and 108 m. The design is an innovative, very shallow suspension bridge with only 2.3 m cable sag over the main span. This is roughly six times shallower than conventional suspension bridges.

The design assessment

A suspension bridge is by nature fairly flexible, and its dynamic response has to be checked. The engineering design followed all the Codes believed to be relevant. Wind tunnel model tests were used to check on aerodynamic stability as well as wind buffeting. The bridges' dynamic behaviour was analysed in all modes – vertical, lateral, and torsional, and believed to be acceptable. The analysis showed the lateral stiffness to be similar to other footbridges. The design was checked against a conservative application of the British Standard 5400 for pedestrian excitation. Tests with a few people on the bridge before it opened confirmed the calculated low accelerations.

The problem

The new bridge was a popular attraction on its opening day, 10 June 2000. Up to 2000 people were on the bridge at one time – close to the maximum that could comfortably walk across. When the most people were on the bridge it developed a sideways movement. The frequencies were a little different for each span but in the range 0.5 to 1.0 Hz. Maximum lateral accelerations were up to $\frac{1}{4}$ g. Although there was no danger of structural damage, the bridge was closed after three days to investigate the problem.

The investigation

Arup carried out detailed investigation into what had happened and why. The dynamic properties of the bridge were measured and showed that the damping and frequencies were close to the initial design analysis. This proved that the bridge movement was due to an unforeseen external driving force.

Very few estimates of lateral dynamic loads from pedestrians were found in the literature. Those that were available showed that the lateral forces were very much smaller than the vertical. Laboratory tests were initiated at Imperial College and Southampton University to measure the lateral forces generated by walking, and how they are affected by horizontal movement of the walkway.

People walk with each foot taking its own path a small distance apart. The body's centre of gravity is between the two paths. When one foot is off the ground, the other has to impart a small sideways load on the ground in order to avoid falling over. Walking produces an alternating load to each side. The typical lateral force was measured to be around 5 lb.

If the bridge starts to sway, the pedestrian has two instinctive reactions. Firstly, it becomes more comfortable to walk in synchronization with the bridge movements. Secondly, in order to maintain balance, the pedestrian walks with legs further apart and thereby creates larger lateral forces on the ground. The stronger the sideways movement, the higher the probability that pedestrians will *lock-in* to the bridge's wobbling frequency. The stronger the sideways movement the larger the additional lateral load – with almost a straight-line relationship. If enough people are on the bridge, this positive feedback overcomes the natural damping of the bridge, and the wobbling rapidly increases. In other words, it resonates.

The Millennium Bridge was instrumented, and the response to the walking of a controlled crowd was measured. Test results on one span are shown in Fig. 1.10.6. The test started at 300 s. More and more people were asked to walk in a path from one end to the other and back again. There was no evidence of the bridge swaying, with up to 150 people walking on the span, but when the numbers reached a critical level of 166 the wobbling took off. The test ended shortly after 1400 s, and the acceleration quickly died down. The phenomenon has been labelled *synchronous lateral excitation*.

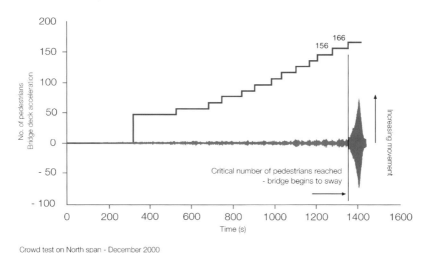

Crowd test on North span - December 2000

Fig. 1.10.6   Test crowd results. Reproduced courtesy of Arup

The solution

One solution would be to limit the number of people on the bridge at any one time to below the critical number that would cause synchronous lateral excitation, for example, 166 for the north span, but this was not acceptable. Various engineering solutions were considered to either stiffen with additional structure or add damping. Additional structure would impact on the appearance of the bridge. So the adopted solution was to install 37 viscous dampers and a number of horizontally acting tuned-mass dampers to suppress the lateral vibration. Additional vertically acting tuned-mass dampers were installed to dampen any vertical or torsional movement that might show up once the horizontal problem had been eliminated. After the dampers had been installed, tests involving 2000 pedestrians proved the solution worked. The bridge was reopened on 22 February 2002.

Is it a new problem?

The publicity surrounding the Millennium Bridge problems brought some previous incidents to light. At least three examples were found in the previous three decades:

- a footbridge over the railway at the National Exhibition Centre (NEC) in Birmingham had vibrated laterally when large crowds crossed for an exhibition or pop concert;
- a 1923 suspension footbridge over the Dee at Chester that experienced sideways movement on one occasion when a large crowd was watching a regatta;
- the Auckland Harbour Road Bridge in New Zealand, normally used by vehicles, vibrated sideways when a large crowd was crossing in a public demonstration.

The NEC bridge was easily stiffened by additional bracing to the foundations. This cured the problem. In the other cases, the behaviour was considered a *one-off* event, and not pursued. The phenomenon was not researched in any detail, quantified or widely publicized.

Conclusions

The research carried out by Arup to tame the wobbling of the Millennium Bridge has now been widely publicized and included in relevant standards. Designers of future bridges have the benefit of this work.

The research shows that any bridge with a lateral natural frequency below about 1.3 Hz is potentially susceptible to excessive swaying because of synchronous lateral excitation. This includes many existing bridges that have not so far exhibited this behaviour. For these bridges, the phenomenon remains a latent issue until the critical number of pedestrians are crossing.

COMMENTS

Was the Millennium Bridge a bridge-too-far? Did the designers ignore evidence from earlier bridge problems? It is easy to quickly blame the designers on both counts – but diving below the surface of superficial judgement presents a different view. The technical innovation of the structural form was not the root of the problem. More conventional bridges with similar modal characteristics would have exhibited similar problems when sufficient people were walking across. The previous examples had never been presented as synchronous lateral excitation. The Millennium Bridge designers could hardly be criticized for missing such weak signals. Designers cannot easily address unknown phenomena.

Tony Fitzpatrick of Arup, who played a huge role in the modification work on the bridge, made an interesting suggestion. He proposed that chartered engineers should be required, in order to keep their status, to report on any issues that arose on their projects. The Institutions could be the forum for making the information available to other engineers. The challenge is to find a way of doing this that would not damage the engineers' reputation and not oblige them to report the slightest detail, while maintaining commercial and legal confidentiality.

The three prior examples had been quickly passed off, or resolved by a simple engineering fix. All three cases reflect Joe McKenna's comment that engineers move on after dealing with a problem – without worrying too much about the cause!

So the Millennium Bridge was not a bridge-too-far. Nor was there prior information on synchronous lateral excitation to alert the designers to this problem. It was a new phenomenon for the profession to consider.

The Millennium Bridge was a popular walkway on opening day. The large number of people intent on walking across, rather than stopping to admire the view created the right circumstances for the first prominent display of synchronous lateral excitation. The designers were unfortunate in having their bridge in a place where the resonance was so publicly displayed. The engineers at Arup are to be congratulated for their thorough investigation – a model of how to solve problems – and for so openly presenting the results of their work **(13)**.

It is interesting to note the media reports. *The Guardian* **(14)** talks about the engineers from Arup who designed the bridge, whereas *Architecture Week* **(15)** designates Lord Foster as the designer.

While the phenomenon is very unusual for large-scale bridges, hikers know it. It is easy to start and amplify sideways movement when crossing a long flexible suspension footbridge on a trail. It is probably easier than starting vertical movement despite the fact that vertical movement has had all the attention. It would have taken a very perceptive individual to consider this experience at the design stage of the Millennium Bridge.

The only rational conclusion is to be humble. A typical answer is to ensure that adequate tests have been performed before opening to the public. But Arup did conduct what was previously considered to be realistic testing before opening. Unfortunately, the tests did not involve sufficient pedestrians to show the latent problem.

Another answer for a very flexible structure is to take a highly conservative approach and say *I don't know where the forcing loads are coming from, but I will increase the stiffness or damping, seeing that both are low* – in other words build in robustness. But the stiffness and damping of the Millennium Bridge were consistent with those of other footbridges of similar span – which have performed perfectly well because they have never carried their critical number of pedestrians. Increasing the frequency of the Millennium Bridge to over 1.3 Hz would have involved a huge addition to the structure and compromised its aesthetic appeal. Had the viscous and mass dampers, that have now been installed, been included as part of the initial construction, they would have added about 10 per cent to the total costs. This raises the conundrum facing engineers that while building in robustness is a rational approach – the added costs are often very difficult to justify.

Fortunately, in the case of the Millennium Bridge, the failure was little more than an embarrassment, and now a new and outstanding footbridge graces the Thames.

1.10.5   Comments on bridges in general

The examples of bridge problems fall into two categories – the maintenance issues with routine bridge works and the dramatic failures that have occurred with some landmark bridges.

Maintenance issues
Shortage of funds
A problem in many jurisdictions is that public works projects are starved of funds for maintenance and repair. What can engineers do? First, the total life of a bridge should be considered as part of the original design. Second, the maintenance aspects should not be treated casually by the designer as a wish list of things to be done by others.

The necessary actions, whether monitoring or maintenance, should be justified as a cost versus the cost of not taking action. Then, there will be a stronger and more compelling case for the recommended actions. Third, the bridge designers should ensure that there will be warning signs of major problems before disaster strikes.

Learning from others

The biggest problem for routine bridges is scour of the foundations. A team on interested parties from the United States took the initiative in 1998 (16) to send a study group to European countries to see how they handled the problem. They found that in some countries, more attention was paid to scour in the early design stage. In general, the placing of large stones around the base of the piers helped prevent the flow from causing the lighter and finer material from being washed away by the turbulence caused by the pier. Scour problems, by their nature, are under the surface – not easily seen and therefore easy to overlook until disaster strikes.

Learning new approaches

Bridge engineers are coping with the shortage of funds by concentrating on computer modelling of bridges to try and quantify potential deterioration and vulnerability. Bridge inspection is moving from visual inspection to installing sensors to monitor for corrosion, loads, and movement.

Whose risk is it?

When politicians and financial managers limit funds for maintenance and repair, they are effectively taking over at least some part of the responsibility of risk management from the engineers. Do they realize this? Do they know the risk they are taking? Are those now responsible for taking the risk clearly identified? The answer in each case is probably no.

Should a few failures be tolerated?

The comments from the US Department of Transport saying that most bridge failures are now due to natural disasters are telling. They appear to consider that no one is responsible for such failures. Maybe it would not be cost-effective to design all bridges, including minor secondary bridges, to withstand the worst conditions from nature. Replacing a few bridges following a major storm might be cost-effective, but this should be a conscious decision not left to chance. This is not an easy task. While statistics are available on the probability of storms, the effect of flood damage is less easy to predict.

It is a moving target, with development continuously taking place in the waterways and flood plains, which affects the hydraulics.

Landmark bridge failures

The examples of landmark bridge failures show a different pattern than the maintenance issue for less prominent bridges. Despite the progress in engineering knowledge, failures still occur. Nearly all these failures seem to be categorized as being steps into the unknown – *overstretching experience*. Despite this criticism, many novel and dramatic bridges have been successfully built.

Do we need to continually stretch our experience? The answer is probably yes. There are still opportunities around the world for longer and longer bridges. The next long bridge is likely to be the Straits of Messina, 65 per cent longer than the current longest bridge and with three times the length of central span of the Golden Gate Bridge. Then there are other locations such as the Bering Straits. Promoters throw a challenge to engineers to come up with designs for these projects.

All the failures rebuilt

One interesting fact is that in all the landmark cases studied (except the Millennium Bridge), a new bridge has been built at the site of the disaster. Obviously, the need for a bridge was justified in each situation, although the initial financing was difficult.

The failures – whose fault?

The landmark bridges were all high-profile projects. The question of whether the failure was due to basic engineering or due to the dominant role and ego of the man in charge can evoke interesting discussion. In the earlier failures, the intuition of the engineer in charge played a more important role than analysis. Now one can question whether the pendulum has swung too far, and the dominance of analysis, led by the use of the inanimate computer, is swamping intuition.

The exception

The Millennium Bridge is an exception in the bridges studied. Despite the millennia of bridge-building experience, a new phenomenon emerged.

Codes and standards – a help or a risk?

The more recent failures have shown inadequacies in Codes and Standards. A false sense of security is generated when Codes and Standards have been followed even when a careful examination would indicate that the design was stepping outside the range covered by the Code. On the other hand, there are no statistics to show how many potential failures have been prevented because engineers followed the Standards. The answer is that Codes and Standards are not panaceas but should be used as an aide mémoire, fully recognizing their limitations. The foundation on which the Code or Standard is based, as well as the underlying science, should be understood.

Heeding warning signals
With the exceptions of the bridge at Ynysygwas and the Millennium Bridge, there were ample warning signals in each example. The signals were not misinterpreted; they were either ignored or overlooked.

The experience of the staff
Ensuring that there are qualified and experienced staff on a project sounds an obvious requirement. In the Québec Bridge Enquiry, one contributing cause was listed as the inexperience of the staff. The young engineers, McClure and Birks, were well-qualified and appeared to have good engineering experience. Was the real problem getting the attention of, and decisions from their authoritative senior engineers?

The climate
There are more aspects of *climate* to bridge construction than the weather. There are the pressures of:

- being a prominent project – in the public's eye;
- working in a team under a prominent builder who has strong opinions;
- dealing with financial and political issues;
- dealing with communication issues between design office and the construction site and;
- being distracted and carried away by the challenge of the project and not spending enough time on objectively assessing what might go wrong.

1.10.6   Lessons learned
The prominent nature of large bridge failures has led others to articulate lessons learned. Some of these from Pugsley, Petroski, and SCOSS are presented in Part 2. Other lessons are below.

Lessons from the maintenance of bridges
- Consider the whole life aspects, including maintenance, as part of the initial design. This should recognize that funding for maintenance may be more difficult to obtain than that for the initial construction.

- Beware of leaving the responsibility vague when others have effectively taken over decisions by limiting finance.

Even solid structures can behave dynamically

Dynamic effects have caused many of the failures. Engineering assessments tend to start with calculating static loads, and dynamic effects can easily be overlooked. As structures become larger and more efficient in the use of materials, the dynamic effects become even more important.

- Make sure you have considered all possible dynamic effects – dynamic loads, dynamic response, and interconnected modes.

Remember history

Most of the failures had precedents. Sometimes the history was in other engineering disciplines or other countries.

- Take time to study the history of engineering and engineering projects.
- Ask the question *Are there any historic failures that might be relevant to my design?*
- Look for past experience in other countries as well as in other engineering disciplines.

Effects of the climate

All engineering work takes place in a climate. The climate may be more influenced by the industry as a whole, or alternatively just the project, or the make-up of the team. The enthusiasm of a large and unique project can spur on the work at a breakneck pace. This may make it more difficult to change tack, critically review the work, and assess how it might fail.

- Take time out to understand all the potential failure modes and make sure that none will result in a failure.

Lessons from human nature

- The best of engineers can make mistakes; no one is immune.
- The psychological aspects can be as important as the engineering. Make sure that your engineering analysis and judgement are not being marginalized by pressure from your ego or from others.

Lessons from the Millennium Bridge

The *wobbly bridge* is a humbling lesson for all engineers. It emphasizes the narrow line that exists between success and failure. Unfortunately, telling the public that this is a new phenomenon, after all the experience with bridges, does not help gain their confidence and support.

- Never be too confident from past experience.
- Recognize that even when you have backed up your confidence with conventional testing, there can still be a latent problem lurking under the surface.

The wobbling of a few bridges before the Millennium Bridge did not attract the attention of the industry to a new phenomenon. It highlights the responsibility engineers have to present newly experienced problems to their colleagues.

- Be prepared to present your experience for the benefit of colleagues and the next generation of engineers. Learned societies can offer a good forum for the exchange and review of knowledge.

References

(1) **Shirhole, A.M.** and **Holt, R.C.** (1991) *Planning for a Comprehensive Bridge Safety Assurance Program*, Transportation Research Record, Washington DC, #1290, 1, 39–50.

(2) www.nationalbridgeinventory.com.

(3) 2002 Bridge Inventory, Better Roads, November 2002.

(4) **Sibly, P.C.** and **Walker, A.C.** (1997) Structural accidents and their causes, *Proc. ICE*, **62**, 191–208.

(5) **Petroski, H.** (1994) *Design Paradigms: Case Histories of Error and Judgement in Engineering*, Cambridge University Press, ISBN 0 52146 108 1.

(6) **Pugsley, A.G.** (1966) *The Safety of Structures*, Edward Arnold.

(7) **Martin, T.** and **Macleod, I.A.** (1995) The Tay bridge disaster – a reappraisal based on modern analysis methods, *Proc. ICE*, **108**, paper 10668.

(8) www.open2.net/forensic_engineering/riddle/riddle_01.htm.

(9) **Middleton, W.D.** (2001) *The Bridge at Québec*, ISBN 0 25333 761 5.

(10) **Billah, K.Y.** and **Scanlan, R.H.** (1991) Resonance, Tacoma Narrows bridge failure, and undergraduate physics textbooks, *Am. J. Phys.*, **52**(2).

(11) Billington, D.P. History and aesthetics in suspension bridges, *Proc. ASCE, J. Struct. Div.*, **117**(10), 103–105.

(12) Scott, R. (2001) In the Wake of Tacoma, *ASCE 2001*, ISBN 0 78440 542 5.

 (13a) Fitzpatrick, A. *et al.* (2001) *Linking London: The Millennium Bridge*, Royal Academy of Engineering, ISBN 1 87163 499 7.

 (13b) Fitzpatrick, A. and **Ridsdill-Smith, R.** (2001) *Stabilising the London Millennium Bridge*, Ingenia, RAEng, London, August 2001.

 (13c) Dallard, P. *et al.* (2001) The London millennium footbridge, *Struct. Eng.*, **79**(22).

 (13d) www.arup.com/MillenniumBridge.html.

(14) Clamcey, J. (2002) *The Guardian*, February 23.

(15) Crosbie, M.J. (2002) London's Bridge Ascendant, Architecture Week, 27 March.

(16) Research Results Digest (1999) *1998 Scanning Review of European Practice for Bridge Scour and Stream Instability Countermeasures*, NCHRP, Washington, DC.

The De Havilland Comet Accidents

In 1954, the accidents to Comet aircraft focussed the engineering community's attention on the danger of high stress at sharp corners. These stress raisers can lead to fatigue failure in components subject to cyclic loading, even for relatively few cycles. But are there other lessons that can be learnt from these disasters?

1.11.1 Geoffrey de Havilland (1882–1965)

Geoffrey de Havilland, a parson's son, had a keen interest in engineering and took a mechanical engineering course. After a brief spell with various petrol engine development projects for cars, he and a friend decided to build an aeroplane. He had read about the Wright brothers' experiences and had once seen an aeroplane in the distance. With £1000 from his grandfather, he designed the engine and aircraft. He had barely left the ground on the Hampshire Downs in 1908 when the wing collapsed. A second aircraft was built and he flew it successfully in 1910. The War Office bought it after a demonstration at Farnborough. So he started a career – on the basis of being a self-taught aircraft designer and self-taught pilot.

He joined the staff of the Royal Aircraft Factory at Farnborough and at the same time became a reservist in the Royal Flying Corps. Shortly before the First World War, he was persuaded to leave Farnborough and become the Chief Designer for the Aircraft Manufacturing Company where he designed DH1 to DH9. The DH4 was the most successful. It was chosen as the design for the American forces when they entered WW1, and 5000 were built in the United States. Post-war, the DH4 contributed to establishing the US Air Mail service.

Following the end of WW1, aircraft companies suffered in the harsh economic times. Eventually, in 1920, de Havilland was able to buy out his company and register it as the De Havilland Aircraft Company. It had a successful history and merged with Hawker Siddeley in 1960. One of de Havilland's most successful designs was the line of Moth aircraft, which were the first practical light aircraft. He flew the first in 1924 and they were still in service in the 1950s.

Another notable De Havilland aircraft was the Mosquito, DH98. It was the first multi-role combat aircraft. Its origins were in the Comet DH88 of the 1930s, developed for racing. The DH88 was the first

British aircraft to combine a retractable undercarriage, flaps, and a variable pitch propeller. R. E. Bishop, who had designed the DH88, designed the Mosquito – with two Rolls Royce Merlin engines and a stressed-skin plywood body. It went straight into production from the drawing board without building a prototype. It entered service in 1941, and eventually 7781 were built in the United Kingdom, Canada, and Australia.

The first jet aircraft designed by de Havilland was the Vampire fighter, DH100, that first flew in 1945. It used a Goblin engine designed by de Havilland.

1.11.2   Origins of the Comet Airliner

It is an amazing fact that as early as 1942, in the midst of WWII, Lord Brabazon chaired a committee to plan the future shape of the British aircraft industry for the post-war era. From the work of the committee came the Bristol Brabazon and Britannia, the Vickers Viscount and De Havilland Dove. Geoffrey de Havilland, a member of the committee, believed that the future of civil passenger aircraft lay with jet engines. He saw enormous commercial benefits for the first to build a jet airliner. The greatest competition would come from the United States. De Havilland saw the US approach as continually making something better. *Our only choice, in contradiction to our old step-by-step policy was to make one great leap and thus gain a lead which would take years to whittle down* (**1**). After exploring various design concepts, the basis for the Comet, DH106, emerged in 1945 under the Chief Designer R. E. Bishop.

1.11.3   The design of DH106 – Comet

De Havilland decided to build the Comet directly off the drawing board without a prototype, in the same way they had pioneered with the Mosquito. To gain the best advantage from using jet engines, the aircraft had to fly at high altitudes of around 40 000 ft to economize on fuel. It should also be as light as possible for maximum speed.

Power controls, hydraulic, and electrical systems were relatively new. Redux bonding of aluminium components followed experiences with adhesives on the Mosquito. The tricycle undercarriage arrangement and having the main wheels on either side of the undercarriage leg were novel. Cruising at high altitudes required a pressurized cabin. All these developments were brought together in the Comet design. This was an aircraft that would fly at twice the speed and twice the height of any airliner then flying (however, a similar aircraft, the Avro Canada Jetliner with four Rolls Royce Derwent engines, was built as a prototype and flew 13 days after the first Comet flight. For a variety of incomprehensible reasons, the Jetliner was never put into production despite the prototype performing well.)

1.11.4   Pressure cabin design

In order to keep the cabin comfortable at cruising altitude, the cabin was pressurized to $8.25 \, \text{lb/in}^2$ (P) – about twice the pressure used in previous aircraft as well as having a larger temperature range. The British Civil Airworthiness Requirements called for a test at 1.33P to show no deformation, and for no failure at 2P. These require-ments were similar to those of the International Civil Aviation Organ-isation. De Havilland, the Air Registration Board (ARB), and other experts believed that a design pressure of 2P would be adequate to avoid fatigue.

De Havilland decided to be conservative and used a design pressure of 2.5 P. They tested a 3 ft square section of cabin wall that included a window. This withstood $20 \, \text{lb/in}^2$ without failing. Two large fuselage sections were tested, one was the front 26 ft from the nose, and the other was a centre 24 ft section. The front section was pressurized 30 times to between P and 2P, then 2000 times to just over P. This was not seen as a fatigue test, but as a series of tests to give confidence in the overall design – which it did.

1.11.5   Fatigue testing to confirm the design

Attention had been drawn to the potential for wing and gust loading to make fatigue failure a possibility during the typical life of an aircraft. As a result, De Havilland carried out fatigue tests of the wings in conjunction with the Royal Aircraft Establishment (RAE) Farnborough.

During 1952, evidence was presented at industry meetings, at which De Havilland had representatives, that fatigue could be a problem on military aircraft with pressurized cabins. The ARB proposed that fatigue tests should be conducted. De Havilland used the nose test section and ran cycling tests. By now, Comets were in service and had exceeded 2500 flying hours and around 800 cabin pressure cycles.

The test section withstood 18 000 pressure cycles before it failed at a defect near a window. This was well above the predicted service life of 10 000 pressurized flights. Also, de Havilland believed that the general stress level was around $28\,000 \, \text{lb/in}^2$, less than half the mate-rial ultimate stress of $65\,000 \, \text{lb/in}^2$. As a result, they were confident that the cabin would survive the service pressure cycling.

1.11.6   Operational experience

There were several early accidents at take-off. They were thought to be pilot error, but after investigation, shown to be aerodynamic design deficiencies, and modifications were made. In 1953, a Comet broke up in mid-air near Calcutta. The failure was attributed to a violent storm with the possibility that the pilot may have overloaded the controls, while trying to fly through the turbulence. Feedback was added to the hydraulic controls to give pilots some feel for the forces they were applying.

In January 1954, a Comet was lost while climbing through 27 000 ft, leaving Rome. The debris came down in flames in the sea near Elba. All Comet flights were suspended and an investigation started. The UK Government requested the Commander-in-Chief Mediterranean, Earl Mountbatten, to use all the Royal Navy resources to recover as much wreckage as possible. The water depth was between 400 and 600 ft. Several vessels were equipped with lifting and search equipment. Underwater television was used for the first time to search for wreckage.

After a detailed review, and while debris was still being recovered from the sea, it was concluded that a fire was the most probable cause of the accident. Several modifications were made, and flights resumed in March. Sixteen days later, another Comet was lost in the sea near Naples, also on a flight out of Rome. All Comets were grounded.

1.11.7   The accident investigation

Following the second crash, the Minister of Supply instructed Sir Arnold Hall, the Director of RAE Farnborough, to make a complete investigation using all necessary resources. In addition, the Government instituted an Inquiry under the Chairmanship of Lord Cohen (2).

1.11.8   RAE

The 'Establishment' at Farnborough had started in 1904 but had its origins in balloons starting in 1878. Prior to the First World War, it was the Royal Aircraft Factory. Following the war, the task of designing and building aircraft was left to private industry, and the title *Royal Aircraft Establishment* adopted. From the 1930s, it grew in stature and capability. Farnborough was a home for mathematicians, aerodynamicists, structural analysts, chemists, metallurgists, mechanical engineers, and system control and instrumentation engineers. They worked in a broad range of laboratories and test facilities. This capability could be devoted to the theoretical and technical assessment of new aircraft. An Experimental Flying Department covered all aspects of flight testing from in-flight flow-visualization to probing supersonic flight and establishing the stability boundary of high-altitude helicopter use. Over the years, RAE had successfully investigated aircraft accidents and had created a separate section for this work. The philosophy for accident investigation had two main thrusts – first, the use of deductive reasoning from evidence, and second, the use of experiments to simulate possible accident conditions. For the first thrust, the collection of evidence was broad, not only covering the wreckage but also collecting data on flying conditions, including meteorology, pre-flight incidents, the aircraft's history and the design and manufacturing history.

Cabin fatigue had seemed incredible so early in service life, especially when the previous test work was considered. But the two failures provided enough doubt to justify a full-scale test. The water tank tests were committed at Farnborough ten days after the second crash.

Water tank fatigue tests

A complete aircraft was filled with water and immersed in a water tank. The wings protruded through rubber sealing flaps. The water pressure inside the cabin was cycled to simulate the pressurized flights. At the same time, the wings were loaded to simulate flight loads. The aircraft cabin failed at the corner of a window after 1930 pressure cycles in the tank. It had previously been in service with BOAC and seen 1230 flights. So the failure had occurred after a total of 3060 cycles. The crack was repaired and now that the location of failure was known, the area was fitted with strain gauges.

A second aircraft was tested in the water tank. It was only cycled with internal pressure, without adding wing loading. It developed fatigue cracks at the corners of many windows after 5248 cycles.

Other avenues of investigation

The RN salvage work was continuing. It had started in January and continued until August. They had been assisted by an established RAE method for predicting the spread of debris from an accident. This was based on calculating the trajectories from the point of failure. When the tank tests pointed to cabin fatigue failure, RAE conducted a test by firing a wooden-scale model from a balloon to help estimate where the debris might be found. The model was constructed so that it would break up in a similar manner to that postulated for the crash. This gave a new pattern of debris in water deeper than where they could use their recovery equipment. The Navy organized local fishermen to trawl in a wider area. The fishermen found the section of cabin roof that was believed to be the source of the failure. Eventually, 70 per cent of the aircraft was recovered from the sea.

Back at Farnborough, all BOAC Comet crews were tested on an embryonic flight simulator, to check if any pilot action might have contributed to the disaster. RAE also test flew another Comet from Farnborough unpressurized, to check other theories.

The recovered wreckage showed that the outer sections of the wings had broken off. Detailed examination looked for clues to establish which piece broke off first. There were imprints on the tail of newspapers and coins from inside the cabin. This evidence suggested that the cabin had failed first.

The engines were recovered and were examined by de Havilland. They had marking consistent with a violent nose-down pitching of the aircraft. A test engine was run in a cradle and rapidly rotated about its horizontal axis. The damage was similar to that on the recovered engines.

Medical evidence from the bodies recovered from the sea was consistent with a rapid depressurization. All the bodies had burns, but these had occurred after death.

Conclusion from the investigation
The RAE report concluded that the crash off Elba had been *caused by structural failure of the pressure cabin, brought about by fatigue.* The most likely location of the initiating crack was at an Automatic Direction Finding window, similar in construction to the regular windows. The Naples crash was presumed to be the same type of failure, as no wreckage was available.

Most parties agreed with the RAE conclusions. However, Mr B. Jablonsky questioned whether the use of adhesives had contributed to the failure. There was no evidence to support his concern. Walter Tye, the Chief Technical Officer of ARB, accepted the conclusions, but had concerns that the two service failures had occurred at much lower life cycles than the water tank tests. He wondered whether another, yet unknown, factor was coming into play. There was no clear answer to his concern. The possibility of overpressurizing the cabin in service was investigated and shown to be remote.

Lord Cohen's Inquiry dealt with responsibility, and he concluded *no suggestion was made that any party wilfully disregarded any point which ought to have been considered or wilfully took unnecessary risks.*

The Inquiry closed off many loose ends and included suggestions for the future direction of various work, including regulation, use of facilities, and research on means of guarding against fatigue failures.

1.11.9   The fatigue results from service and test

A detailed analysis is given in the book *Engineering Catastrophes* (**3**).

Cycles

The aircraft that crashed near Elba had seen 1290 pressure cycles, the one at Naples had seen 900 cycles. The full simulation in the first tank test at Farnborough failed after a total of 3060 cycles – part in service and part under test. The second tank test showed fatigue cracks after 5248 cycles, but this had no wing loads added to the cabin stresses. Dr P. B. Walker, head of the Structures Department at RAE, told the Inquiry that the full simulation probably lacked some loading, such as from turbulence and buffeting. He estimated that the 3060 test cycles might be the equivalent of 2500 service cycles. The aircraft that failed in service were probably those at the lower end of the spectrum of failures. Dr Walker pointed out that a scatter of 3:1 was small. A typical scatter of fatigue results could be in the range of 9:1.

Stress

The highest stress measured by the strain gauges on the tank test was $40\,800\,\text{lb/in}^2$ close to the edge of the 3 in radius in the corner of the window. Extrapolating to the edge of the window gives an estimated maximum stress of $47\,700\,\text{lb/in}^2$. The edge of the window has a sealing and reinforcing strip joined by Redux adhesive. But the window frame was attached to the cabin skin by two rows of $\frac{1}{8}$ in rivets. The rivet holes would increase the local stress levels by roughly three times the general stress level. These stresses were clearly high enough to initiate fatigue failures in relatively few cycles.

Critical crack growth

Fracture Mechanics was in its infancy at the time of the Comet crash. A. A. Wells from the British Welding Research Association (now TWI) was seconded to the Naval Research Laboratory in Washington, who were the leaders in the field. Fracture Mechanics can predict when a crack becomes unstable and run at high speed. Wells tested the material from the Comet and found a critical crack length of 6.3 in. This was precisely the length of crack found in the tank tests when failure occurred. More recent Fracture Mechanics analysis confirms the assessment (**4**).

It was observed that by the time the cracks were $\frac{1}{4}$ in long, 95 per cent of the fatigue life had been used. The growth from 1 in to the critical length of 6 in only took 200 cycles.

The second tank test showed 16 different cracks developing at the corners of 11 windows. All cracks started at rivet holes near the corners of windows.

The de Havilland fatigue tests
The de Havilland fatigue tests were conducted on a section that had previously been loaded up to 2P. This overpressure test had taken the material in the corner of the windows into the plastic range. When the test pressure was removed, this material would have been forced into compression locally. This reduced the tensile stresses during fatigue testing and gave a longer life.

1.11.10   De Havilland versus RAE
The investigation soon surfaced radical differences of view between De Havilland designers and RAE staff. The differences were so marked and well known that they have been addressed in Lord Cohen's Report. The designers' approach was to make broad calculations of the general stress level in their designs and then carry out tests to confirm whether the design was adequate. RAE staff, coming from a more scientific background, considered that detailed stress calculations were essential and should have been part of the design. Lord Cohen saw that it was natural for them to approach the *safe life* from different points of view. He noted that the differences of opinion diminished in the course of the Inquiry as greater mutual understanding developed.

COMMENTS
Traditional design versus analysis
Geoffrey de Havilland had very clear views. He believed that good designers *are born and not made*. They possess the gift of being able to create a clear mental picture and have the ability to visualize the loads acting on a component. A deep insight into mechanical engineering is essential as well as practical experience – lots of it. His philosophy was to have good designers working to well-proven practices and then have their designs confirmed by test.

The staff of RAE were not designers. They based their investigations on exploring the background science – aided and abetted by tests. The title of a paper on the work at Farnborough (**5**) is telling – *The scientific investigation of Aircraft accidents.*

Which view is correct? The de Havilland approach could miss scientific facts that were not in the designers' experience base. The RAE scientific view could miss practical problems coming from experience. Probably both views are too extreme, and the ideal is a combination of each. In the case of the Comet, more detailed calculations at the design stage should have shown the high stresses at the corners of the windows. The corner stress of $47\,700\,\text{lb/in}^2$ was well above the general stress of $28\,000\,\text{lb/in}^2$ that de Havilland believed was present.

The Comet accident happened at a time when aircraft were becoming more complex in the approach to supersonic flight. This was an era when the size of design teams was increasing by about a factor of five each decade. The extra design staff were largely in the technical areas of stressing and aerodynamics. So, natural evolution was already bringing together the two views of *design by tradition* and *design by analysis*. Unfortunately, there has been a tendency for the two areas to be handled by different staff, which introduces an extra interface.

Geoffrey de Havilland liked to wander into the design office and chat around a drawing board. This created an atmosphere where ideas were openly discussed and explored. An atmosphere that allowed staff to put forward their ideas and give the designer the opportunity to gain from the experience of others. This atmosphere has been lost with larger design teams, the separation of drawing office from technical office and formal review processes.

A myth

The author Nevil Shute wrote his novel *No Highway* in 1948. His full name was Nevil Shute Norway. As well as being an author, he was a well-experienced aeronautical engineer. While writing had become his full-time career, he still kept current with aeronautical developments. He had read Pugsley's RAE report on the potential for fatigue in aircraft. He based *No Highway* on the fatigue failure of an aircraft due to tailplane flutter. The myth created in the mid-1950s was that the Comet crashes could have been avoided if the De Havilland designers had paid as much attention to the RAE report as had Nevil Shute. The explanations in this chapter show that de Havilland did pay attention to fatigue. The fatigue concerns raised by Pugsley were related to aerodynamic forces and gust loading on wings, and de Havilland had worked with RAE on avoiding wing fatigue. In his autobiography, Sir Geoffrey de Havilland expresses his view that *fatigue had been associated with high periods of vibration – and jets didn't vibrate like piston engines.*

Confidence from testing

The Comet problems can be summed up as – proved by test but failed in service. It is an example of incorrect and inappropriate confidence coming from an apparently successful test. Tests have to be realistic and must replicate service conditions. Because of the previous overpressure tests of the cabin section, the fatigue test results were misleading. Another issue is to ensure that the boundary conditions do not influence the test results. In the 3 ft square cabin section test, the edge effects from clamping and sealing the section probably affected the results by artificially strengthening the section.

Scatter of fatigue results

The traditional method of design was, and to some extent still is, to conservatively estimate the loads and ensure that they are a reasonable amount less than a conservative estimate of the strength. This tends to be a static calculation where the difference between the two may be a factor of two or three – the safety factor. It was natural to apply this same thinking to fatigue strength.

De Havilland thought that a single fatigue test showing failure after 18 000 cycles was adequate for a service life of 10 000 cycles. The Chief Technical Officer of ARB was concerned that there might be some other effects resulting in the life of the crashed aircraft being only one-third that of the tank tests. Even Nevil Shute's book is based on getting near to the actual number of cycles predicted for failure. So while Dr P. B. Walker told the Inquiry that a range of 9:1 could be expected, this fact was not widely understood at the time.

Stress concentration factors and fatigue

The likelihood of fatigue failure of shafts with sharp changes of section and subject to a high number of cycles was known since Wöhler's experiments in the late nineteenth century. But the rivet holes in the 3 in radius in the corner of the window of the Comet were not understood to be a similar stress concentration. The high-cycle fatigue failures of machine components were not seen to be related to the large, relatively low-pressure, aircraft cabins that were subject to a much smaller number of cycles. This is a classic case of not being able to visualize from one experience to another in a different field and in a different context.

Stopping cracks

The Inquiry explored the practice in De Havilland and other aircraft companies of 'repairing' small cracks by drilling a small hole at the end of the crack. The presence of the cracks was logged in the records of the aircraft and the cracks checked in service to ensure that they did not grow. This was a practical way of removing the sharp notch at the tip of the crack. The development of Fracture Mechanics was eventually able to provide a mathematical basis for this practice, a case of science eventually proving that the methods developed from practical experience were sound. The science has now enabled the critical crack size to be calculated as well.

Expert help

De Havilland was stepping well outside their previous experience. The advice they had from experts and regulators like ARB and ICAO (International Civil Aviation Organisation) gave them confidence. While it is good to ask for the advice of others, when you are the one taking the big step forward, then you are on your own. You cannot allow yourself to become complacent just because others, who are not taking your big step, cannot envisage the problems that you may face.

How big a step should you take?

Igor Sikorsky's career had several similarities to Geoffrey de Havilland. Sikorsky first flew a fixed-wing aircraft in Tsarist Russia on 3 June 1910. He flew successfully for eight minutes before he stalled and crashed when the engine was not powerful enough to provide the lift in a turn. His real interest was in helicopters, but he put these ideas in a bottom drawer until the engine designers could come up with an engine with the necessary power/weight ratio. Sufficiently powerful engines were eventually available, and Sikorsky turned his energy into pioneering helicopters in the United States. Like de Havilland, Sikorsky was a self-taught pilot and designer.

Sikorsky was a strong believer in *engineering common sense*. He saw it as an extremely valuable source of guidance – although he recognized that it could, on occasions, mislead. But with training and experience, *engineering common sense* usually provided the best guidance. After a lecture in London in the 1950s, Sikorsky was asked what was his design philosophy. His answer was to *never agree to build anything that was larger than a factor of two over proven experience – in any one dimension, i.e. twice the speed or twice the weight but not both.*

The Comet was designed for twice the speed, twice the height and with twice the cabin pressure of previous airliners!

The use of government facilities

Lord Cohen's comments in his report read like a plea to the aircraft industry to utilize the expertise and facilities at RAE. It seems remarkable today that there were 16 different companies designing and manufacturing aircraft in the United Kingdom at that time. Their commercial outlook was different from the thrust of RAE, which was a government organization answering to the Minister of Supply. RAE was a highly competent establishment that made significant contributions to the aircraft industry and indeed to a much broader area of engineering and academia. But how do you measure the worth of such an organization? This is a question that still bedevils decision-makers. The decision on whether to open or close a national laboratory, and on the level of funding, tend to be made by political philosophy. Having said that, Lord Cohen's advice to industry is still appropriate today – make sure that all national and international expertise is used, even if it exists in an unfamiliar environment. Lord Cohen would be surprised to find that neither de Havilland nor RAE exists today, but the lessons from his Inquiry are still relevant.

The effect of climate

Simon Bennett, a sociologist at Leicester University, has used the Comet history as a case study (6). He sees several climate issues surrounding the aircraft's history.

First is blaming the pilot for the crash on take-off. The Comet was seen as the pinnacle of aeronautical achievement. When one crashed, it could not be the fault of the aircraft – it must be pilot error. But the Comet was underpowered, and early jet engines were slow to accelerate. The aerodynamics of the wing was optimized for high-speed flight rather than for take-off. If the nose was a few degrees higher than the designed angle, then there was insufficient power for take-off. A crash occurred at night and in rain, when it was not easy to assess the aircraft's attitude. For crashing such a prestigious aircraft, the Captain was banished to flying piston-engined York freighters. De Havilland issued clear instructions that the nose should not be raised higher than the designated angle. Their test pilots had no problem in meeting this instruction. This raised the issue as to whether all airline pilots needed to be as skilled as test pilots. Or should test pilots explore the broader boundaries into which commercial pilots might stray? Eventually, after a second crash, De Havilland recognized the need to modify the wing for take-off and regulators required better margins for pilots.

Another aspect was the political and economic times under which the Comet was developed. Post-war Britain was looking for success – politically, economically, and technically – something to boost the economy, boost morale, and be a source of pride. This influenced everybody involved. Did these ambitions lead to a less critical environment for assessing all the decisions that were taken – in design, testing, and regulation?

Final comments

Geoffrey de Havilland was knighted in 1944 after a distinguished career. The Comet crashes were a heavy burden for him, and he personally followed the whole investigation. He had experience of the risk of flying from his early attempts to fly, but also by having lost two sons in test flying accidents. His cousins, Olivia de Havilland and Joan Fontaine, had less stress in their successful careers in Hollywood.

Eventually, De Havilland produced the Comet IV, which flew successfully. But by now, the market had been taken by the Boeing 707.

In his autobiography, Sir Geoffrey sums up:

> *The human error in the case of the Comet was the general lack of deeper knowledge of the problem of metal fatigue. Although much was known about metal fatigue, not enough was known by anyone anywhere.*

1.11.11 Lessons learned

The Comet crashes highlighted for a whole generation of engineers the importance of stress concentrations and their effect on fatigue life. No doubt this helped avoid many other failures – but we will never know for sure.

- Make sure that the test conditions properly represent service conditions. In particular, do not pre-condition test samples by an overpressure test and redistributing the stresses.
- Tests can be misleading if you do not understand the underlying science affecting the results.
- Recognize the large scatter in fatigue results.
- Analysis and experience should go hand-in-hand.

- When you are doing something that has never been done before, get the advice of experts, but recognize that you are on your own.
- Recognize how far you are moving outside current experience.

References

(1) **De Havilland, G.** (1961) *Sky Fever – The Autobiography of Sir Geoffrey de Havilland CBE*, ISBN 1 84037 148 X.

(2) Report of the Inquiry into the accidents to Comet G-ALYP on 10th January 1954 and Comet G-ALYY on 8 April 1954, HM Stationery Office, 1955.

(3) **Lancaster, J.** (2000) *Engineering Catastrophes: Causes and Effects of Major Accidents, 1996*, Second edition, Woodhead Publishing, ISBN 1 85573 505 9.

(4) **Withey, P.A.** (1997) Fatigue failure of the de Havilland Comet 1, *Eng. Fail. Anal.*, **4**(2), 89.

(5) **Walker, P.B.** (1964) The scientific investigation of aircraft accidents, *Proc. IMechE*, **179**, 997.

(6) **Bennett, S.** (2001) *Human Error – By Design?* Perpetuity Press, ISBN 1 89928 772 8.

SECTION 1.12

The Danger of Not Knowing

There is nothing worse than experiencing the starting stages of a disaster while not having the information to know what is going on. The most prominent example of this is the accident at the Three Mile Island nuclear power station. Neither the operators nor the experts had the information to be sure that they knew what was going on inside the containment building. In this chapter, two aircraft incidents are reviewed where the flight crews lacked information. Fortunately, neither resulted in any fatalities or serious injury.

1.12.1   Example 1. The Gimli Glider

On the evening of 23 July 1983, an Air Canada Boeing 767 en route from Ottawa to Edmonton ran out of fuel over Manitoba and made a successful emergency landing at a disused airfield.

The incident

Cruising at 41 000 ft and at 469 knots, a warning light came on, indicating a fuel pump problem. The pump was isolated as the engine has duplicate pumps. Shortly afterwards, a warning came on for the second fuel pump, and Captain Bob Pearson decided to divert to Winnipeg. Then the left engine quit, and Pearson prepared for a single engine landing. Meanwhile, the Captain and First Officer thought it might be a computer problem and tried to cross feed between the fuel tanks to correct the problem. The Engine Indication Caution and Alerting System then indicated both engines were out – and things went very quiet. The Auxiliary Power Unit (APU) designed as an emergency power source could not start as it uses the same fuel tanks as the main engines, and all three tanks were now empty. The glass cockpit panels all went blank. Fortunately, basic instrumentation is still available. The final power back-up is the Ram Air Turbine (RAT), which uses the airstream to drive a turbine and provide limited hydraulic power. This allows operation of the main control surfaces.

Pearson opted for his best guess at a gliding speed and flew at 220 knots. First Officer Maurice Quintal searched the manuals for guidance. Loss of power meant that the aircraft identification signal stopped working. Fortunately, Winnipeg Air Traffic Control had an old system and was able to manually track the plane's course. They and the flight crew soon recognized that the aircraft would not make it to Winnipeg. Closer by was a disused RCAF airfield at Gimli. The navigation books gave no details for this long-disused airfield. But coincidentally, First Officer Maurice Quintal had, years previously, been stationed there for flight training when he was with the RCAF.

There was not enough hydraulic power to lower the undercarriage. A gravity drop was attempted. The main wheels locked down but the nose wheel would not lock in place. The aircraft was high for a direct approach. Pearson was a keen and experienced glider pilot. He tried a glider manoeuvre and threw the 132 ton aircraft into a sideslip to lose height. It worked – although disconcerting for the passengers. The sideways motion limited the airflow into the RAT, and hydraulic pressure started to reduce, making the aircraft more difficult to control. Pearson crossed the runway threshold at 180 knots – only 30–50 knots higher than a normal landing. On landing, he stood on the brakes, and two tyres exploded. The nose wheel gear collapsed, and the aircraft ground along the runway throwing up a curtain of sparks. It stopped at 3000 ft. A small fire broke out in the nose, and smoke started to enter the cockpit. All on board exited via the emergency chutes. A few passengers using the tail chutes sustained minor injury because the chutes were steep with the aircraft down on its nose (Fig. 1.12.1).

Fig. 1.12.1   The *Gimli Glider.* **Photo by Wayne Glowacki, reprinted with permission from the Winnipeg Free Press, 24th July 1983**

The Winnipeg Sports Car Club was using the airfield for a rally.
They scattered when they saw the plane approaching. It stopped less
than 100 ft from campers. The Sports Car Club used their fire extin-
guishers to put out the small fire in the nose of the aircraft.

The aircraft was relatively undamaged. The tyres were changed, a
few items repaired, fuel added, and it was flown out two days later.
It was taken to a maintenance depot for thorough checking.

Why did flight 143 run out of fuel?

The flight started in Montréal. The ground crew had found that the
Fuel Quantity Indication System computer was not working properly.
Errors were made in calculating the fuel to be loaded in Montréal.
It has popularly been described as confusion between imperial and
metric measurements – but was it?

Problems with the Fuel Quantity Indication System. (FQIS)

The FQIS (Fuel Quantity Indication System) is a computer system
aimed at simplifying the management of fuel for the flight crew as
well as providing the signal for the fuel gauges in the cockpit. Boeing
specified a highly reliable system and incorporated dual channels with
Built-In-Test-Equipment (BITE). With redundancy and self-testing,
the system should provide adequate information even if one channel
fails. Honeywell supplied the system to Boeing.

This aircraft, C-GAUN, had problems with the FQIS at Edmonton
on 5 July, 18 days before the forced landing. One of Air Canada's
ground staff, Conrad Yaremko, checked the system and found that
the cockpit fuel gauges would only work if channel 2 was disabled.
He logged it as a deviation. When the flight arrived at Toronto, the
system was checked. It now appeared to be working correctly and
was reset.

On 14 July, en route from Toronto to San Francisco, this aircraft
lost all fuel gauges. A new processor was recommended for installation in San Francisco. But when the original processor was taken
out and reinserted, it worked. It was rechecked in Toronto and met
specification.

Back in Edmonton on 22 July, during a routine check, the fuel
gauges went blank. Yaremko did not realize this was the same aircraft,
but remembered what he had previously done to correct this type of
fault. He disabled channel 2, the fuel gauges worked, and he put
a warning tape around the disabled circuit breaker. He also entered
a brief note in the aircraft logbook. The rules say that if only one
channel is working, the fuel quantity should be checked by a drip

check. This check is effectively a float stick built into each fuel tank. The stick appears through a small sealed hole under the fuel tanks. This procedure was followed in Edmonton before Captain John Weir flew the aircraft to Montréal.

In Montréal, the ground crew of Messrs Ouellet and Bourbeau were assigned to prepare and fuel the aircraft. Ouellet noted the log entry and, while waiting for the fuelling truck, decided to investigate. He remembered that there had been similar snags that seemed to be repeating. He could not understand why one channel had been deactivated. He had read that the system automatically went to one channel if there was a problem with the other channel. So he reactivated the breaker and tried a BITE test. The fuel gauges went blank. The fuel truck arrived and as he had been instructed to carry out the drip check, he left the cockpit – and forgot to deactivate channel 2. So the aircraft now had blank fuel gauges.

The calculation
Captain Pearson and First Officer Quintal worked with Messrs Ouellet and Bourbeau to carry out the drip checks and calculate the fuel to be loaded. The drip sticks on the Boeing 767 are calibrated in centimetres. Tables in the aircraft manual convert these measurements into litres – showing 7682 litres on board. But the fuel gauges and fuel loading on the B-767 are in kilograms. The flight to Edmonton via Ottawa needed 22 300 kg of fuel. The First Officer asked the ground crew what the specific gravity was, and was told 1.77. So the calculation they made was:

$$(22\,300 - 7682 \times 1.77) \times 1/1.77 = 4916 \text{ litres to be loaded.}$$

Both flight and ground crews participated in the calculation. Ouellet had an idea that the figure should be divided by roughly 50 per cent to get kilograms. He did not have the exact figure and left it to the others. Captain Pearson rechecked the hand calculations on his computer.

The factor 1.77 was in fact the conversion factor from litres to lbs, although it was often incorrectly referred to as the specific gravity. It was the correct conversion factor for the rest of Air Canada's fleet that had fuel load measured in lbs. The correct calculation, using the specific gravity of 0.8 for aviation fuel is:

$$(22\,300 - 7682 \times 0.8)/0.8 = 20\,193 \text{ litres to be added.}$$

So the aircraft took off short of 15 277 litres of fuel. In the brief stop in Ottawa, Captain Pearson requested a recheck of the drip measurements. The ground crew in Ottawa gave the conversion factor as 1.76.

Minimum Equipment List (MEL)

With various systems replicated, it is permissible to fly with some components not functioning. There is a Minimum Equipment List (MEL), which specifies the minimum equipment that must be operable. For the FQIS, the MEL states that if only one channel is working, then a drip check must be carried out to confirm the fuel on board.

Either the ground crew or the Captain can decide whether an aircraft is below the MEL level and ground the aircraft. In this case, at Montréal with no fuel gauges, the aircraft was below the MEL and should not have been flown. However, it was widely believed in Air Canada that there was a Master MEL that allowed greater freedom. Also, before 1970, the MEL had only been a guide for pilots and not mandatory.

Captain Pearson had seen Captain Weir in the car park, and while chatting, Captain Weir had told him that there was a problem with the FQIS. When Captain Pearson arrived in the cockpit, he saw that the fuel gauges were blank. He also read the note in the logbook. However, he believed that this was the state in which the aircraft had arrived from Edmonton and that with the drip procedure, he could take off below the minimum MEL. In fact, he was mistaken on both counts. The fuel gauges had worked on the incoming flight, and the MEL rules forbade operating without fuel gauges.

Fuel loading – whose job is it anyway?

Medium- and long-haul passenger aircraft used to have three flight crew – Captain, co-pilot, and engineer. President Reagan had a Commission look at flight deck manning, and they recommended that two crew members are adequate. One of the tasks of the flight engineer was to monitor fuel loading and use. When that position disappeared, the responsibility was not clearly reassigned – to ground or flight crew or both jointly.

Normally, aircraft carry a fuel logbook. On the short haul DC 9, with a two-man flight crew, the logbook was dispensed with in order to reduce the paperwork load on the crew. Air Canada followed the same practice on the B-767 – so there was no fuel logbook.

What was the reason for the FQIS problems?

Honeywell tested and inspected the FQIS processor from C-GAUN. They found that the failure was a 'cold solder' joint between an inductor coil wire and the terminal post. The terminal post was pre-tinned – the coil wire was not, and adhesion was poor.

This hardware fault exposed a design fault. The voltage drop created by the poor joint prevented the switchover to the good channel. This explains why Yaremko was able to get the fuel gauges working by isolating channel 2 and why Ouellet was not able to get them working with both channels energized.

Corporate responsibility

The Government Inquiry undertaken by the Hon. Justice George Lockwood (1) saw that, on top of Human Factors, errors and equipment failures, there were corporate deficiencies by Air Canada.

Metrication

Air Canada was under pressure from the Canadian Government, which at the time was its owner, to metricate. The Boeing 767 was the first metric aircraft in the fleet – but only partially. Fuel was in metric – altitude and speed were still in feet and knots. However, Justice Lockwood was clear that the problem was not confusion between metric and Imperial but was by using the wrong conversion factors from metric volume to metric weight. He criticized the failure by management to alert all parties to the changes that are necessary when a new type of aircraft is added to the fleet.

Decision making and control

Justice Lockwood criticized Air Canada for being top heavy and having a complex decision-making process. Responsibility for fuel load calculations had not been assigned. The requirements of MEL had not been enforced even though the corporate headquarters were aware of deviations through their morning briefings. Deviations were occurring at a rate of about 2/month.

Training – flight crew

Over the years, the underlying principles for training pilots has changed. In the past, pilots were expected to learn how everything worked. With the advent of more complex aircraft, this was not easy. The basis then changed to limiting the training to those items where there was a need to know. As a result, they were not taught about the intricacies of the FQIS.

Justice Lockwood talked to many witnesses, including those from other airlines and countries. One was Captain Last, Principal VP from the Airline Pilots Association (IFALPA) who was also a Boeing 757 Captain with British Airways. Captain Last answered Justice Lockwood's general question on training as follows:

> *You asked what the answer is, and I am afraid the answer is a rather unpalatable one to an awful lot of the industry. This situation is generated by one thing and one thing alone, and that is money. The fact of the matter is that the major manufacturers have no interest in being in the business of building airplanes other than to make money: the more they can sell the better off they are. And the easier they can sell them to operators, the better off they are. The operators are in the business of reducing their costs to the maximum possible extent. Training is a cost which does not show up as having any positive benefits on the balance sheets. It does not produce revenue unless you can sell your training to someone else, and you will sell it to someone else if you can sell a five-day course instead of an eight-day course.*
>
> *So the whole thing is loaded in the direction on minimising costs, which means minimising information, minimising time, et cetera, et cetera, and there is nobody on that side of the industry, either the operators' side or the manufacturing side who is going to gain any benefit out of rectifying the situation. So whoever you are talking to, whether it be a management pilot or a manufacturer's pilot or whatever, his main concern, regardless of what he actually believes deep down – and a lot of them are saying informally that they do not agree with what is happening – but they are not in a position to do anything about it.*

Captain Last reiterated that the only answer was to allow the line pilots, through their pilots' associations, to participate more in the design and philosophy of training courses. He went on to say that with older aircraft and the older type of pilot training, if something went wrong with an aircraft, the margin of safety did not drop dramatically. On the other hand, with modern complex airplanes, 99 per cent of the time there are no problems, but when they do occur, the drop in the overall level of safety is much more severe. Designs are being pushed closer and closer to the limits to squeeze out the last ounce of performance. The knowledge given to crews to deal with the edges of the envelope is that much less. *We have removed a crew member from the cockpit on the basis that none of the things around the edges would ever go wrong. But they still go wrong.*

Training – ground crew
A similar issue occurs with the ground crew. The new approach is to use the computer-based checking equipment to identify which piece of equipment to replace. The attitude is *You do not need to know how the system works.*

Inadequate corporate communication within Air Canada
The Inquiry concluded that Air Canada was too compartmentalized and somewhat bureaucratic. Despite having highly qualified and dedicated individuals, it can be hard to overcome the hurdles in the corporate structure. One example was the difficulty in disseminating historic data on aircraft problems between aircrew and maintenance personnel.

Another was that manuals were unclear. Neither the manuals nor other documents identified who was responsible for the work. The Air Canada Boeing 767 MEL was customized from the Federal Aviation Administration (USA) Master MEL but was *neither clear nor concise.* In general, those writing documents and manuals did not communicate or interact with the users. While updated documents were sent to the ground crew at the ramp offices, there was no check that they had been properly posted and filed.

Justice Lockwood congratulated Air Canada Maintenance on an effective daily check on outstanding deviations in the fleet. Unfortunately, this only takes place every weekday, and the Gimli incident occurred on a Saturday.

Spare parts
Recommendations for spare parts inventory come both from the manufacturers and from an assessment by the airline of component reliability. The answer is only an estimate, and the Director of Safety Services with BA had a pithy comment:-

> *When you buy a new airplane and you decide on the spares, one thing that you can be absolutely certain of: you are going to get it wrong.*

Air Canada decided to buy one spare fuel processor for their eventual fleet of 12 Boeing 767s. But earlier in 1983, they had sent it to Aerospatiale, a member of Airbus in Toulouse for use in developing a computer repair programme.

Flight safety
Justice Lockwood noticed the lack of a sophisticated and well-developed safety organization within Air Canada. In reviewing the

approach of other airlines, he was impressed with the BA Board Air Safety Review Committee. This committee has the airline Chairman as its Chairman. As a result, there is a clear understanding and communication of safety issues from those directly involved right up to the Board level.

1.12.2   Example 2. The day the Azores were in the right place

A Canadian charter airline, Air Transat, flies a regular route from Toronto to Lisbon. On 24 August 2001, the experienced crew of Captain Robert Piché and First Officer Dirk DeJager flew a two-year-old Airbus 330 with 304 people on board on the overnight flight to Lisbon. While still dark over the Atlantic, they got a warning signal of fuel imbalance between fuel tanks. After five minutes' investigation, it appeared that they might be short of fuel, so they decided to divert to Lajes airfield on Terceira Island in the Azores. Seven minutes later they declared an emergency. After a further 25 minutes, the right engine flamed out while still at 39 000 ft, but 135 nautical miles from the Azores. The left engine quit 13 minutes later. They glided the next 70 nm and made a fast, hard landing that burst eight of the ten main tyres and started a small brake fire. Fortunately, by now dawn had broken and Lajes airfield, a military airfield with a long runway, was fully operational. A few people were injured using the emergency chutes but the plane was evacuated in 90 s.

Why did flight 236 run out of fuel?

The investigation quickly showed that the fuel line to the right engine had failed where it rubbed against an hydraulic line.

The right-hand Rolls Royce Trent 772 engine had been replaced five days before the incident. The replacement engine did not have a pump fitted, but the Air Transat mechanic was able to fit a pump from a later model. A year earlier, Rolls Royce had issued a service bulletin addressing the problem of clearance between the fuel and hydraulic lines near the pump. The advice was not mandatory, and Air Transat had not completed the work recommended by Rolls Royce.

The Fuel Control and Monitoring System has two dedicated computers. Fuel imbalance is normal, and fuel is routinely transferred between tanks. In this case, it seems that the computers directed fuel into the cracked and leaking pipe and hence out of the aircraft. About 20 te of fuel was lost. The instructions in the manual for dealing with a suspected fuel leak are not simple – largely due to having to cover a wide variety of possible leak paths.

The aftermath from the incident

The immediate reaction from the passengers was to praise the flight crew for their excellent handling of the emergency. A little later, the mood changed and a $30 million class action lawsuit was filed against the flight crew, Air Transat, Airbus, and Rolls Royce.

The airline was fined $250 000 for improper maintenance. This was influenced by a report that an Air Transat mechanic had warned that this particular aircraft was not ready to fly. The supervisor had overruled the mechanic. For a while, Air Transat had its Extended Twin Operations Standard (ETOPS) licence, allowing twin-engined aircraft to operate over large bodies of water, limited to 60 min from any airfield.

An All Operations Telex requiring inspection for interference between fuel and hydraulic lines on Trent Series 700 engines was issued by Airbus five days after the incident.

The French regulator recommended that flight crews should carry out a verification of fuel status every 30 min. They should take the computer calculation of fuel burnt, away from the fuel on board at the ramp and compare this with the indication of fuel remaining in the tanks.

Another feature of the incident has worried air regulators. The flight data and cockpit voice recorders stopped working when both engines failed. This should not have happened.

COMMENTS

While both incidents had different issues, they had a common thread. Both lacked essential information, which if known early enough could have prevented the emergencies. This highlights the importance of providing good information – good instrumentation and good manuals.

First Officer Quintal was searching the manual to find out how to lock the nose undercarriage while helping to calculate the glide path and initiating all the other tasks on the checklist to prepare for an emergency landing. The information on locking the undercarriage was not in the obvious place in the manual and was also not in the index. Quintal gave up and concentrated on the other tasks.

Similarly, there are reports (2) that the fuel management instructions in the Airbus 330 manual are complex and confusing. In the case of the Air Transat flight, one fuel pipe leak should not have caused the aircraft to run out of fuel. The principle of redundancy should have allowed some fuel to be retained in the other tanks. Then there would have been the possibility of a single engine landing rather than gliding to a dead-stick landing. Whether this was the fault of the computer system, confusing manuals, or flight crew error will be difficult to establish with the class action lawsuit pending.

The benefit of additional skills – and luck
Both incidents had a sequence of small issues that eventually led to the emergencies. Both had fortunate coincidences. At Gimli, they were that the Captain was a skilled glider pilot, that the First Officer knew Gimli and that there was still daylight to see the abandoned runway. In addition, the fact that the nose wheel collapsed added to the friction with the runway and helped bring the aircraft to a halt before it ploughed into tents and campers. For Air Transat, the coincidence was having the Azores within gliding distance – the only land within 1600 km!

While flight crews practise regularly on simulators, a dead-stick landing is not in need-to-know training. Fortunately, many pilots explore unusual situations on a simulator. This broadens their understanding for use when faced with new problems.

Both emergencies were outside the envelope of problems considered as part of design or operating analysis. Computers can only help in situations that have been foreseen. It takes human skills to analyse and take decisions when faced with the unknown. Having said that, it would seem simple to get the computers to compare the fuel used against the fuel remaining in the tanks and be able to give a warning of a possible leak.

What was the cause of the Gimli crash?
Justice Lockwood did a very credible job in exposing the multitude of factors that led to the Gimli crash. But can one of these factors be established as THE cause? Probably not, but it does show how, in different types of inquiries, the blame can be placed on one of a variety of causes. Then the other faults may continue uncorrected. During the Gimli Inquiry, some parties representing specific interests proposed 'the cause' that was clearly biased towards their interests.

Justice Lockwood states that the conversion from imperial to metric measurements was not a cause. While this is strictly true, the error in calculation was caused by using a conversion factor from metric volume to Imperial weight when the specific gravity should have been used. Many would like to blame the Canadian Government for putting pressure on Air Canada to convert to the metric system. But Air Canada should, like other international airlines, be familiar with both systems as both are used at the different destinations they serve.

The nature of the faults

The faults that Justice Lockwood tabled went all the way through the spectrum from the cold-soldered joint to Corporate Management issues. He even criticized the regulator, Transport Canada, for certifying manuals that were *neither clear nor concise.*

It is a reasonable generalization that equipment problems tend to be more clear and precise, whereas the management faults tend to be vague and general. That does not mean that they are any less important. It is probably the reason some investigations prefer to stick to the equipment faults, which are more 'black-and-white'. In between, there are the human errors. Pilot error has tended to be an easy 'catch-all' excuse for investigations, but deeper insight into the nature of decision-making and the influence of environment has questioned this simplistic answer.

Manuals and documents

In both incidents, it was not easy or quick to find the relevant information in the extensive manuals. More complex aircraft lead to more complex manuals. Restricting the training to need-to-know does limit the ability of the crew to know where to look for information on unusual incidents that are not in their need-to-know envelope.

Justice Lockwood criticized the manuals for not saying who should do the tasks. However, with organizational changes more common than the reissuing of manuals, it is sensible to separate how-to-do manuals from who-does-what instructions. These can best be handled in separate documents.

A valid criticism was made, which is common in many circumstances, that those who wrote the manuals did not communicate with the users. People can be classified into those comfortable in two dimensions (2D) and those happier with three dimensions (3D). Office workers are more comfortable with 2D – for example, paperwork. Others hate that environment and want to get out into the 3D world. Pilots and ground crew are in the latter category, while writers of manuals are in the former. All players have to communicate and work together.

Communications

The Gimli investigators exposed many instances of poor or mistaken communication. It needs care and attention to be concise and precise with the English language. Using language that cannot be misinterpreted requires skill and practice. This applies to both written and verbal communication. Space or time often limits the ability to expand on an explanation. The logbook entry by Conrad Yaremko was brief and terse. It was written in shorthand used by maintenance staff. It was less comprehensible for the flight crew, and Captain Pearson misunderstood the details. In fact, some other ground crew also misunderstood it. It is useful, after writing, to stop and read what you have written. And to see if it communicates the information you want to transmit – clearly and unambiguously.

Design faults
The design fault with the B-767 fuel processor resulted in one failure losing the whole of a supposedly redundant system. The B-767 had been in airline service for about a year and had first flown nearly two years earlier. This shows how, even with extensive testing by both the manufacturers and the FAA (Federal Aviation Administration) that there is always the possibility of yet another design fault lurking in the details. Documents from Air Canada showed that between October 1984 and March 1985, two other B-767s in their fleet had 48 snags recorded with the fuel system, and blank fuel gauges on 20 occasions. Why did it take so long to cure what appeared to be a simple solder connection problem?

The Gimli glide highlighted another design weakness. Without the power supply, the basic instruments were airspeed indicator, altimeter, artificial horizon, and magnetic compass. The compass was in a position where neither the Captain nor the First Officer could read it accurately while they sat in their seats. In part of the descent, there was some light cloud cover, and the aircrew were being given directions by Winnipeg Air Traffic Control. Captain Pearson resorted to making course corrections by estimating the change in angle against features in the clouds. And a rate-of-descent meter would have avoided the crew having to manually calculate the glide path.

Design of fuelling procedures
The review of the practice of other airlines showed that most have a more rigorous procedure for establishing and checking fuel loads. Many have either a logbook or a computer record that tracks fuel use. This gives a cross-check on fuel loading and provides a measure of redundancy. While this only gives an approximate check, it would have highlighted the major discrepancy at Montréal. Procedures should be designed to give the same reliability and redundancy as in the design of equipment.

Intermittent faults

The Gimli aircraft had intermittent faults. In hindsight, it seems sensible to change any item showing any fault, even if intermittent, and not reinstall it until it has gone through a more thorough investigation to understand the intermittent fault. The Gimli aircraft had a string of similar problems, which should have sounded warning bells.

Spares

Both incidents could have been avoided if spares had been readily available. Sending the only spare fuel processor for their B-767 fleet to France seems incredible, especially after the history of problems with the system. Similarly, according to the statement of claim in the class action suit, Air Transat did not have the pump for the model of engine nor the correct hydraulic hose in store. No doubt this fits with Captain Last's comments – spares are seen as a non-productive asset and best kept to the minimum.

A thought on basic numbers

An independent check could have been made on the conversion factor used in the Air Canada incident. Going back to basics, a litre is the volume of one kilogram of water. Aviation fuel floats on water therefore weighs less – say 80 per cent. Therefore, the specific gravity should be about 0.8 – not 1.77.

In Chapter 10 page 20.1 of the aircraft manual, the specific gravity of Jet A fuel is given as 0.8. But in forms used by ground crew, the 'specific gravity' is listed as 1.77 without any qualifying units. As the Inquiry Report highlights, there was an amazing lack of understanding of the meaning of specific gravity.

1.12.3 Lessons learned

Accidents happen from the coincidence of apparently trivial events. Some basic lessons from these two emergencies are as follows:

- provide good information and instrumentation;
- make sure that manuals are clear and concise. Check that the time taken to look up any information is consistent with the time available in an emergency. Talk to the users;
- recognize that with complex equipment even the most thorough testing may not expose a hidden problem. Always be ready to hear the problems from field experience;
- keep a record of problems – an event/trend analysis;

- intermittent faults can create confusion and misunderstanding. Fault reporting should watch for problems caused by intermittent faults;
- failures do not usually have a single cause. Corrective actions are usually needed throughout the organization – from top to bottom;
- common mode failures can occur with humans as well as with equipment;
- with evolving technology, watch out for issues that might quietly become more of a concern as practices and technology changes. In particular, while pursuing efficiency and cost effectiveness, watch the impact on safety;
- public opinion can change quickly from praise to lawsuits. Are you prepared?

References

(1) Final Report of the Board of Inquiry into Air Canada Boeing 767 C-GAUN Accident – Gimli, Manitoba, 23 July 1983, ISBN 0 660 11884 X, April 1985.
(2) Air Transat Flight 236 at www.iasa.com.au/folders/Safety_Issues/ others/AirTransatA330Flameout.html; http://aviation-safety.net/ database/2001/010825-0.htm.; details of the accident and the claims are presented in the legal Statement of Claim on www. flight236.com/documents/StatementofClaim_OCT3101.pdf.; a review of the computer issues by Peter B. Ladkin is in Risks Digest at http://catless.ncl.uk/Risks/21.94.html.

A lot has been written about the Chernobyl accident on 26 April 1986 and its aftermath. A selection is in Ref. (**1**). This chapter briefly summarizes the background history and the accident. Some features not usually emphasized are highlighted.

1.13.1 Science in Russia

Peter the Great's vision led to the expansion of scientific studies in Russia. He created the Academy of Sciences in St Petersburg in 1724. Prominent international scientists worked at the Academy, including Euler and Bernoulli. In 1869, the Russian chemist Dimitri Mendeleev, simultaneously with the German Lothar Meyer, established the Periodic Table of the elements.

In 1918, Abram Ioffe, a member of the Academy and the leading light in Soviet physics, founded the Leningrad Physical Technical Institute. His aim was for post-revolution Soviets to keep up with international developments. Igor Kurchatov studied physics and engineering in the Crimea in the early 1920s before being fascinated by the atomic physics work of Rutherford and Soddy in the United Kingdom. In 1932, Ioffe set up a nuclear group with Kurchatov as the key player. They followed the progress in atomic research at the Cavendish Laboratory in Cambridge where several Russians were studying. In 1937, the first cyclotron in Europe was built at the Radium Institute in Leningrad. Georgii Flerov and Konstantin Petrazhak demonstrated spontaneous fission of uranium in 1939. This reconfirmed the work of Fermi, then at Columbia University, and Hahn and Strassman in Germany.

In 1934, the Kremlin decided to move the Academy of Sciences to an elegant palace built in Catherine the Great's era in Moscow. It is still located there today. The German invasion of Russia disrupted the scientific work in Leningrad, and key staff moved further east. In 1942, one of Kurchatov's colleagues, Goergii Flerov, wrote to Stalin, proposing an atomic bomb project. Stalin and Beria were suspicious of scientists but appointed Kurchatov to lead the bomb project under Beria and provided him with large resources. Kurchatov established a laboratory in the Moscow suburbs – now known as the Kurchatov

Institute. The first Russian reactor started there in 1946. The first Russian atomic bomb exploded in August 1949.

1.13.2   A good fit – nuclear power and communism

Lenin linked electricity with Communism. Electricity would provide power to the workers. But there was little fuel in European Russia, and transmission distances were long from the coal mines in the Urals.

Kurchatov had gained both power and respect from his successful weapons programme. He addressed the 20th Party Congress in 1956 and outlined a grand programme of nuclear-generated electricity, as well as other peaceful nuclear projects. He refocused the work of his institute onto the peaceful uses of nuclear science. Nuclear could provide the power source for electricity in European Russia and fulfil Lenin's vision.

1.13.3   Choosing the reactor for power generation

By 1950, various concepts for power reactors were being considered in the different institutes that had grown up during the nuclear weapons activities. Kurchatov asked Nikolai Dollezhal to design a reactor to produce steam for a turbine. Dollezhal had hands-on experience of building chemical plant and had also built the weapons production reactors. He designed a reactor that heated water in pressure tubes and used a graphite moderator. At other institutes, Anatolii Aleksandrov promoted a high-temperature, helium-cooled graphite reactor, and Aleksandr Leipunskii proposed a liquid metal breeder reactor. The decision was made to choose Dollezhal's design and build a 5 MW unit at Obninsk. The reactor went critical in 1954 and produced its first electricity on 26 June 1954. The first atomic generated electricity lit a light bulb at the Experimental Breeder Reactor in Idaho in December 1951. The first Calder Hall reactor produced 50 MW into the UK grid in 1956.

1.13.4   Competition during the Cold War

While there was a competition for weapons during the Cold War, there was also a competition for leadership in other areas. President Eisenhower initiated the *Atoms for Peace* program. The first Geneva Conference on the Peaceful Uses of Nuclear Energy was held in 1955. Scientists at that conference were astounded at the announcement of electricity production from the Obninsk reactor.

Following the conference, the British Government invited the Soviet scientists to visit Harwell. Kurchatov led the group, unexpectedly accompanied by Khrushchev and Bulganin. Kurchatov gave a lecture outlining the broad scope of Russian nuclear activities and stressed the goal of fusion research.

The second Geneva Conference in 1958 was again a coup for Soviet scientists. Papers showed progress on several large and experimental reactors as well as the first nuclear icebreaker – the **Lenin**. They announced that two experimental sodium-cooled breeder reactors had started.

1.13.5   Fast expansion of the nuclear programme

Kurchatov got the resources to build a rapidly expanding nuclear programme. He died in 1960, and Anatolii Aleksandrov became Director of the Kurchatov Institute – a post that he held until he retired. Aleksandrov had developed the VVER reactor, similar to the American Pressurised Water Reactor, and the nuclear icebreaker. He was also President of the Academy of Sciences from 1975 until 1986, and a member of the Central Committee of the Communist Party from 1965–1986.

Aleksandrov was a large man with a commanding personality. He changed the nuclear programme from scientific innovation into a top-down controlled plan with emphasis on building a large nuclear power programme to achieve a Communist Utopia. The nuclear power programme was based jointly on the RBMK pressure tube reactor and the VVER pressure vessel reactor.

The fast expansion spawned many institutes and factories to provide standardized designs, components, and construction. The scale of the plans overtook the ability to deliver. Trained staff was stretched thinly around the facilities.

1.13.6   The RBMK reactor
Basic design

The two per cent enriched uranium fuel is contained in vertical zirconium pressure tubes. Water at 280°C boils in the pressure tubes. The steam is separated in a steam drum and delivered directly to the turbines. The pressure tubes are arranged in a square lattice in the graphite moderator. Some nuclear heat is generated in the graphite, which operates at around 700°C.

Evolution of the RBMK

The RBMK (Fig. 1.13.1) evolved from the Obninsk reactor and was extrapolated at Dollezhal's Design Institute, NIKIET, to 1000 MW, 1500 MW, and with plans to go to 2000 MW. The large RBMK reactors have on-load refuelling, which means that they could stay on-line

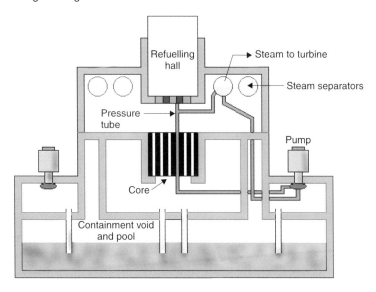

Fig. 1.13.1   Diagram of RBMK reactor

for long periods. The RBMK uses modest-sized components that can be mass-produced, as opposed to the special heavy pressure vessels required for the VVER.

Programme

Aleksandrov and Dollezhal proudly described the RBMK-1000 at the fourth Geneva Conference in 1971. The first four units were built at Sosnovy Bor, just outside Leningrad between 1973 and 1981. Four more RBMK-1000 units were built at Kursk, between 1976 and 1985; four at Chernobyl between 1977 and 1983 and three at Smolensk between 1982 and 1990. In addition, two RBMK-1500 units were built at Ignalina in what is now Lithuania.

By the time of the Chernobyl accident, there were about 100 reactor-years of RBMK operating experience. The units had a good record for electricity production, despite a few serious incidents.

Safety characteristics

Control and safety of any reactor depends on managing the production of neutrons. In the RBMK, increased steam production produces voids that absorb fewer neutrons than the water, a +ve void coefficient. At full power, the −ve fuel temperature coefficient is larger than the +ve void coefficient, and together they produce a −ve power coefficient, in other words, a stable configuration. Below 20 per cent full power, the reactor becomes unstable because the control system is not fast enough to handle the power fluctuations from the +ve void coefficient.

Reactor size

In nuclear physics terms, the RBMK-1000 is about 50 per cent larger than other reactor designs of the same output. Large size means greater susceptibility to local criticality, which complicates the distribution of instrumentation and control devices.

The RBMK-1000 has 1661 channels, each containing two fuel elements. Criticality occurs with only around 100 fuel elements. This created concern after the accident to ensure that no assembly of fuel could become critical again.

RBMK reactor control

The RBMK reactors are unique in relying on operators to drive the reactors rather than using computers. The difference in philosophy could be due to having less-developed computer technology in the USSR. In this era, the United States classified advanced computer technology and limited its sale. Also, the USSR had a different attitude to human effort. In the 1960s, Professor Timoshenko revisited Russia and compared engineering education in the USSR and United States. He noted that both were good at the leading institutions. However, while the engineer in the United States was trained to become a member of a team, in the USSR, he was expected to be able to be sent to Siberia to build a bridge and not come back until it was finished. This puts more reliance on the skill and capability of the individual. With the rapid expansion of engineering projects in the USSR and limited funding, the standards of Russian engineers have not matched the high standards noted by Timoshenko.

In Western nuclear philosophy, the control system is kept separate from that required for safety. The RBMK design mixed them. The RBMK-1000 had 211 control rods, with 24 for emergency protection and 24 under local automatic control. The bulk of 139 rods were under manual control. The RBMK had no fast scram capability.

1.13.7 The test plan

The accident occurred while trying to conduct a safety test on unit #4. Following the Three Mile Island Accident, safety analysts worldwide had concentrated on demonstrating the adequacy of emergency core cooling after a reactor trip. The heat to be removed from the core following a reactor trip comes from two sources, the first depends on the speed of shutting down, and the second depends on the decay heat. Decay heat reduces exponentially but is still two per cent after 30 minutes. This may sound small but is a large amount of heat to remove. For example, the RBMK-1000 would be producing 64 000 KW, 30 minutes after shutdown. This heat can be removed by inherent characteristics such as heat sinks, passive systems such as natural convection, and active systems such as forced cooling. Inevitably, some form of forced cooling is necessary for long-term cooling. Forced cooling requires power. Safety analysts study a variety of situations, including remote scenarios such as assuming no electric power is available – even in a power station. This was the scenario for the Chernobyl test. The test aimed to find out whether the inertia of the turbine, once tripped, could power the pumps until the emergency diesel generators could be started.

The write-up for the test was poor, and the approval by the station management was perfunctory.

1.13.8 Events leading up to the test

The test was planned to take place at 30 per cent power as the reactor was being shut down for maintenance. At 1:00 a.m. on April 25, power was reduced to 50 per cent ready to start the test. Grid control asked the station to hold at that level, to provide electricity to the local area. Eventually, at 12:28 a.m. on April 26, steps were taken to start the test. However, an operator error dropped the power to one per cent. The operators then tried to increase power so that the tests could start.

A nuclear poison, xenon, is produced in the fuel during normal operations as a by-product of the nuclear reaction. The xenon production takes time to decay. Hence, when a reactor reduces power, it still sees the xenon poisoning from the earlier full power operation. The operation at 50 per cent power and the drop to one per cent led to the xenon distorting the reactor flux, as well as making it more difficult to raise power.

Over the next hour, the operators tried to raise power by pulling more control rods out of the reactor and adjusting the water flowing through the core. The rules said that operative reactivity margin should be not less than 15 effective rods. Nevertheless, the operators pulled out all except six to eight rods.

Eventually, at about seven per cent power, conditions seemed to have stabilized, and the tests started. The tests were directed by the electrical engineers from Moscow and not the nuclear group. The electrical engineers were under pressure from Moscow to complete the tests, despite known problems. The RBMK-1000 has two 500 MW turbine-generators. One was shutdown the previous day, the second was used for the test. For some unexplained reason, possibly to repeat the test, the operator disabled an automatic reactor trip that would have operated when the turbine was disconnected from the reactor.

1.13.9   The accident

Now the reactor was at low power with a distorted flux, with nearly all the control rods withdrawn further than the approved limits and with water in the core almost boiling – but not quite. As the turbine speed dropped, the coolant pumps it was driving also slowed, and the water started to boil. The +ve void effect from boiling started to increase the reactor power. Twenty seconds later, the operator pushed the button to shut down the reactor by driving the control rods into the core. In the next four seconds, the reactor power surged to 100 times full power. The fuel disintegrated, the water turned to steam, the pressure tubes burst, and the reactor exploded.

The explosion lifted the 1000 te concrete shield 14 m into the air. This broke the remaining pressure tubes, jammed the control rods, and exposed the reactor core to the atmosphere. The shield settled back at an angle over the core. A second explosion followed. This might have been a chemical explosion caused by hydrogen generated in the overheated pressure tubes. The explosions threw active debris into the air that landed on the turbine hall roof and adjacent areas, causing fires.

1.13.10   Why did the power surge?

Many explanations say the power rose because of the +ve void coefficient. This is only partially true. An analysis showing what actually happened was made by Atomic Energy of Canada (AECL) **(2, 3)**. AECL had, from an earlier prototype reactor, computer codes that were able to calculate the thermal and nuclear conditions at various points up the pressure tube. This sophistication is necessary to see the detailed interaction of the nuclear and thermodynamic changes on each other.

The flux was distorted because of the xenon poisoning, and instead of being a cosine shape, was in two humps (Fig. 1.13.2). The control rods were made of boron carbide – a good absorber of neutrons. But they had been fitted with graphite followers. This displaced water from the pressure tubes, which would otherwise have been absorbing some neutrons. Graphite has around 70 times smaller absorption cross-section than water. In the run up to the tests, the control rods had been withdrawn so far that the bottom of the graphite follower

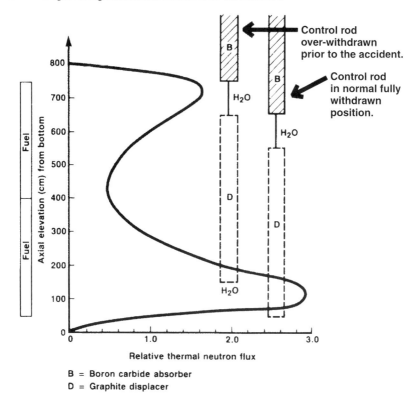

B = Boron carbide absorber
D = Graphite displacer

Fig. 1.13.2   Cause of the power increase. (From Ref. 3) Reproduced courtesy of AECL

was above the lower flux peak. As the rods lowered, the graphite increased reactivity in the bottom of the core, leading to an uncontrolled increase in power.

While the power started to increase because of the +ve void coefficient, the real cause of the explosion was the reactivity insertion when the graphite follower entered the lower flux peak. On its own, the power increase from the +ve void coefficient, while much larger than in other reactors, could have been controlled by the control rods if they had been partially inserted as required by the operating instructions.

1.13.11   Role of Valeri Legasov

Academician Valeri Legasov was appointed First Deputy Director of the Kurchatov Institute by Aleksandrov in 1975. Legasov was a chemist by training and was a strong supporter of the Party. His father had been in the secretariat of the Central Committee.

Involvement with the events at Chernobyl

Within an hour of the accident, local Party members as well as the power station management informed Moscow that a serious nuclear accident had occurred. The nature of the accident was not immediately apparent in Moscow, but by the afternoon, it was realized that it was a major disaster. The government set up a commission under Deputy PM Boris Scherbina and sent it to site to take charge. Legasov was on the commission and left that evening for Chernobyl. He flew over the reactor in a helicopter and determined that the fission process had stopped. However, the graphite was burning, and radioactive particles were being thrown into the air. Local fire fighters had considered dumping water on the reactor, but the commission realized that the resultant steam could blow apart the fragile reactor. Instead, they decided to smother the reactor with sand containing lead, boron, and dolomite. The lead absorbed neutrons and acted as shielding. Boron is a strong nuclear absorber and would stop any future fission. The dolomite would break down at high temperature into calcium and carbon dioxide. The carbon dioxide would act as an inert gas and slow the graphite fire. In the next few days, 5000 te of sand was dropped onto the reactor by helicopters.

The sand appeared to dampen the reactor, but for ten days, radioisotopes were being released into the atmosphere at various rates. The sand acted as a blanket, keeping heat in the reactor. The next idea was to inject liquid nitrogen into the core. This had only just started when, for reasons unknown, the discharge of radioisotopes reduced dramatically.

The next concern was that molten material could fall into the water pools below the reactor and cause another explosion. With dramatic effort, the water was removed. Later, lava-like material was found in various chambers below the reactor. The largest was called the *elephant's foot* because of its appearance. The lava material was silicone oxide from the sand blended with molten fuel and structure.

The presentation in Vienna

The assessment of the accident involved a large team – many of whom had been involved with the RBMK programme at either the technical or the political level. Gorbachev promised that the cause of the accident would be disclosed at a conference organized by the International Atomic Energy Agency (IAEA) in Vienna in August – four months after the accident. The Politburo managed the drafting of the presentations so that they were ideologically acceptable. They chose Legasov to lead the Russian delegation. He tabled the main results – in a five-hour marathon presentation to the 600 international delegates. There was mixed reaction among the delegates. Some saw the detailed information as unusual from Russia, others saw that not enough had been said about the reasons for the accident. Later, in Russia, Legasov admitted that he had tabled *what* had happened – but not *why*.

Legasov's personal views following the accident

Legasov was shocked by the accident. Prior to the accident, he had fully supported Aleksandrov's view that the reactors were all safe – there was no possible danger of a serious accident. After the accident, he contemplated what he had experienced, and its cause. He was concerned at the lack of preparedness for any accident. He became concerned that his presentation in Vienna had placed most of the blame on the operators. He could not understand how the operators *seemed to have lost all sense of danger.*

While he still believed in the competence of the leading players and the staff at the Kurchatov Institute, he saw that the top scientists had become disconnected from the day-to-day work in the power stations they had invented. The detail designers did not fully understand the basic principles of the systems. He saw a generation of engineers who were now in design, operations, and maintenance – but who *have been ignoring the moral principles, the role of our history and our culture.* This new generation was morally indifferent to the impact of their actions – they accepted inefficiency and incompetence. Design problems were not addressed and quietly ignored. Furthermore, the designers did not foresee the types of errors that the operators could make, resulting in an unsound man–machine interface. Even those in high places and with high responsibility had grown to accept slipshod work.

Actions taken and proposed by Legasov

In the era of glasnost and perestroika, Legasov believed that he had a duty to tell the results of his analysis. He tabled proposals on how institutions should be broken into units that are more manageable and the funding redirected. He anticipated that his ideas would be well received, but they were rejected or ignored. He proposed a new Institute of Industrial Safety to the Academy of Sciences. The idea was turned down. Shortly afterwards, on 27 April 1988 – two years and one day after the accident, Legasov committed suicide. His grave is close to that of the aircraft designer Tupolev as well as Khrushchev. In the same cemetery lie Molotov, Prokofiev, Shostakovich, Eisenstein, Gogol, and Chekhov.

1.13.12   Role of Evgeny Velikhov

Five days after the accident, PM Nikolai Ryzhkov decided that the first team sent to Chernobyl from Moscow was exhausted. He nominated a replacement team lead by another Deputy PM, Ivan Silayev. On this team were Academician Velikhov, then a Deputy Director of the Kurchatov Institute, and Prof. Eugene Ryzantzev, head of the Kurchatov Nuclear Reactor Institute.

Legasov was expected to return to Moscow but stayed on at Chernobyl. While he was a colleague of Velikhov, they had different views and styles.

Velikhov made many helicopter flights over the reactor. He proposed installing a heat exchanger under the core to remove heat. In his six weeks at Chernobyl, he received about 35 times the maximum allowable annual dose for nuclear workers.

Velikhov began his graduate work at the Kurchatov in 1958 on magnetohydrodynamics. His prime interest is fusion. Russia had pioneered the Tokamak design of fusion reactor, from the 1950s concept of Andrei Sakharov. Lev Artsimovich, who studied at the Ioffe school, led the fusion research for decades, and Velikhov was one of his students.

Velikhov was appointed the Director of the Kurchatov when Aleksandrov retired. From 1977 to 1996, he was a VP of the Academy of Sciences. In 1989, he was elected a Deputy in the Supreme Soviet – in the first democratic elections since the Revolution. This gave him a central role for promoting science in the political arena. He was the scientific advisor to several Russian Presidents. In particular, he played a key role in Mikhail Gorbachev's arms reduction proposals. In the 1980s and 90s, he worked actively through the Pugwash organization to promote an end to the arms race.

In the era of limited funds, it has been difficult to advance fusion research. Velikhov persuaded Gorbachev to promote international cooperation in fusion and is a strong supporter of the International Thermonuclear Experimental Reactor (ITER) project.

In 1989, the Academy of Science asked Velikhov to organize a Nuclear Safety Institute (IBRAE). He established an institute, wide-ranging from nuclear safety analysis and research to training and emergency response planning.

The main lesson that Velikhov learned from Chernobyl is that, despite it being a major catastrophe, it is nothing compared to what a nuclear bomb attack would be like. Hence, scientists worldwide should work together to end the nuclear arms race.

1.13.13   Aftermath of the accident

The most serious aftermath is the health issue – from both radiation and the psychological trauma. A summary is in the next chapter.

The sarcophagus

The Government Commission decided to entomb the reactor. Construction started in May and was completed by November. It was important to build the concrete shelter, the sarcophagus, quickly. Now the concern is with its long-term stability, and a new structure is needed.

Finding someone to blame

It was inevitable that there was pressure to blame someone. The public prosecutor in Kiev laid charges and conducted a trial. The station manager and five of his staff were found guilty and sentenced to between two and ten years in prison. They were given early release.

Several other investigations took place. One worthy of note was started by Deputy PM Scherbina. His review was wide-ranging and concluded that *in the design of the reactor there were insufficient technical measures to ensure its safety*. While he saw the real cause being the failure of staff on the station, he blamed the Ministry of Energy for permitting the test at night and tolerating the shortcomings of the RBMK. Even the leaders Aleksandrov and Dollezhal were implicated, but they were now in their 80s and losing their power. The State Committee for Safety in the Atomic Power Industry came in for criticism. They were accused of not ensuring adequate supervision or regulation and of not fully using their statutory rights – and their leaders were indecisive.

International activities following the accident

The IAEA had established an International Nuclear Safety Advisory Group (INSAG) in 1985. This group of international experts has produced reports, guidelines, and advice for anyone running or embarking on a nuclear power programme. INSAG carefully studied the Chernobyl accident.

The peer review concept of INPO was extended, and a new international organization, WANO, was established (see Part 2). This ensures international peer review of all power reactors – including all those in Russia.

What has happened to the RBMKs?

Eventually, the three remaining Chernobyl reactors have been permanently shut down. The other 13 units are still in operation. Changes have been made to control rods, fuel, procedures, and operator training that bring them within the minimum safety standards. What appears to be the final word on RBMKs is in a wide-ranging paper (**4**) by Velikhov and his colleagues at the Kurchatov when they reviewed nuclear power for the twenty-first century. They say *RBMK's were unable to confirm the parameters of safety and economics declared once by their developers.*

COMMENTS

The Chernobyl accident follows a familiar pattern for disasters – with problems in three areas: technical, operational, and institutional. In addition, the technical problems had lain dormant and ignored for decades. No new scientific phenomenon was involved.

James Reason (**5**) has studied the human errors involved in the Chernobyl operations. He sees seven. One was a mistake – an operator error in allowing the power to drop to one per cent. The other six he labels *misventions*. He defines *misvention* as behaviour that involves both a deviation from appropriate safety rules and errors, leading to unsafe outcome. The sequence of errors at Chernobyl successively stripped away the limited safety that was built into the rules.

The not-invented-here syndrome

Immediately after the accident, spokespeople in other countries were quick to say that their reactors could not have similar accidents. The arguments varied, depending on the type of reactor being used. Countries using Light Water Reactors (LWRs) argued that they would not allow a +ve void coefficient, that their reactors used water as a moderator, that water does not burn like graphite, and that they have containment. In the United Kingdom, with Magnox and AGR reactors using graphite and without containment, the arguments had to be different. The United Kingdom claimed a superior safety culture and that the graphite operated at much lower temperatures and was in an inert gas. Canada argued that CANDU has two fast shutdown systems to cover all credible accidents. Ultimate safety is more dependent on these systems than on operator action.

In the United States, little was said of the fact that the RBMK was similar to the weapons plutonium-production reactors at Hanford. They had water-cooled fuel in pressure tubes within a graphite core, with no conventional containment and a +ve void coefficient. The first of the nine reactors at Hanford started in 1944. The last, the N reactor, also produced electricity and was shut down in 1987.

Many of the explanations needed to be more detailed to tell the whole story. For example, the claim that Chernobyl had no containment was only partially true. The designers considered failures of pipes and pressure tubes. They used a system of venting into water-filled spaces beneath the reactor (Fig. 1.13.1) to condense any steam and contain any fission products. The design did not consider a large explosion within the reactor, which could lift the reactor lid. Similarly, LWR containment designs do not cover all accidents but only contain realistic accidents and assume that the reactor vessel cannot fail – as it is manufactured to high quality and has regular in-service inspection.

The argument on coefficients also needs deeper explanation. At low power, the RBMK had much larger +ve void coefficients than international safety regulators would tolerate. Small +ve coefficients can be handled by control systems. Large coefficients, either +ve or −ve, can create problems when trying to change reactor power or deal with sudden events like tripping the turbine. The ideal is to have near neutral coefficients to allow both safe operation and easy manoeuvring. Riding a bicycle is a simple analogy. A bicycle is unstable on its own, but with a rider, the combination can be controlled safely. Making a bicycle more stable by adding a wheel to convert it into a tricycle may appear to improve stability. It is stable when at rest – but try cornering at speed, and it can be argued that the bicycle is easier to control and safer.

Some commentators have tried to draw conclusions between the Communist system and Capitalism. Aleksandrov claimed the Three Mile Island accident was the result of profit-hungry capitalists. Western commentators tried to blame Chernobyl on Communism. The familiar pattern to this disaster (technical + operational + institutional) shows that these claims were not true. More relevant were the harsh economic conditions at that time in Russia. One difference that is worth noting is the fact that scientists and engineers are much more involved in the top political process in Russia than in the West.

No one has a monopoly on latent design weaknesses, on operational errors, or on institutional mindset and bureaucracy.

An earlier test?

An attempt had been made to run the test on Chernobyl unit 3. The voltage had fallen off too rapidly for powering the pumps. New electrical equipment was being tried for the repeat tests on unit 4. Maybe the fact that the earlier test had taken place without incident had added to the casual handling of the repeat test.

Secrecy

With the power and military programmes being intertwined, there was a blanket of secrecy on all nuclear activities in the USSR. While the pronouncements of Aleksandrov that their reactors were safe and just simple steam producers were well known, the knowledge of incidents at other reactors was not made available. There had been problems on other RBMK reactors that might have made operators at Chernobyl more alert to the consequences, had they known about them.

Getting advice accepted

A UK nuclear delegation had visited Russia in 1976. Following the visit, they wrote a report listing their concerns with the safety of RBMK reactors (6). Their report did not appear to get much attention. It is probably a generalization that unsolicited advice from outside is ignored or rationalized away – no matter how good the advice. The greater the cultural differences, the less likely the advice will be acted upon.

Getting modifications installed

During commissioning of the Kursk unit 1 in 1976, short control rods were installed in the bottom of the core and included in the scram control. These were then fitted to all RBMKs except Chernobyl, where they had not found time to install them.

Impact of Chernobyl on the nuclear power industry

While all the Chernobyl reactors have been shut down, the Ukraine still generated 51 per cent of its electricity in 2003 from its VVER reactors.

   The Chernobyl disaster has impacted the worldwide nuclear industry. There have been fewer new nuclear power stations built. From a casual reading or watching of the media, one could assume that the nuclear industry is dead. However, the annual worldwide production of electricity from nuclear power has increased 66 per cent since the Chernobyl accident. This has mainly been due to better operation of existing units. There has been new construction. In 2002, six new nuclear power stations entered service, and construction started on seven, while four small old units were shut down. At the end of 2003, there were 441 power reactors in operation generating about 24 per cent of the world's electricity.

Chernobyl has encouraged anti-nuclear campaigners to lobby against all nuclear projects regardless of the design. This makes political decision-makers cautious about committing new nuclear programmes. The advice of Sir Bernard Ingham to all those in the UK nuclear industry who bemoan its state is to point to their own failings. He cites the nuclear industry as a prime example of industries and organizations that do nothing to meet attacks by their enemies or try to correct false impressions deliberately created about them. He adds:

> *The British nuclear industry is a classic example of one which abdicated the field to its enemies. Its scientists, engineers and administrators should stop whining about its misunderstood state and do something about it. They should carry the fight to the enemy with fact and expose the fiction.*

1.13.14   Lessons learned

The prime lesson learned by Academician Velikhov, related to nuclear weapons, is the most dramatic. One's thoughts are sharply focused when experiencing a disaster first hand.

Academician Legasov saw the need to have a person, with knowledge of all aspects, of history, and the background of a design, available throughout the process, from first design through to operation. He also noted the need for everyone, from the most senior down to the working level, to reject sub-standard work and take personal responsibility.

Sir Bernard Ingham highlights the need for engineers to stand up for their technology and fight opponents.

Lessons for design

- Take time to get the concept right – it is nearly impossible to correct conceptual weaknesses in the detailed design stage or in operation.
- Detail designers should be familiar with the overall concept and principles, rather than just concentrating on the detail they are involved with, for example, the impact of control rod followers on reactor control. Know how the detail impacts the whole.
- Designers have to anticipate mistakes and errors by operators and design so that these do not result in catastrophe.
- Do not make it easy to isolate or override safety systems.

Lessons for design and operations

- Know the safety limits of your plant.
- Make sure that everyone is aware of problems and near misses at similar plants.

Lessons for the owners operators and regulators

- Make sure that there are emergency procedures in place to deal with the worst accident.
- In hazardous industries, it is essential that designers have an effective safety culture. Safety weaknesses will show up – one day.
- Take action when a product is not able to meet acceptable standards – even if it has been in operation for some time.

References

(1a) *INSAG Summary Report on the Post-Accident Review Meeting on the Chernobyl Accident*, August 30–September 5, 1986, GC (SPL 1.) IAEA, Vienna, 24 September 1986.

(1b) **Collier, J.G.** and **Davies, M.L.** (1986) *Chernobyl, The Accident at Chernobyl Unit 4 in the Ukraine*, April 1986, CEGB, Barnwood.

(1c) **Josephson, P.R.** (1999) *Red Atom, Russia's Nuclear Power Program from Stalin to Today*, W.H. Freeman & Co., New York, ISBN 0 71673 044 8.

(1d) **Read, P.R.** (1993) *Ablaze – The Story of the Heroes and Victims of Chernobyl*, Random House, New York, ISBN 0 67940 819 3.

(1e) **Mould, R.F.** (2000) *Chernobyl Record*, Institute of Physics Publishing, Bristol and Philadelphia, ISBN 75030 670 X.

(1f) **Vargo, G.J.** (2000) *The Chernobyl Accident: A Comprehensive Risk Assessment*, Battelle Press, Columbus, Ohio, ISBN 1 57477 082 9.

(2) **Chan, P.S.W., Dastur, A.R., Grant, S.D., Hopwood, J.M.** and **Chexal, B.** (1987) The Chernobyl accident: multidimensional simulations to identify the role of design and operational features of the RBMK – 1000, *ENS/ANS Topical Meeting on Probabilistic Risk Assessment*, Zurich, August 30/September 4, 1987.

(3) **Howieson, J.Q.** and **Snell, V.G.** (1987) Chernobyl – A Canadian technical perspective, *Nucl. J. Can.*, **1**(3), 192–215.

(4) **Velikhov, E.P.** *et al*. *Concepts of Nuclear Power of the Future*, Paper for World Federation of Scientists International Study – World Energy in the XXI Century. Accessible at http://info.nucinfo.dk/denmark/Landevis/Russia/floating NNP/concept.htm

(5) **Reason, J.** (1987) The Chernobyl errors, *Bull. Br. Psychol. Soc.*, **40**, 255–256.

(6) NNC Report, March 1976.

Radiation is an issue for all engineers, not just for those in the nuclear industry. Nuclear fuel travels by rail, and radioactive isotopes are carried on the roads. Every hospital uses nuclear medicine and has to deal with its nuclear wastes. Engineers use isotopes and X-rays for non-destructive testing and examination. Some domestic fire alarms contain americium 241. Radon gas builds up in buildings, particularly in basements. Some building materials such as granites emit radioactivity. Are engineers sufficiently knowledgeable of the effects of radiation to deal with these situations?

1.14.1 Background

Professor Röntgen discovered X-rays at Würzburg in 1895. Doctors quickly saw X-rays as a potential treatment for cancer. As early as 1911, they found it was a two-edged sword: X-ray therapy could itself induce cancer, after a long latent period. This dampened their enthusiasm. On the other hand, irradiation therapy of patients suffering from gas gangrene reduced their mortality rate from 50 to 25 per cent (**1**).

Becquerel discovered radioactivity in 1895, and the Curies continued studies with radium. Madame Curie suffered burns and lesions on her hands from handling active materials. She died in 1934 at the age of 66 from a failure of her bone marrow caused by radiation.

Starting in the late nineteenth century, radiation was seen to have beneficial properties. Health spas at radium springs became popular in Europe and North America. This was part science but mainly promotional marketing, sometimes by charlatans.

A new realization of the power of radiation came when Fermi started the first nuclear reactor in 1942, and when the atomic bombs were dropped on Hiroshima and Nagasaki in 1945. With conventional bombs, the concern had only been with explosive power. The deaths in Japan were mainly caused by the explosive blast, fire, and burns – but the public associated atomic bombs with radiation deaths. Nevertheless, it was also realized that the atom could be used for peaceful purposes, and nuclear-generated electric power, nuclear medicine, and the use of radioactive isotopes emerged around the world.

1.14.2   Radiation safety standards and regulation
Early advice for workers
By the 1920s, the need to give advice to workers on the safe limits of radiation was seen. The initial advice was to prevent the obvious and immediate signs of damage, such as loss of hair and skin burns. Limits were proposed from 200 to 800 rem (see note at end of chapter for description of units of radiation).

In the 1930s, what is now known as the International Commission on Radiological Protection (ICRP) set a limit of 0.2 roentgens/day (~44 rem/year). The United States established the National Committee on Radiation Protection (NCRP), which recommended 0.1 roentgen/day. These recommendations stood until around 1950.

Studies at the start of the nuclear age
The US Academy of Sciences established the Biological Effects of Ionising Radiation Committee (BEIR). The United Nations set up the Scientific Committee on the Effects of Atomic Radiation (UNSCEAR). Both BEIR and UNSCEAR base their studies heavily on analysis of the effects of radiation on the survivors at Hiroshima and Nagasaki. They conclude that the data on deaths versus high dose received shows a linear relationship. Double the dose gives double the risk of death. They extrapolated the high-dose data to show zero risk at zero dose. Consequently, BEIR and UNSCEAR promote a linear no-threshold hypothesis.

One implication of the linear no-threshold hypothesis
The linear no-threshold hypothesis has a mathematical consequence. A dose will have the same number of fatalities no matter how it is distributed. For example, at low-dose rates, the risk proposed by UNSCEAR 2000 is 4.0×10^{-2}/Sv. The linear no-threshold hypothesis means that if this dose is spread over 10 000 people during a year, giving an average dose of 0.25 rem/year, then one person will die. If it is spread over 100 000 people at an average dose of 0.025 rem/year, then still one person will die.

The regulators
The use of nuclear energy for power production led to the formal establishment of regulatory agencies such as the NRC in the United States, the HSE/Nuclear Safety Directorate and the NII in the United Kingdom, and the CNSC in Canada. These bodies have consistently set their standards on the basis of the studies of BEIR and UNSCEAR and the recommendations of ICRP. The principles adopted by ICRP

since the 1970s are that no practice shall be adopted unless it produces a net benefit, and all exposures shall be kept as low as reasonably achievable (ALARA), economic and social factors being taken into account.

The nuclear regulators have all adopted the linear no-threshold hypothesis. At the same time, they have put limits on exposures for radiation workers and members of the public. The regulations give different limits for various parts of the body, but the basic whole body limit is 2 rem/year for radiation workers. The limit for members of the public is 0.1 rem/year. These limits only apply to doses from nuclear facilities, and not from medical examination or therapy. The limits only apply to doses above the natural background.

Changes to the regulations over the years

The nature of the rules and regulations has changed over the decades. The early advice was to protect those handling X-ray devices and radium by minimizing immediate health problems. After the atomic bombs were dropped, the rules were aimed at protecting the public with the view that this would automatically protect the individual. The rules were based on limiting cancer risks over the long term, that is, a lifetime. Since the 1990s, the emphasis has changed to protecting the individual, whether a member of the public or an industry worker.

The yearly limits have significantly reduced from 800 to 200 rem in the 1920s, 44 to 22 rem from the 1930s to 1950s, 5 rem from the 1960s to the 1990s, and now 2 rem for industry workers.

1.14.3   Data from the atomic bomb survivors

The bombs dropped on Hiroshima and Nagasaki took a horrendous death toll of 200 000. A Life Span Study is monitoring 86 572 of the survivors, about half of those who were within 2.5 km of the blasts. The study is ongoing, as cancers have a long latency period. The cancer statistics for the survivors, up to the end of 1997, are that 9335 cancer deaths had occurred among the survivors with significant exposures (22.6 per cent of the total of all deaths among the survivors). Of these, it is estimated that only 450 were the result of radiation exposure from the bombs (**2**).

The projections for when all the survivors have died is that about 800, or 0.9 per cent of the survivors, will have died as a result of cancer from the radiation they received from the bombs.

The view of the joint Japanese/US study group Radiation Effects Research Foundation (RERF) is that it is more informative to show loss-of-life expectancy rather than counting cancer deaths (**3**). Their

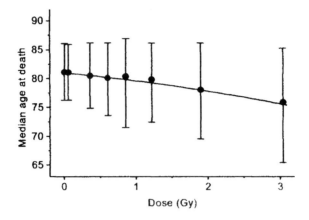

Fig. 1.14.1   Japanese atom bomb survivors, life expectancy with dose

data are shown in Fig. 1.14.1, reproduced with kind permission of RERF.

Radiation has been shown to produce genetic damage in mice subject to high radiation. Genetic changes are continuously occurring in all living organisms, including humans. Studies in Japan have shown no genetic effects caused by radiation in the children born to the survivors **(4)**.

1.14.4   Challenges to the radiation regulations
Different conclusions from the Life Span Study of survivors
The results of the Life Span Study, and hence the regulations flowing from it, have been challenged. Calculations of the estimated dose rate and the potential for survivors to have been partially shielded by buildings are debated. It has been suggested that only the fit and healthy survived the blast, and so the survivors are not representative of an average population.

The studies predict the total cancer death rate and therefore have to be extrapolated to full life span. By 2000, 44 per cent of the survivors were still alive. There are different models of how the data should be extrapolated to full life span. Consequently, there is a range of estimates. The BEIR Committee (1990) estimated the risk as 7.8 excess deaths per 10 000 person rem. The UNSCEAR 2000 report estimates 11 excess deaths per 10 000 person rem at high and medium doses.

Dr John Gofman was the founder and first Director of the Biological Division at the Lawrence Livermore National Laboratory. Later, he became the Professor of Molecular and Cell Biology at Berkeley and founded the Committee for Nuclear Responsibility. Dr Gofman has extensively studied the data and concludes that at low doses, the risk is 27 excess deaths per 10 000 person rem.

The UNSCEAR 2000 report is based on extensive expert opinion and after a mathematical analysis of all the data. It estimates that for low doses and low rates, the mean value for the total cancer risk is four excess deaths per 10 000 person rem.

Every dose a risk?

Dr Gofman believes that every extra man-made dose received by members of the public or a worker is criminal. Dr George Wald, Nobel Laureate and Professor of Biology at Harvard, says *Every dose is an overdose* (**5**). Dr Helen Caldicott says *all it takes is one cell and one radioactive decay for the possibility of cancer, or genetic defect* (**6**). These comments have been well-publicized. They are made by well-qualified people, although neither Wald nor Caldicott is an expert in radiation biology. With their credentials, it is even more important that they demonstrate that their comments are supported by a sound scientific foundation. The growing body of scientific data appears to show that these are unsupported views and do a disservice to the public.

Are regulations too restrictive?

The French Academy believes that a threshold exists, below which radiation is not harmful to humans. They consider the linear no-threshold hypothesis at low doses to be without scientific validity. In their view, the hypothesis has only been adopted because it is convenient to administer.

The Health Physics Society in the United States, with 6000 members, issued a Position Paper in January 1996 and reaffirmed it in March 2001. Their position statement *recommends against quantification estimates of health risk below an individual dose of 5 rem in one year or a lifetime dose of 10 rem in addition to background radiation.* They point out that there is substantial scientific evidence that the linear no-threshold model is oversimplistic, particularly in the low-dose range, and ignores biological repair mechanisms.

The American Nuclear Society (ANS) issued a Position Statement on the 'Health Effects of Low-Level Radiation' in June 2001. It claims that there is *insufficient scientific evidence to support the use of the linear no-threshold hypothesis in the projection of the health effects of low-level radiation.* Its statement references many scientific studies and concurs with the position taken by the Health Physics

Society. The ANS recommends further multi-disciplinary research and monitoring.

The view from the IAEA

The United Nations Agency, the International Atomic Energy Agency (IAEA, separate from UNSCEAR), promotes nuclear technology and produces standards. They have a *no regrets* approach to low-level radiation risk (7). The IAEA stands by its reliance on the linear no-threshold hypothesis, coupled with a small permissible dose. However, the IAEA does recognize that there are anomalies with its approach, that it can foster an *exaggerated fear of radiation*, and that its approach to cancer induction is simplistic.

1.14.5   Sources of radiation from nature and man-made sources

Natural radiation

To get a perspective on the scale and risk of man-made radiation, it is worth relating it to the natural radiation dose. Everyone is continuously exposed to radiation from nature. Cosmic rays come from space and, because of the shielding effect of the atmosphere, are weaker at lower altitudes. The Earth's crust holds many radioactive nuclides, well-spread throughout the land and oceans and present in building materials. These are slowly reducing because of the radioactive decay process. When life began on Earth, the natural radiation levels were three to five times greater than current levels. The cosmic and earth-bound radiation creates external exposure to the body. Internal exposure comes from the ingestion of potassium-40, as well as natural uranium in water and foods. Potassium is an essential ingredient for health. Radon gas, released from the ground and breathed in, gives another internal exposure.

Every person on Earth is struck by about 15 000 natural particles of radiation each second. Radiation particles and rays come in a variety of forms; α-, β-, γ-, X-rays and neutrons. They vary in size, speed, and energy, which affect their ability to penetrate the body and deposit energy.

The average worldwide annual radiation dose from natural sources, quoted by UNSCEAR, is 2.4 mSv (0.24 rem), with a typical range in different parts of the world from 1 to 10 mSv. Lifetime natural radiation doses in the United Kingdom are the lowest in Europe at an average of 100 mSv, and the highest is in Finland at 500 mSv. There are great variations within countries. Cornish granites can produce doses from 10 mSv up to 100 mSv per year.

Man-made radiation

In everyday life, there are additional man-made radiation doses: from medical and dental use of X-rays and radioactive isotopes, airborne pollution from atmospheric testing of nuclear weapons from the 1940s–60s, from nuclear accidents such as Chernobyl, and from nuclear power. The UNSCEAR 2000 report estimates diagnostic medical examinations give an average yearly dose of 0.4 mSv, with a range from 0.04 to 1.0 mSv. Atmospheric nuclear testing, mainly from the 1950s, decays with time and is currently around 0.005 mSv, having decreased from 0.15 mSv in 1963. Similarly, the Chernobyl accident contributes 0.002 mSv/year in the Northern Hemisphere, down from 0.04 mSv in 1986. Nuclear power adds 0.0002 mSv/year. These man-made exposures add an average of 0.41 mSv/year, or 17 per cent to the average natural background.

1.14.6   Low-dose radiation models

The main hypotheses being debated for the relationship between low doses of radiation and risk to people are shown in Fig. 1.14.2.

Dr Gofman proposes the superlinear hypothesis. The linear relationship is the current basis used by regulators. The French Academy, and others, postulates the threshold model. The sublinear hypothesis is promoted for dealing with leukaemia, and Luckey **(8)**, Pollycove **(9)**, and Cohen **(10)** promote the hormetic hypothesis.

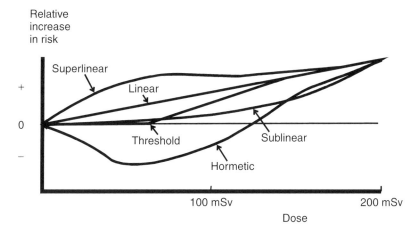

Fig. 1.14.2   Risk models at low dose

1.14.7   Epidemiology

A typical engineering approach to assess competing hypotheses is to look at the statistical data. In the case of radiation effects, this has been done repeatedly, and a typical example from the Registry of 120 000 radiation workers in the United Kingdom, NRRW, is shown in Fig. 1.14.3.

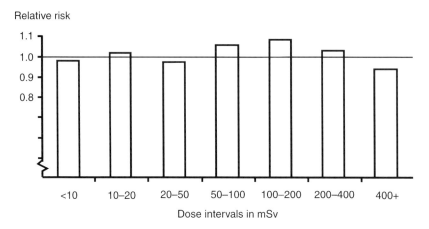

Fig. 1.14.3   Relative risk for UK radiation workers

The conclusion from Fig. 1.14.3 is that over the range experienced, there is no significant additional risk from the radiation received. Note that this covers a range of up to four times the average lifetime background radiation dose in United Kingdom. When the population is large, a variation of a small percentage in the risk can impact a large number of people. This means that the epidemiological approach does not provide a sufficiently precise answer to use for regulatory decisions.

There have been several examples of clusters of health effects near nuclear facilities. These have all been extensively studied. None has shown an increase in local risk due to radiation, but rather all have shown that a variety of extraneous factors can lead to apparently anomalous clusters.

1.14.8   DNA damage
A model
If the epidemiological data cannot give sufficient help, then what can the underlying science provide? The supposition is that the main impact of radiation is to damage the DNA at the centre of the nucleus of a cell. A possible model of the DNA damage and repair mechanisms and cancer is shown in Fig. 1.14.4 (based on (11)).

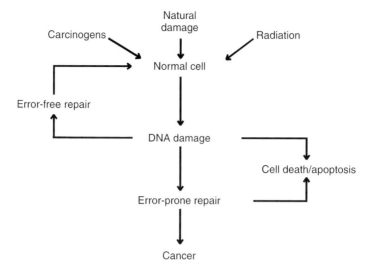

Fig. 1.14.4   DNA and cancer model

DNA damage and the linear no-threshold hypothesis

The linear no-threshold hypothesis is based on a single particle of radiation damaging a single DNA molecule in a single cell and thereby being the potential initiator of cancer growth. The more the particles, the greater the risk. The hypothesis also means that with more cells, the risk is greater. In other words, the greater the mass, the greater the risk. But cancer risks appear to be similar for mice and men – of very different masses **(10)**.

DNA damage and repair

DNA damage occurs naturally from a variety of processes other than radiation. A model of DNA shows that this occurs a million times a day for every cell. DNA comprises of a double-stranded helix. The repair mechanism repairs 99.99 per cent of single-strand breaks and 90 per cent of double-strand breaks. On average, one mutation per cell gets past the immune system each day **(9)**. The accumulation of mutated cells with ageing gradually impairs DNA repair processes and leads to the increased probability of cancer with age.

Are there benefits from low-level radiation?

There is extensive evidence showing that low-level radiation stimulates the DNA repair process. The body has a process of killing off unwanted cells – apoptosis – and this process is also stimulated by low-level radiation. In addition, radiation can alter the timing of cell division. All these effects can result in reduced risk from low-level radiation. At higher dose levels, the repair and rejection rates are overcome by the bombardment.

A tentative summary from DNA modelling

As Professor Cohen summarizes (10), the principle effect of radiation is to modify the biological defence mechanisms rather than to provide the initiating event.

Recent research suggests that it is RNA, a cousin of DNA, which has a role in the smooth division of cells and may be the link to understanding cancerous growth (12).

1.14.9   Studies of hormesis and other work at low doses

Hormesis

Hormesis is defined as an adaptive and stimulatory response to low levels of exposure. Many substances are beneficial at low doses but become toxic at higher doses. Examples are essential minerals, salt, and vitamins. In 1990, scientists from a variety of backgrounds founded a group (BELLE (13)) to study the biological effects of low-level exposure to chemical agents and radioactivity. Their May 2002 newsletter says that the majority of those involved in industrial hygiene believe that toxic effects can have thresholds. *But what we believe and what we have reasonable power to assert or prove with the data and scientific certainty are quite different matters.* They see the need for research in an area that is paradoxical, where at low levels there is wide uncertainty and where statistical assessment is limited.

The hormesis effect of low-level radiation was first noticed by Atkinson in 1898 and reported in *Science* Vol. 7 (7). Gofman's view on hormesis is that *the idea is manifestly absurd* – an emotional statement that summarily rejects, without explanation, a mounting body of findings.

Clinical studies using low radiation doses together with chemotherapy

Two Harvard trials gave patients with non-Hodgkin's lymphoma low doses of radiation prior to chemotherapy. The survival rate was 70–74 per cent, compared to 40–52 per cent for patients having only chemotherapy.

Trials in Japan have given similar results – 84 per cent survival versus 50 per cent.

Health surveys in areas of high natural background radiation

While UNSCEAR assesses the world average natural background as 2.4 mSv and a range of 1 to 10 mSv, there are significant populations with much higher background radiation. Around 100 000 people live in the Kerala coastal region of India with an annual background of up to 35 mSv. At Ramsar in Iran, lifetime doses of 17 Sv have been recorded (240 times current limits for the public). Even across the United States from the Gulf coast to the Rockies, the background varies from around 1 mSv to 12 mSv.

All surveys of populations living in high natural background radiation show no increased cancer mortality. If anything, there appears to be less risk, although this could be due to other effects such as lifestyle and less atmospheric pollution. A study of the population in Ramsar showed a higher level of chromosome damage but no increase in cancer mortality (**14**).

Studies of workers exposed to radiation

A US study of nuclear Navy dockyard workers showed that the cancer mortality rate for those exposed was 85 per cent of that for those not exposed (**15**).

A study of 95 673 monitored radiation workers in the United States, the United Kingdom, and Canada showed that there were no excess deaths over the 3830 normally expected from traditional mortality statistics. The risk is reported as $-0.07/\mathrm{Sv}$, with 90 per cent confidence. In other words, the natural, spontaneous risk reduces with dose at low levels (**16**).

A study of UK radiologists who were working between 1955 and 1979 has shown that they were half as likely to contract cancer as the general population (**17**). Some have attributed this to getting better health care and monitoring, but this effect is unlikely to account for all the benefit.

Adaptive response studies

Studies at Chalk River Laboratories in Canada on human skin cells showed that prior low doses resulted in less damage from a subsequent high dose. Tests have shown that low doses increase the error-free DNA repair ability. This improved repair ability has been observed for all damage, not just that caused by radiation (**18**).

Other tests on human skin at Chalk River have shown that individuals have different sensitivity to radiation. Further tests studied the effects of chemical carcinogens. People with sensitivity to radiation did not necessarily have the same sensitivity to chemicals, and vice versa. Each individual appears to have their own pattern of sensitivities to different carcinogens. In mice that were either normal or genetically prone to developing cancer, a low dose of radiation delayed the onset of both natural and radiation-induced cancer (**19**).

Experience from radiation accidents

An explosion in Russia in 1957 at a nuclear waste facility in Siberia exposed local villagers to radioactivity. A study of 7852 exposed people showed a cancer mortality rate 28 per cent less than similar non-exposed villagers nearby (**20**).

An accidental melting of a cobalt-60 source into recycled steel used in apartment construction in Taipei 20 years ago led to the occupants being exposed to radiation. About 10000 people were exposed for 16 years. Local statistics predicted 232 cancers from natural causes. Linear no-threshold hypothesis (ICRP Model) estimates an additional 70 deaths. However, only seven cancer deaths have occurred in total (**21**). The conclusion from this academic study is that doses of around 5 rem/year, that is, 50 times the maximum allowable dose for the public, greatly reduces cancer mortality. Also, current radiation protection policies are inappropriate and should be re-evaluated.

Luminous watch dial painters

Between 1910 and 1930, watch faces were painted with luminous paint containing radium. Young women using small paintbrushes did the painting. They moistened the brushes with their mouth, and this led to ingesting radium as well as getting high doses to the jaw and head. After a latent period, many developed cancerous growths and many died. A detailed study (**22**) showed cancers developed above a dose of 10 Gy. Below that level, there were no cases. There appeared to be some benefit from small doses. On the basis of linear theory, five cancers were expected in the 1391 patients who received less than 10 Gy, and none was diagnosed.

How hazardous is radon?

Radon hazards were studied in the BEIR IV report using the linear no-threshold hypothesis. On the basis of this work, the US Environmental Protection Agency (EPA), in its June 2003 report, estimated that there are between 8000 and 45 000 deaths per year from radon in the United States. The EPA has proposed actions that effectively aim to regulate exposure from natural radiation. It recommends that houses be tested for radon. If the levels are over 4pCi/l, it suggests that homeowners spend $800 to $2500 to reduce radon in their homes. The EPA also proposed reductions in radon in drinking water and estimated that it would cost \$180 million to improve water plants. The American Water Works Association estimated that it would cost \$2.5 billion. In the United Kingdom, the NRPB recommends action if the radon levels are above $200\,Bq/m^3$ (5.4 pCi/l). However, it does say that the increased risk of lung cancer at this level is small.

Professor Cohen carried out an extensive study of the effects of radon exposure in buildings and outside, for about 90 per cent of the US population (23). The results, Fig. 1.14.5, show a negative risk, that is, a benefit with dose rather than the risk postulated by the linear no-threshold hypothesis.

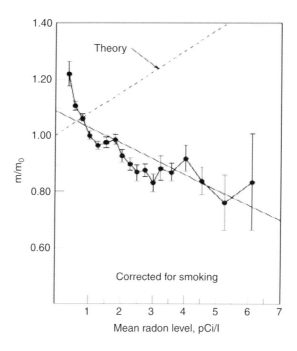

Fig. 1.14.5   Radon exposure risks in the United States. Reproduced with kind permission from Professor Cohen

The EPA disputes Cohen's results, suggesting that they are due to problems of correlating the effects of smoking. They do say that *the controversy continues.*

1.14.10   Effects of radiation from Chernobyl

The accident at Chernobyl in 1986 released 100 per cent of the noble gases (Krypton and Xenon, etc.), ~50 per cent of Iodine, ~30 per cent of Caesium, and three to four per cent of the fuel into the atmosphere as a result of the explosion and subsequent fire.

There were 600 workers at the site when the accident happened. Of these, two died immediately, one from the explosion, and one by a coronary. 134 received acute doses and suffered radiation sickness. 30 died in the next few months despite international medical efforts. These deaths were consistent with previous dose/mortality data.

The main radioactive exposure to people outside the plant came from Iodine and Caesium. Iodine is inhaled from the air and ingested from drinking local milk and eating local green vegetables. Caesium enters the body both by ingestion, and also gives an external dose from deposits on the ground.

Appendix J to the UNSCEAR 2000 report is a 151-page report on the radiological consequences of the Chernobyl Accident. The report draws on 476 references. The summary states:

> *There have been about 1800 cases of thyroid cancer in children who were exposed at the time of the accident, and if the current trend continues, there may be more cases during the next decades. Apart from this increase, there is no evidence of a major public health impact attributable to radiation exposure 14 years after the accident. There is no scientific evidence of increases in overall cancer incidence or mortality or non-malignant disorders that could be related to radiation exposure. The risk of leukaemia, one of the main concerns owing to its short latency time, does not appear to be elevated, not even among the recovery operation workers. Although those most highly exposed individuals are at an increased risk of radiation associated effects, the great majority of the population are not likely to experience serious health consequences as a result of radiation from the Chernobyl accident.*

Many cancers have a long latency period, and cancers from the accident may still surface. The projected excess deaths for solid cancers for full lifetime, based on the rate for low-level low-dose rate in the UNSCEAR 2000 report and (24), are shown in Table 1.

The predicted excess solid cancer deaths for the recovery workers are just under two per cent of the normal background cancer deaths. With the variations in normal background cancer numbers, particularly with the major changes occurring in the former USSR, it will be very difficult to detect this predicted increase from statistics. These predictions are based on the linear no-threshold hypothesis. The

Table 1   Predicted extra deaths from radiation from Chernobyl

Population	Number	Average dose mSv	Normal expected cancer deaths	Predicted excess cancer deaths
Recovery workers	200 000	100	41 500	800
Evacuees	135 000	10	21 500	54
Residents of control zones	270 000	50	43 500	540
Other residents of contaminated areas	6 800 000	7	800 000	1940?

dose to the 6.8 million people in the contaminated areas is only three additional average years' natural background radiation in a lifetime, and the predicted excess deaths are questionable.

Dr Gofman has predicted 470 000 deaths from Chernobyl (25). Of this number, 244 786 are in Europe, where the maximum dose was around half of one year's background dose. The normal background dose varies by a factor of five between countries in Europe, with no related difference in cancer mortality. Hence, this prediction appears to be a play on statistics and not a real phenomenon. This is the type of exaggerated extrapolation that the American Health Physics Society was concerned about when it issued its Position Paper.

Childhood thyroid cancer

Thyroid cancers are normally rare at around one per cent of all new cancers. The sudden increase in childhood thyroid cancers around Chernobyl was not expected, as the latent period for this type of cancer is around 10 years. The peak thyroid cancer incidence appeared in 1995. By that time, three children had died from cancer.

More work needs to be undertaken to fully understand this situation. There were few reported childhood thyroid cancer cases before the accident. Hence, there is some question as to how many of the reported cancers are the result of better examination. There was known iodine deficiency in many of the children in the region. However, the predominant type of thyroid cancer post-Chernobyl is not similar to that seen in other countries.

A statistical study of the childhood thyroid cancers gives the best estimate for the total number who may contract the disease as 4400 (26). Thyroid cancer can now be treated very successfully. The study concludes that there is a realistic chance that the total number of deaths will be less than 1000. This may be the bulk of the total death toll attributable to the accident.

Summary

The Japanese atom bomb survivors experienced one short burst of concentrated dose. The Chernobyl survivors experienced a distributed dose over days, weeks, and years. As a result, they can provide complementary data for assessing the effects of radiation doses on humans.

The Chernobyl accident has been devastating for those people living in the immediate vicinity who have had their lives disrupted. However, the vast majority need not live in fear of the health consequences from radiation. It has been suggested that the use of the linear no-threshold hypothesis led to more people being disrupted than necessary as well as giving exaggerated fear of radiation.

There have been fewer than 50 radiation deaths from Chernobyl to date, including those from acute radiation of workers and childhood cancer. Based on the no-threshold hypothesis, more deaths may be expected in the future and may reach the low thousands as shown in Table 1. Alternatively, based on the hormetic model, there will be fewer cancer deaths overall than if the Chernobyl accident had never happened. Zbigniew Jawarowski, a past Chairman of UNSCEAR, has reviewed the lessons from Chernobyl and notes: *the incidence of all cancers appears to have been lower than it would have been in a similar but unirradiated group* (**27**). In this reference he shows that in the immediate reaction following the accident, decision-makers in various countries implemented radionuclide concentration limits, which varied by up to a factor of 50 000. His overall conclusion from the Chernobyl accident is that: *it is unlikely that* any *fatalities were caused by radiation among the public*.

COMMENTS

It is not easy to make engineering decisions when there is such a wide range of views and dissent in the underlying science. Does one take the consensus view or ignore the controversy? One approach is to simply work to the rules set by the regulator, despite the questions.

Working to the rules

Working to the rules sounds a simple answer. But the rules contain an inherent conflict by setting permissible limits and at the same time calling for ALARA. With pressure growing to question the appropriateness of the linear no-threshold hypothesis at low doses, it has been pointed out that ALARA has qualifications. These are that economic and social factors be taken into account. Therefore, it is back to the engineer to make the case that ALARA is not economic or socially acceptable.

Evidence is mounting to show that the linear no-threshold hypothesis is not correct at low doses. Consequently, applying ALARA may increase the risk by preventing the activation of the defence systems, thereby increasing both natural, spontaneous cancer risk and the risk associated with an accidental high exposure. This is not consistent with the cautionary approach advocated by the regulators.

What will be the permissible dose in the future?

An apparently plausible argument can be made to extrapolate the current permissible dose into the future based on history. There has been a continuous reduction over the last century. Will this trend continue? The current permissible dose to the public is 42 per cent of the average background radiation, while the range from minimum to maximum is over ten times. It is hard to see a case to reduce it further – although there are people trying to do just that.

Can the rules lead to a waste of resources?

The scientific consensus from BEIR, UNSCEAR, and ICRP is based on caution and not changing the rules until there is a proven benefit. Cohen's study of radon in the United States shows that vast amounts of money could be spent unnecessarily by homeowners and water works. Fortunately, no significant funds have been spent. However, in other non-nuclear fields, Hahn and Hird (**28**) found that regulations for occupational safety and health in United States have cost around $9 billion with negligible benefits.

One future issue where the rules can have a big cost impact is the decommissioning of nuclear facilities and storage of nuclear wastes.

In addition, the French Academy of Medicine is concerned that proposed European directives could restrict work in medical radiology by setting tighter dose limits. This in turn could limit health benefits for patients.

Logical anomalies

Applying the nuclear regulations logically over a wider range of situations produces some amusing aberrations. If extra radiation, no matter how small, can result in additional risk, then is it wrong to ask someone to move from London to the high radiation of Cornwall?

It would be worse still to ask them to relocate to Finland. Should Ramsar be evacuated, despite the longevity of its citizens?

Grand Central Station in New York, the Capitol Building, and the Library of Congress in Washington are all impressive granite buildings. The measured dose rate inside these buildings is three to five times the maximum allowed to the public from a nuclear facility. Should we evacuate these buildings? Should we stop building with granite?

Judging the science

The linear no-threshold hypothesis is unproven at low doses. It is based on the postulation that a single particle can result in cancer. There is consensus for this view among the many experts involved with BEIR, UNSCEAR, and ICRP. However, the proof of science should be based on facts, not consensus.

The alternative view is that low-level radiation stimulates the repair and rejection mechanisms. The statistics for the public are that one in four deaths is caused by cancers. Any cell damage from low-level radiation will only increase this by a very small amount. The real effect is the stimulation from the low-level radiation. This reduces the risk from all cancers. It is the biological response to the damage, not the damage itself, that governs the risk.

Cancer has a long latency period that will make scientists cautious at making changes to rules without a long checking time. There are around 200 different forms of cancer and only some seem sensitive to induction by radiation. In addition, there are various forms of radiation. It may be that some types of cancer exhibit a threshold for a particular form of radiation while others do not.

The fact that many scientists have spent their entire career believing in the linear no-threshold hypothesis makes it more difficult to change. One can only hope that these conflicting views can soon be reconciled.

The media and the public

The media has the same problem as engineers – how do lay people reconcile the different scientific views? Giving equal time or column inches is hardly the answer. The extravagant claims of some are hard for the media to resist. On 22 April 2000, the BBC said that 15 000 people had been killed by Chernobyl and that 50 000 were left handicapped in the emergency clean up. The *Guardian* on 26 April 2000 reported that there had been 50 000 new cases of thyroid cancer as a result of Chernobyl.

The public is not in a position to assess the scientific data and judge between the hypotheses. Even the reasonably established facts are drowned in a sea of conflicting comments. It is too easy to take a small risk and multiply it by a large population and produce a large number of victims. Should a calculated loss-of-life expectancy of days or weeks count as a radiation cancer death or not?

The net result from these confusing data are to create *radiation-phobia* in large segments of the community. The public feels that there is no one to turn to for advice because they suspect the motives of the various experts. This is especially true for the people living in the areas affected by Chernobyl. With such a large proportion of the population dying of cancer, it is easy to attribute a portion of this to one cause or another by using disputed science or doubtful statistics. Despite the prevailing view, the fact is that radiation is a weak carcinogen.

One approach taken by regulators and politicians is to open the debate to the public. The two-and-a-quarter-year Sizewell B, Hinkley Point C and Sellafield Inquiries have provided debate, as well as platforms for the protest industry. The controversy has not been resolved. Have the Inquiries provided any comfort or increased understanding for the public?

A pragmatic approach
Despite the controversy, a pragmatic course can be taken. First, recognize that radiation can be dangerous at high levels. The difficulty comes in assessing low doses, below $\frac{1}{2}$ Sv.

The regulations have to be followed but the evidence shows little risk at the allowable dose limits. Therefore, there appears to be good reason to challenge ALARA.

It would not seem to be reasonable to spend large sums of money or other resources to limit radiation doses to levels below the variations in background level. In terms of dealing with long-term nuclear wastes, it might be prudent to spend only limited funds on temporary storage until there is more clarity in the risks involved.

A final comment
Going back to the opening question of radiation hazards – *Are engineers failing the public?* – the most reasoned answer would be no. The failure in this case is not having a confirmed scientific basis on which to judge low-level radiation risks. With low-level radiation all around us, it is prudent for engineers to be knowledgeable on the

subject. There is a need to understand the risks. There is also a responsibility to ensure that the public has access to the facts – known as well as unknown.

1.14.11 Lessons learned

- It is not easy to make engineering decisions when the underlying science is in dispute.
- It is difficult to work in an area where the regulations do not appear to be consistent with the evidence. The social and economic qualifiers to ALARA move the debate from regulation based on science to perceived cultural standards.
- There is an obligation to look beyond the rules and regulations and understand the risks.
- There is a need to try and understand the different sciences and technologies that impact on engineering decisions – one cannot just ignore or walk away.
- If there is no communication with the public and they are not kept well-informed, then there is every prospect for misinformation and phobia developing, which will make any decision more difficult. The public includes the politicians acting on behalf of the public.

Radiation units
Exposure

Roentgen –   a measure of X-ray–and γ-ray–induced ionization in air. Produces 83 ergs of energy/gram of dry air.

Absorbed dose

Rad   – radiation absorbed dose. 0.01 joule/kg. 100 ergs/g.
Gray*   – 100 rad.

Equivalent dose

Rem   – roentgen equivalent man. Defined to take account of the effects of different radiation on a human.
Sievert* – 100 rem. (Note: 1 milli Sievert = 1mSv = 0.1 rem)

* SI units.
Equivalent dose = Relative Biological Efficiency × Absorbed dose.

Relative Biological Efficiency = 1 for X-rays and γ-rays, 10 for fast neutrons, and 15 to 20 for α particles.

References

(1) **Cuttler, J.M.** (2002) Disinfecting wounds with radiation, *Annual Conference on CNS*, Toronto, June 2002.

(2) **Pierce, P.A., Shimizu, Y., Preston, D.L., Vaeth, M.** and **Mabuchi, K.** (1996) Studies of the mortality of A-bomb survivors, Report 12, Part 1, *Rad. Res.* **146**, 1–27.

(3) **Cologne, J.B.** and **Preston, D.L.** (2002) Radiation risks, longevity and the impact of the comparison group on low-dose risk estimates, *RERF Update*, **13**(1) 17–22 and Facts and Figures, RERF Update, **14**(1), 24, www.rerf.or.jp.

(4) UNSCEAR 2001, Report to the General Assembly UN, Vienna, Austria.

(5) **Wald, G.** (1972) *Introduction to Secret Fallout*, Sternglass, McGraw-Hill, ISBN 0 07061 242 0.

(6) **Caldicott, H.** (1994) *Nuclear Madness*, W.W. Norton, New York, ASIN 0 39303 603 0.

(7) **Rosen, M.** (1996) *Radiation and Society: A No Regret Approach to Low Level Radiation Risk*, Uranium Institute, London.

(8) **Luckey, T.D.** (1991) *Radiation Hormesis*, CRC Press, ISBN 0 84936 159 1. Accessible at www.giriweb.com/luckey.htm.

(9) **Pollycove, M.** and **Feinendegen, L.E.** (2003) Radiation-induced versus endogenous DNA damage..., *Hum. Exp. Toxicol.*, **22**(6).

(10) **Cohen, B.L.** (2002) The cancer risk from low level radiation, *Am. J. Roentgenol.*

(11) **Mitchel, R.E.J.** and **Boreham, D.R.** (2000) Radiation Protection in the world of modern radiobiology, *Proceedings of the IRPA 10th Meeting*, Japan, May 2000.

(12) **Couzin, J.** (2002) *Science*, **298**, (2296–2297).

(13) www.belleonline.com.

(14) **Mortazavi, S.M.J.** (2002) *High Background Radiation Areas of Ramsar*, University of Kyoto, Kyoto, Japan.

(15) **Matanoski, G.M.** (1991) *Health Effects of Low-Level Radiation in Shipyard Workers*, DOE Report DE AC02 79 EV10095.

(16) **Cardis, E., Gilbert, E.S., Howe, C., Kato, I., Armstrong, B.,** and **Beral, V.** (1995) Effects of low dose and low dose rate of external ionizing radiation; Cancer mortality amongst nuclear industry workers in three countries, *Rad. Res.*, **142**, 117–132.

(17) **Berrington, A., Darby, S.C., Weiss, H.A.,** and **Doll, A.** (2001) 100 years of observation on British radiologists: mortality from cancer and other causes 1897–1997. *Br. J. Radiol*, 507–519.

(18) **Mitchel, R.E.J.** (2001) Radiation biology of low doses, *Rad. Sci. Health*, Nov.

(19) **Mitchel, R.E.J., Jackson, J.S., Morrison, S.M.** and **Carlisle, S.M.** (2003) Low doses of radiation increase the latency of spontaneous lymphomas and spinal osteosarcomas in cancer prone, radiation sensitive Trp53 heterozygous mice, *Rad. Res.*, **159**.

(20) **Kostyuchenko, V.A.** and **Krestina, L.Y.** (1994) Long term irradiation effects in the population evacuated from the East Urals radioactive trace area, *Sci. Total Environ*, 119–125.

(21) **Chen, W.L.**, *et al.* (2004) Is chronic radiation an effective prophylaxis against cancer? *J. Am. Phys. Surg.*, **9**(1), 6–10.

(22) **Thomas, R.G.** (1994) The US radium luminisers, A case for a policy of 'below regulatory control', *J. Rad. Prot.*, **14**, 141–153.

(23) **Cohen, B.L.** (1995) Test of the linear no threshold theory of radiation carcinogenesis for inhaled radon decay products, *Health Physics*, **68**, 157–174; Update, *Technology* (7), 2001, 609–631.

(24) **Cardis, E.** *et al.* (1996) Estimated long-term health effects of the Chernobyl accident, *IAEA Conference*, April.

(25) **Gofman, J.W.** (1990) *Radiation Induced Cancer from Low Dose Exposure: An Independent Analysis*, Committee for Nuclear Responsibility, San Francisco.

(26) **Thomas, P.T.** and **Zwissler, R.** (2003) New predictions for Chernobyl childhood thyroid cancers, *Nucl. Energy*, **42**(4), 203–211.

(27) **Jaworowski, Z.** Lessons from Chernobyl, with particular reference to thyroid cancer, Australasian Radiation Protection Society, Newsletter No. 30, April 2004, also in *21st Century Science and Technology*, Washington DC, ECR May 7, 2004, 58–63.

(28) **Hahn, R.W.** and **Hird, J.A.** (1991) The costs and benefits of regulation, *Yale J. Regul.*, **8**(1), 233–278.

PART 2

Part 2 includes views of experts and information that provides background to the failures discussed in Part 1.

There are valuable insights and advice on failures in papers and books by prominent engineers. This chapter presents the succinct thoughts of a few selected engineers that are worth repeating.

2.1.1 Sir Alfred Pugsley (1903–1998)

After graduating from Battersea Polytechnic in 1923, Alfred Pugsley became a civil engineering apprentice at the Royal Arsenal, Woolwich. From 1926 to 1931, he was a Technical Officer at the Royal Airship Works, Cardington, where he worked on the structure of the R101. Following the demise of the R101, he moved to the Royal Aircraft Establishment (RAE) at Farnborough, where he became the Head of the Structural and Mechanical Engineering Department. He was Professor of Civil Engineering at Bristol University from 1945 until his retirement in 1968.

Pugsley was both an engineer and a scientist. He explored basic scientific principles and studied the history of how structures had evolved. Then he used mathematical analysis to test his understanding – and followed that with laboratory tests to confirm his hypothesis. When he was satisfied with his understanding of a problem, his next step was to present the results in a simple and easily understood manner for general design use. His approach brings together history, science, mathematics, and engineering judgement to provide practical advice.

Aeroelasticity

Pugsley's first major contribution to engineering science was on aeroplane wing flutter in the mid-1930s (**1**). A crop of accidents and crashes in the 1920s initiated this work. Pioneering work took place at RAE Farnborough and at the National Physical Laboratory (NPL). The review of the accidents, mathematical analysis, and tests led to an understanding of how the aerodynamic forces, elastic forces, and inertia forces could interact. This showed four possible failure modes: wing divergence (potentially breaking the wing), aileron reversal (loss of control), wing flutter (potential fatigue failure), and wing-aileron flutter (fatigue and loss of control).

The mathematical analysis of these instabilities was complex but was summarized in a simple diagram showing the limiting boundaries. This research led to increasing the torsional stiffness of the wings of the Hurricane and Spitfire, which prevented them from crossing the wing flutter boundary at high speed. Pugsley saw this as an example of research preventing failure – *without the need for an accident to teach the lesson.*

Fatigue

In his work at Farnborough, he became concerned that the evolution of aircraft design to higher speeds, using stronger materials and higher stresses, could lead to fatigue problems. His approximate analysis showed that the structural life could be of the same order as the service life. He presented his concerns publicly in a paper to the Royal Aeronautical Society in 1947 (**2**), while pursuing the issue.

He promoted the installation of V–g (Velocity–acceleration) recorders in aircraft on routine flights to gather data on the variable loads. V–g recorders had been developed by NACA (National Advisory Committee for Aeronautics) in 1937, and Pugsley saw them as ideal tools to gather statistical data for design.

From his work on fatigue, he saw that all structures had a limited life and that the designer should consciously consider the life as a routine part of design analysis and not only the life from repeated loading but also the effects of the environment on corrosion and the relationship between corrosion and fatigue.

Accidents

At Bristol, Pugsley turned his attention to bridges and other structures (**3**). He noted that about 1 in 14 suspension bridges had failed within their normal expected lifespan (**4**). The failure rate for major unconventional bridges was around 1 in 10, for ordinary road bridges 1 in 10^2, for railway bridges 1 in 10^3, and for small beam bridges 1 in 10^4. For dams, the figures varied from 1 in 50 to 1 in 250, depending on the design. Fortunately, bridge failures have not resulted in a large loss of life, but some dam failures have led to large numbers of deaths.

Pugsley concluded that the failure rates were too high. The main cause was a lack of understanding of all the possible failure modes. His solution was to improve design thinking to get a broader understanding, rather than just using larger safety factors. In each case, he saw a failure mode that had been latent, and secondary, in earlier designs become the first mode of failure as designs were built on a larger scale. Following a string of successful projects, the designer *perhaps a little complacent, simply extended the design method once too often.* He noted that all the failures had been preceded by incidents on other projects, usually on a smaller scale. If these warnings had been properly interpreted, they could have prevented the subsequent failure.

Pugsley noted that after accidents *structural engineers wish they could have foreseen such troubles before they occurred. This sort of pious hope. . . has always dogged engineers. . .*

He saw the need for an accident-risk philosophy for design. He wrote *A philosophy of aeroplane strength factors* (**5**) in 1942 and *Concepts of safety in structural design* (**6**) in 1951. Both are classics on design philosophy for preventing failures.

Risk and safety

The study of the probability of structural failure inevitably leads to reviewing risk. Pugsley studied risk, and the perception of risk by the public. He saw the difference in risk for passenger aircraft carrying over 100 passengers and for fighter aircraft. The aim for passenger aircraft should be to have a large-enough safety factor to prevent structural failure during service life. If a large safety factor were applied to the fighter, it would make it unnecessarily heavy and less manoeuvrable. This would be a greater risk to the pilot in combat. Hence, a greater structural risk, or smaller safety factor, is appropriate for the military aircraft.

Pugsley believed that designers should be aiming for efficiency and economy in all structures. He recognized that it was not possible to achieve perfect safety – there would always be some degree of risk. What risk is the public prepared to take? He concluded *it would be unwise not to draw attention to the fact that mankind seldom wants or aims for perfect safety; pioneering work and risk are inseparably bound up, and are the essence of human behaviour.* He saw a hazy borderline between success/safety and disaster/failure. He is often quoted as saying *A profession that never has accidents is unlikely to be serving its country efficiently.* Taken out of context, this view sounds cavalier. However, he promoted redundancy in structures so that there would be a warning of failure.

Safety factors

Safety factors have evolved via different routes. Simple safety factors emerged from early engineering failures. If it failed, just make it a bit stronger and standardize on that safety factor. This resulted in safety factors based on the ultimate load to failure. Other approaches were to define the maximum allowable stress or to define a safety factor based on the load to cause deformation or instability.

Pugsley saw that neither the loads nor the strengths have a single numerical value. A more realistic approach is to recognize this – gather statistics on the variables and adopt a statistical approach.

He saw the need for a broad set of safety factors to assist designers where a project was not covered by a code of practice or to check on established historical practice. He chaired a committee of the Institution of Structural Engineers in the 1950s to review safety issues. They identified five parameters that fitted into two groups:

Group X, factors influencing the probability of collapse:

(A) materials, workmanship, inspection and maintenance;
(B) loading, having regard to control of use;
(C) assessment of strength, having regard to the accuracy of analysis, experimental information, and kind of structure.

Group Y, factors influencing the seriousness of the results of collapse:

(D) danger to personnel;
(E) economic considerations.

The ultimate load safety factor is the factor from group X multiplied by the factor from group Y. Pugsley favoured a probabilistic approach, but agreed with the committee that an adequate answer could be derived from a simple qualitative judgement on each parameter. This resulted in the factors in Tables 2.1.1 and 2.1.2 (**6**).

Similar work was taking place in the United States. This was based on a more mathematical probability approach. Pugsley reviewed their results and found that they gave similar safety factors. Rather than seeing this as competition, he saw the advantage of having confirmation from an alternative approach, which gave added confidence to the results.

Test results

Testing specimens gives credible data on which to base a design. In most cases, it is not possible to test enough samples to establish the statistical distribution. Pugsley promoted the practice that if only one sample were tested, then the measured failure load should be divided by 2.0 to give the failure load used in design. If three samples were tested, then the lowest failure load should be divided by 1.5. These factors were based on an engineering judgement of the shape of the statistical distribution.

Table 1

VG = Very Good		GROUP X G = Good	F = Fair	P = Poor
A	B		C	
		VG	G	F	P
VG	VG	1.10	1.20	1.30	1.40
	G	1.30	1.45	1.60	1.75
	F	1.50	1.70	1.90	2.10
	P	1.70	1.95	2.20	2.45
G	VG	1.30	1.45	1.60	1.75
	G	1.55	1.75	1.95	2.15
	F	1.80	2.05	2.30	2.55
	P	2.05	2.35	2.65	2.95
F	VG	1.50	1.70	1.90	2.10
	G	1.80	2.05	2.30	2.55
	F	2.10	2.40	2.70	3.00
	P	2.40	2.75	3.10	3.45
P	VG	1.70	1.95	2.20	2.45
	G	2.15	2.35	2.65	2.95
	F	2.40	2.75	3.10	3.55
	P	2.75	3.15	3.55	3.95

Table 2

NS = Not Serious	GROUP Y S = Serious	VS = Very Serious
D		E	
	NS	S	VS
NS	1.0	1.0	1.2
S	1.2	1.3	1.4
VS	1.4	1.5	1.6

Codes of practice, standards, and safety rules

Pugsley promoted and contributed to many codes of practice and standards. Yet at the same time he recognized that, because they are based on past practice, they tend to become out of date. While they help pass on the experience of pioneers and help avoid errors, they can be restrictive and result in uninspired pedestrian design. Codes are *no substitute for a clear understanding by the designer.*

Standards evolve as materials change. Revisions to rules are often conservatively made by simply scaling one property, such as strength, rather than making comprehensive changes to recognize the full

potential of the new material. Pugsley used the example of Lloyds'
Rules for shipping, as they changed from timber to iron and then steel.

Pugsley saw that the usual sequence was:

$$Accident \rightarrow investigation \rightarrow safety\ rules.$$

He would rather see the sequence:

$$Safety\ rules \rightarrow investigation \rightarrow prevention\ of\ accident.$$

Climate

Pugsley noticed the impact of what he called the *climate*, where
there are time and money pressures, lack of attention by staff and
management, and shortage of skilled people. This is where the seeds
of failure are sown (**7**). The *climate* affects construction as well as
design and increases the risk of failure.

As projects get larger, there is likely to be more government
involvement through finance, ownership, or regulation. Civil servants
are responsible to ministers whose tenure is usually far shorter than
the project. Decision-making is subject to changing political pressures
and tends to be a slow process, further affecting the *climate*.

General lessons

Pugsley listed the main lessons he drew from studying accidents. His
list, summarized from his book on structures (**6**) published in 1966 is:

1. *Be wary of treating an extrapolation of successful past practice
   in a routine manner. Consider it pioneering work, and put fresh
   minds to work.*
2. *When using codes of practice, particularly in pioneering
   work, try to quantify any qualitative advice, by calculation
   or experiment.*
3. *Make sure that project staff are experienced and that they fully
   understand the principles of the design they are building.*
4. *Pay attention to combinations of load, and the interactions of
   load and deflection.*
5. *Do not consider severe natural hazards as acts of God. Design
   for them.*
6. *Do not treat structures as rigid bodies. Always consider transient loads like wind and earthquakes.*
7. *Beware of remote control during construction. Have close communication between all players, including design teams and
   the customer.*

8. *On completion of preliminary design, reassess the dead weight and repeat all relevant design calculations.*

9. *Do not make do with inadequate data. Consider getting data from experiments.*

10. *Beware of gaps in supervision, for example, from illness.*

11. *Be careful when there are undue pressures of time, money, or politics.*

12. *There is natural pressure from people outside a project to get it completed quickly. Do not let this pressure lead to taking short cuts.*

13. *Particularly in pioneering work, beware of strong personalities dominating. Have sparring partners to ensure thorough debate and review.*

14. *Do not let overenthusiasm by a company or profession lead to taking unreasonable risks.*

15. *Look for relevant experience in other fields.*

16. *Review, study, and learn from all major accidents.*

As Pugsley pointed out, most of these general lessons cannot be resolved by larger safety factors. Neither do they appear in codes of practice or safety regulations. Yet they are important. He cautioned *Much has yet to be done!* A lot of progress has been made in structural engineering and engineering in general since this list was drafted.

COMMENT
A lot of sage advice, but much has yet to be done to fully apply these lessons!

References
(1) **Pugsley, A.G.** (1937) Control surfaces and wing stability problems, *J. R.Ae.S.*, **41**.

(2) **Pugsley, A.G.** (1947) The behaviour of structures under repeated loads, *J. R.Ae.S.*, September 1947.

(3) **Pugsley, A.G.** (1972) The engineering climatology of structural accidents, *Proc. Int. Conf. on Structural Safety and Reliability*, Pergamon Press.

(4) **Bulson, P.S., Caldwell, J.B.** and **Severn, R.T.** (Editors) (1983) *Engineering Structures Developments in the 20th Century: A Collection of Essays to Mark the 80th Birthday of Sir Alfred Pugsley*, University of Bristol Press, ISBN 0 86292 105 8.

(5) **Pugsley, A.G.** (1942) *A Philosophy of Aeroplane Strength Factors*, R&M 1906.

(6) **Pugsley, A.G.** (1951) Concepts of safety in structural engineering, *J. I.C.E.*, **36**(5).
(7) **Pugsley, A.G.** (1966) *The Safety of Structures*, Edward Arnold.

2.1.2   Alfred M. Freudenthal (1906–1977)

Alfred Freudenthal was born in the Carpathian region of what was then the Austro-Hungarian Empire. His degree in Civil Engineering and doctorate were from Prague. His doctorate was on the study of plasticity. He emigrated in the mid-1930s and became the Resident Engineer for building the new port of Tel Aviv. During WWII, he was in charge of building two minesweepers for the Royal Navy in Tel Aviv.

In 1947, he presented a paper on the statistical aspects of fatigue to the Royal Society. With recognition from this, he was invited to move to the United States. He was appointed first as Professor at the University of Illinois and then at Columbia before becoming the Director of the Institute for the Study of Fatigue and Structural Reliability. At the Institute, he encouraged Emil Gumbel and Waloddi Weibull to join in his studies.

Freudenthal was both a scientist interested in understanding material science and an engineer dedicated to studying fatigue and statistical concepts for design. He pioneered reliability engineering, and in recognition of his contribution, the American Society of Civil Engineers published a *Civil Engineering Classic – Selected Papers by Alfred M. Freudenthal* (**1**). Two aspects of his work are briefly highlighted here – the safety of structures and comments on fatigue.

The safety of structures – a case for statistics

In the mid-twentieth century, design practice was to either apply a safety factor between the load and strength or design within an allowable stress. Freudenthal recognized that a true understanding of risk could only be quantified by using statistical methods. Variations and uncertainties occur in two areas – factors affecting load, and the capacity of a structure to carry that load. Every physical factor used in design has a statistical distribution. In his landmark paper on the Safety of Structures, Freudenthal presented the concept shown in Fig. 2.1.1.

The probability of occurrence of load-carrying capability is plotted against the probability that the structure can resist failure. This allows the curves of equal probability of failure to be drawn, which in turn allows the safety factor to be quantified – for a specific probability of failure. Freudenthal recognized the difficulty of establishing the complete probability distribution that takes into account all factors. For example, inspection can theoretically reject all material below a

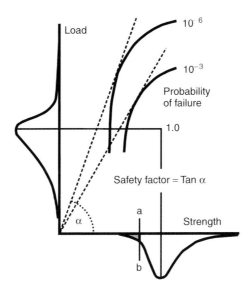

Fig. 2.1.1   Curves of equal probability of failure based on Fig. 2.1.4 of Ref. (2). Reproduced with permission from ASCE

certain strength – as shown in the shaded area to the left of the line a–b in Fig. 2.1.1. But there is always the possibility that some low-strength material will get past inspection. To get a true quantification of risk, the probability of this event needs to be included.

Freudenthal saw a psychological barrier that needs to be overcome before engineers are comfortable with a probabilistic approach to design. In a 1954 paper **(3)**, he said *the concept of safety is deeply rooted in engineering design, whereas the notion of finite (no matter how small) probability of failure is repulsive to a majority of engineers.*

Fatigue test interpretation – the limitations of the s–n curve

Freudenthal's study of fatigue started at the sub-microscopic level. He saw that materials are neither homogeneous nor isotropic. Damage at the molecular level starts with a statistical distribution. The traditional presentation of fatigue test results is a curve of stress (s) versus cycles to failure (n). A relatively small number of tests are conducted at several stress levels. Then a best-fit line is drawn through the results. This form of presentation gives no indication of the possible scatter of results.

Freudenthal presented an alternative approach in a paper to the Royal Society in 1953 **(4)**. He proposed plotting the probability of survival at a given stress level as shown in Fig. 2.1.2.

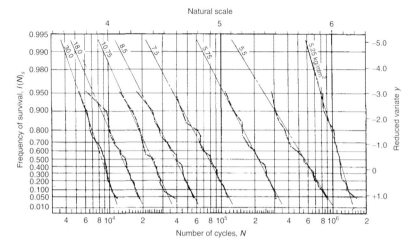

Fig. 2.1.2   Survivorship functions for annealed aluminium wire. Figure 4 from Ref. (2). Reproduced with permission from ASCE

Freudenthal recommended that at least 20 specimens should be tested for each stress level. If the test results do not follow the expected pattern, then there is a problem with the consistency of the test method or samples. With consistency, statistically significant conclusions can be drawn for use in design.

COMMENTS
Alfred Freudenthal has contributed to the understanding of risk in engineering by pioneering the concept of reliability engineering. His papers show a keen interest in the science of materials and a comfort with mathematical and statistical analysis. While attempting to quantify risk, he was always looking for an economic answer – aiming to improve engineering by removing ignorance from the judgement of risk versus economy. He searched for a scientific and statistical understanding of failure as a means for developing logical and economic engineering designs. The span from an understanding of science to practical application, using insight from mathematical and statistical analysis, is still a necessary skill for an engineer.

References
(1) **Freudenthal, A.M.** (1981) *Civil Engineering Classics*, Selected papers by Alfred M. Freudenthal, ASCE, New York, ISBN 0 87262 263 0.
(2) **Freudenthal, A.M.** (1947) The safety of structures, *ASCE Trans.*, **112**, 125–180.

(3) **Freudenthal, A.M.** (1954) Safety and probability of structural failure, *ASCE Trans.*, **121**, 1337–1397.

(4) **Freudenthal, A.M.** and **Gumbel, E.J.** (1953) On the statistical interpretation of fatigue tests, *Proc. R. Soc. A*, **216**, 309–332.

2.1.3   Henry Petroski

Since 1980, Henry Petroski has been at Duke University, where he presently is the Aleksandar S. Vesic Professor of Civil Engineering and also a professor of history. He has written many books on engineering, which have become popular with the public as well as engineers. Through his books, and articles in the magazine *American Scientist*, Dr Petroski has brought an understanding of engineering to a large audience.

One interesting example of Petroski's writing is a section in the *Reference Manual on Scientific Evidence* (**1**). This volume, published by the Federal Judicial Center, provides background to judges. It is distributed to all federal judges. Dr Petroski's section on Engineering Practice and Methods is a well-presented summary of the nature of engineering and the roles of design and analysis. He also describes the differences between science and engineering and the role of failure in engineering.

Petroski's books (**2–6**) have covered the design of a range of items from paperclips and zips to aircraft and bridges. Underlying all the examples is a focus on history, failure analysis, and design theory.

History

History provides lessons from past failures. It also allows us to be aware of the roots of the profession. Petroski points out that history gives a perspective on how engineering interrelates with society. History is an important part of the design process.

Failure analysis

> *Form follows failure.*

Petroski encourages engineers to always keep failure lessons in mind and focus on avoiding the features that made previous designs fail. Make failure analysis part of the design process from the start, rather than the traditional approach of conducting failure analysis after the design is complete.

Design theory

Petroski views disasters as ultimately being failures of design. He defines design as *the avoidance of failure*. Designers should be motivated more by disaster avoidance than cost saving. Nevertheless, there is always some risk and uncertainty in design. Judgement and decision-making are key to the design process. Back-of-envelope calculations are effective in giving a broad check and understanding throughout the design process. Detailed calculation can come later to confirm the design.

Case studies can teach a lot about design and the design process. Even familiar cases can be retold from a fresh perspective and alert designers to pitfalls.

Some notable quotations

- *Engineering is inextricably involved with virtually all aspects of society. No engineering problem is without its cultural, social, legal, economic, environmental, aesthetic or ethical component – and attempts to approach an engineering problem as a strictly technical problem will be fraught with frustration. While engineering is grounded in maths and science, it serves society in very human ways.*
- *Tragedies are rooted in two human characteristics: the cultural drive to build ever-bolder structures and the hubris of master builders and engineers in their attempts to do so.*
- *Engineering design and analysis are only as good as the knowledge upon which they are based.*
- *Engineering is the art of compromise – and there is always room for improvement in the real world. To stay within the status quo would be irresponsible. Every Dollar that goes unnecessarily into strength may be a Dollar that is not available for social (needs), R&D, education, etc.*
- *Despite maths, science, and the computer, engineering best practice still needs a good deal of common sense reasoning.*
- *It is inevitable that errors are going to be made in design. Some conceptual designs are just bad ideas from the start, and it is the self-critical faculty of the designer that must be called into play to check him- or herself and to abandon bad ideas on the drawing board.*
- *Possibly the greatest tragedy underlying design errors and the resultant failures is that many of them do indeed seem to be avoidable.*

References (a selection of books by Henry Petroski)

(1) *Reference Manual on Scientific Evidence*, Second edition, 2000, Federal Judicial Center, http://www.fjc.gov.

(2) **Petroski, H.** (2003) *Small Things Considered: Why There is No Perfect Design*, Alfred A. Knopf, New York, ISBN 1 40004 050 7.
(3) **Petroski, H.** (1994) *The Evolution of Useful things*, Vintage Books, New York, ISBN 0 67974 039 2.
(4) **Petroski, H.** (1998) *Inventions by Design: How Engineers get from Thought to Things*, Harvard University Press, ISBN 0 67446 368 4.
(5) **Petroski, H.** (1994) *Design Paradigms: Case Histories of Error and Judgement in Engineering*, Cambridge University Press, ISBN 0 52146 649 0.
(6) **Petroski, H.** (1992) *To Engineer is Human: The Role of Failure in Successful Design*, Vintage Books, New York, ISBN 0 67973 416 3.

2.1.4 Trevor Kletz

Dr Kletz's first career was with ICI, where he gained experience with chemical plant and became committed to improving safety. He was appointed a safety advisor to the Petrochemical Division. In his second career, he is the Visiting Professor of Chemical Engineering at Loughborough and Texas A&M Universities. He is still pursuing process industry safety. Through his books **(1–10)** and lectures, his advice and experience is available to a wide audience. A brief summary of his views covers safety, history, human error, design, and reviewing failures.

Safety

It might seem to an outsider that industrial accidents occur because we do not know how to prevent them. In fact, they occur because we do not use the knowledge that is available.

Most senior managers genuinely want fewer accidents but their actions do not always produce the required effect. *Exhortation is often thought sufficient – but it is not.*

After construction and before commissioning, do a walk-around inspection. Check for potentially hazardous situations. See if any changes are necessary and whether the information and signage are adequate and easily understood. Try to identify any hazards before they occur.

History

Over time, the importance of standard practices and rules is forgotten. Staff become complacent; and staff change. Memory is with the staff – the organization has no memory, and the reason that accidents happened is forgotten. As time passes, conditions become ripe for another accident – and history repeats.

Human error

Human beings are actually very reliable, but complex. Nevertheless, people do make errors. Accept that people make mistakes, but blaming human error is a path that leads nowhere. Looking at the reasons for human error provides the best way of dealing with them. Dr Kletz sees four types of error:

1. ***Mistakes*** – *incidents preventable by better training or instructions. Sometimes the result of poor communication. What we do not say is as important as what we do say.*
2. ***Violations*** – *incidents that occur because some one knows what to do but decides not to do it.*
3. ***Items out of reach*** – *incidents that occur because people are asked to do tasks beyond their physical or mental capabilities,*
4. ***Errors*** – *due to a slip or momentary lapse of attention.*

In general, the solution is to try and change the situation, not the people. The most effective actions we can take are to design our plants and methods of working so as to reduce the opportunity for error or minimize their effects.

Design

Safety by design should always be our aim. Dr Kletz found in reviewing accident case histories that some features appeared frequently. These were all details – rust, insulation, plugs, operating temperature variations, and drains. He speculated that these details get less attention than the heartland of the plant. Design engineers give scant attention to items such as drains – they consider that they did not get an engineering degree to become a glorified sewerman.

Chemical plant design has to recognize hazardous material. Design practice should be to:

- *use less hazardous material – intensify the use;*
- *use safer materials – by substitution;*
- *use hazardous material in the least hazardous form;*
- *recognise that very little energy is needed for a spark. Despite all precautions, it is impossible to remove all sources of ignition; and*
- *make sure that there is more than one line of protection so that leaks are contained. Additional lines of protection should have effective instrumentation and warnings so that it is known that a leak has occurred.*

In terms of equipment design:

- *simplification is key – complexity means more equipment that can fail and more opportunities for human error;*
- *limit the effect of hazards, not by adding on protective equipment but by equipment design, by changing reaction conditions, by limiting the energy available, or by eliminating hazardous phases, equipment, or operations;*
- *avoid knock-on or domino effects;*
- *make incorrect assembly impossible;*
- *make the status of equipment clear; and*
- *use equipment that can tolerate poor operation or maintenance.*

And, as general reminders:

- *take the time to make a sound assessment of project cost and duration before you start;*
- *always ask yourself to what extent the project is a development project, and if it is, allow for it in both costs and time; and*
- *make sure that design changes during commissioning and operation get full reviews.*

Reviewing failures

A lot can be learned from reviewing failures. But we often miss the opportunity to learn all the lessons because:

- *we find only a single cause, often the last one in the chain;*
- *we find only immediate causes. We should also look for ways of avoiding the hazard and for weaknesses in the management system;*
- *we list human error as a cause without saying what sort of error. Yet different actions are needed to prevent those due to ignorance, those due to slips or lapses of attention, and those due to non-compliance;*
- *we list causes we can do little about;*
- *we do not help others to learn as much as they could from our experiences;*
- *we forget the lessons learned and the accidents happen again;*
- *we need better training, by describing accidents first rather than principles, as accidents grab our attention. We need discussion rather than lecturing, so that more is remembered. We need databases that can present relevant information without the user having to ask for it.*

Dr Kletz has summarized the difference in effort usually expended on establishing causes of accidents with their relative importance in a simple chart, Fig. 2.1.3.

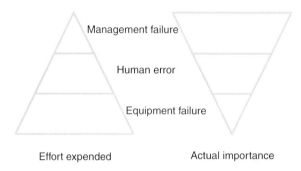

Fig. 2.1.3   **The effort expended on different causes of accidents and their relative importance**

COMMENTS

While Dr Kletz is primarily addressing engineers in the chemical industry, most of the lessons are relevant to all engineers. Any engineer dealing with plant fittings such as valves and piping should read his books. They will help recognize potentially hazardous situations – and provide practical advice on how to avoid those hazards.

A quotation from Fred Astaire is appropriate to Fig. 2.1.3. *The higher up you go, the more mistakes you are allowed. Right at the top, if you make enough of them, it's considered your style.*

References (a selection of books by Henry Petroski)
Some of Dr Kletz's books are listed below:

(1) **Kletz, T.** (2001) *Learning from Accidents*, Gulf Professional Publishing, Elsevier, ISBN 0 75064 883 X.

(2) **Kletz, T.** (2001) *An Engineer's view of Human Error*, Taylor & Francis, ISBN 1 56032 910 6.

(3) **Kletz, T.** (1999) *Hazop and Hazan – Identifying and Assessing Chemical Industry Hazards*, IChemE, ISBN 1 56032 858 4.

(4) **Kletz, T.** (1998) *What Went Wrong? – Case Histories of Process Plant Disasters*, Gulf Professional Publishing, Elsevier, ISBN 0 88415 920 5.

(5) **Kletz, T.** (2003) *Still Going Wrong!: Case Histories of Process Plant Disasters and How They Could Have Been Avoided*, Gulf Professional Publishing, Elsevier, ISBN 0 75067 709 0.

(6) **Kletz, T.** (1997) *Dispelling Chemical Engineering Myths*, Taylor & Francis, ISBN 1 56032 438 4.

(7) **Kletz, T.** (1993) *Lessons from Disaster – How Organisations have No Memory and Accidents Recur*, Gulf Professional Publishing, Elsevier, ISBN 0 85295 307 0.

(8) **Kletz, T.** (1998) *Process Plants: A Handbook of Inherently Safer Designs*, ISBN 1 56032 619 0.

(9) **Kletz, T.** (1995) *Computer Control and Human Error*, Butterworth-Heinemann, ISBN 0 88415 269 3.

(10) **Kletz, T.** and **Harvey-Jones, J.** (2000), *By Accident: A life Preventing Them in Industry*, PFU, ISBN 0 95384 400 5.

2.1.5   Hyman G Rickover (1898 or 1900 [uncertainty]–1986)

Admiral Rickover is the *Father of the US Nuclear Navy* (**1**). He was born near Warsaw and immigrated to the United States with his family when he was a small child. Hyman Rickover graduated from the Naval Academy at Annapolis in 1922, and later earned an MS in Electrical Engineering from Columbia University.

In the 1930s, he had sea-going experience on warships, including submarines. He spent most of WWII on engineering work, repairing and improving vessels. This gave him experience of managing relationships with contractors – as well as helping develop his own views on management.

Rickover was assigned to the Manhattan Project at Oak Ridge in 1946, and saw the potential for nuclear power in submarines. He came away from Oak Ridge with a vision for a Nuclear Navy: and the drive to achieve that vision. By 1948, he had command of the Nuclear Power Branch of the US Navy. He remained in charge of the Nuclear Navy until his retirement in 1982.

Admiral Rickover's achievements

Admiral Rickover was one of the greatest engineers of the twentieth century. But what did he really do? He did not design the reactors – although he had significant input and provided critical review. He did not design the submarines – but again had essential input. For example, he insisted that the only novel feature of the first nuclear submarine, *Nautilus,* would be the nuclear power plant. On the project side, he applied the new Critical Path Method in the form of PERT to great effect and set new standards for quality control in the marine field. His real achievement was to create a complete nuclear navy – a navy that met all the requirements of the Cold War;

a navy built on his principles, which still dominate today, decades after he established them.

His achievements sound simple, but are the most difficult to achieve comprehensively and are worthy of more detailed examination.

The Naval Nuclear Propulsion Program

Rickover grew the program into a venture that included:

- the operation of the fleet of surface and subsurface nuclear-powered vessels;
- the R&D, specification, design, acquisition, procurement, construction, testing, commissioning, maintenance, operation support, refuelling, and ultimate decommissioning of the nuclear plants;
- the selection of staff, and the operation of schools and consistent training programs;
- the provision and operation of training reactors for the live training of all personnel who will operate, maintain, and supervise nuclear powered vessels;
- the supervision of dedicated laboratories;
- the direction of shipyards and dedicated prime contractors and suppliers.

This comprehensive program is run by the Director of Naval Reactors, a post that has decision-making authority in both the Navy and the Department of Energy (DoE). The Director's rank has to be a four-star admiral, and Deputy Administrator in DoE. The Director has a HQ staff of fewer than 400 to manage the program. He manages as many reactors as the whole US civil electrical power industry. The program has 5100 reactor years of safe operating experience. To put this in context, the world's total civil nuclear power station experience is 9200 reactor years.

The scope of the Director's responsibilities, including safety, and the stature of the position are enshrined in Public Law. The program's philosophy is to have centralized technical control of all aspects. This requires an in-depth technical understanding of the work of the program at all levels.

Achieving performance

Rickover expected high performance in all aspects of the program. He was annoyed and disappointed when people asked him for his formula for success. He told them *any successful program functions as an integrated whole of many factors. Trying to select one aspect as the key does not work. Each element depends on all the others.*

The current Director of Naval Reactors is Admiral F.L. 'Skip' Bowman. He summarized the work and philosophy of the Naval Nuclear Propulsion Program at a meeting of the House Science Committee, who were reviewing the CAIB(Columbia Accident Investigation Board) Report**(2)**. Bowman, while agreeing with Rickover that there is no magic formula, nevertheless summarized his core values in four headings: People, Formality and Discipline, Technical Excellence and Competence, and Responsibility.

On people
Rickover knew the importance of having the right staff. His approach was simple, select the best, select them yourself, make sure they have integrity, and a sense of purpose – then continually train and challenge them. He probably unfairly rejected many good engineers with his dogmatic and dictatorial approach – but those he did select were good.

On formality and discipline
Engineering is a discipline – and engineering decisions have to be formal and disciplined. All activities with reactors, whether in ships or at training facilities, must be followed to the letter. Changes and decisions need to be recorded. The Director, following Rickover's practice, still receives a copy of every recommended action prior to issue.

Rickover was adamant of the need for this discipline, yet at the same time he decried some aspects of traditional Navy discipline. He preferred wearing a suit rather than his uniform. Tradition looks to develop leadership on the sports field. In Rickover's view, the players are selected by the coach and the rules confine the play. How could this prepare for the real world? Tradition can be used to manipulate and pressurize officers into conforming to the official line – whatever that is.

Although Rickover required discipline from his staff, he made his own decisions. He went to sea on the commissioning trials of most of the nuclear submarines. On a couple of occasions, he instructed the Captain to repeat some tests, which placed the submarines in dangerous situations. In the case of the ***La Jolla***, she dived 240 ft below the planned depth, though within the design limits. His ego had taken charge. Fortunately, none of the incidents resulted in damage – except to his image.

On technical excellence and competence
This starts with choosing the right people, but the true level of skill will only come with education and training. Education and training

are continuing activities – not one-off. Training should be led by the most experienced people.

It is crucial that the people making decisions understand the technology they are managing and the consequences of their decisions. This understanding comes partly from having all staff operate live reactors during their early training. Rickover stated his criteria for managing technical excellence as:

- *the person must feel that he owns the job and will remain on that job indefinitely;*
- *only one person can be truly responsible for the work;*
- *if the boss is not interested in details, the subordinates will also not consider them important. Most managers would rather focus on lofty policy matters;*
- *establish simple and direct means to find out what is going on in detail. Most managers avoid keeping up with details, instead they create 'management information systems';*
- *face the facts – resist the temptation to hope that things will work out.*

One means of providing technical excellence and competence is redundancy and multi-layered defence. Redundancy philosophy can be applied to people as well as equipment.

On responsibility
Admiral Bowman quoted Rickover to the House Committee:

> *Responsibility is a unique concept: it can only reside and inhere in a single individual. You may share it with others, but your portion is not diminished. You may delegate it, but it is still with you. You may disclaim it, but you cannot divest yourself of it. Even if you do not recognize it or admit its presence, you cannot escape it. If responsibility is rightfully yours, no evasion, or ignorance, or passing the blame can shift the burden to someone else. Unless you can point to the person who is responsible when something goes wrong, then you have never had anyone really responsible.*

Safety
The success of the Naval Reactor Program has been built on a culture of safety. Admiral Bowman has summarized the philosophy as:

> *Safety is the responsibility of everyone at every level in the organization. Safety is embedded across all organizations in the*

Program, from equipment suppliers, contractors, laboratories, shipyards, training facilities, and the Fleet to our Headquarters. Put another way, safety is mainstreamed. It is not a responsibility unique to a segregated department that then attempts to impose its oversight on the rest of the organization.

While safety is everyone's responsibility, there is a small Reactor Safety Analysis group. They provide:

- liaison with other safety and regulatory agencies to ensure that 'best practices' are known and followed;
- an independent check on safety practices; and
- the database of safety issues, which effectively provides a 'corporate memory' of past problems.

COMMENTS

The CAIB saw that the most difficult area to deal with, where failures can germinate, is in the overall infrastructure and corporate bureaucracy. Rickover and his successors have grasped this issue and provided effective answers.

Politicians are bound to try to control a program as large as this. Rickover recognized that unless he could manage this aspect, then his whole vision was at risk to political intervention. He was skilled in stroking the egos of politicians. Yet at the same time the politicians knew that Rickover was 'his own man'. Despite being under the direction of Congress and the Navy, he would only do what he believed to be right. Politicians accepted this power balance because Rickover delivered on his promises.

Rickover valued courage – courage to table the facts and be completely candid. He told a Congressional Committee that *a certain measure of courage in the private citizen is necessary to the good conduct of the State, otherwise men who have power through riches, intrigue, or office will administer the State at will and ultimately their private advantage.* This was courage and candour that politicians are not used to!

In his early days, Rickover answered to both the Atomic Energy Commission and the Navy. He found this a useful arrangement to avoid being directed by just one corporate view. Most engineers understand the benefit of redundancy in equipment design. Some recognize the benefit of redundancy in human control and management systems. But Rickover is probably unique in engineering a situation where he had a redundancy of bosses! With this dual line of command, as well as the job description of the Director cast into National law, it has provided a strong barrier against political pressure. He truly *engineered* the Nuclear Navy in all its aspects.

References (a selection of books by Henry Petroski)
(1) **Polmar, N.** and **Allen, T.B.** (1982) *Rickover*, Simon and Schuster, New York, ISBN 0 67124 615 1.
(2) Statement of Admiral Bowman before the House Science Committee, Washington DC, 29 October 2003. Accessible at www.house.gov/science/hearings/full03/oct29/bowman.pdf.

Background – Placing Engineering into Perspective

2.2.1 Science and engineering

Science and engineering are words frequently used in conversation, often interchangeably. What do they mean, and what is the difference between them? Some years ago in their annual review of science, the *Economist* described a scientist as someone who loves ignorance – because the driving force for scientists is to remove ignorance, to establish the facts and laws of nature. Compared to this definition of a scientist, an engineer is someone who has to work within existing knowledge, or lack of knowledge, to make and operate things. Engineers do not have the time or money to explore all the science involved and remove all the unknowns. Engineering is goal-oriented. Science seeks an understanding – science does not guess.

Science is synonymous with learning and knowledge. The scientific method is to observe, develop a hypothesis, make deductions, and then verify and check the hypothesis by experiment. It is a repeating process as hypotheses are challenged, improved, and revised. Hypotheses are concepts of the mind, useful aids on the road to searching for the Holy Grail – the fundamental laws of nature.

Engineering has been defined as a science. Alternatively, it has been defined as an art – for constructing and using machinery. So engineering can be both a science and an art.

Engineering is sometimes described as the application of scientific knowledge to create practical things for the benefit of society. Few engineering products meet this definition. Most rely on empirical data and judgement, backed up by limited testing. It would be preferable to rely on established scientific data but they seldom exist in a form, and to the extent, needed. The general laws of science underpin engineering, but the detailed scientific data are sparse.

Just as there are overlapping definitions of science and engineering, there are similar variations in the work done by scientists and engineers. Engineers do scientists' work and vice versa. There is synergy between the two. Many eminent engineers and scientists have had a foot in each discipline, often following a pattern – spending time *finding out*, followed by *applying* the new knowledge.

Scientists and engineers need to work closely together and understand each other's role. When engineers have to use empirical data, they are helpful if scientists can then explore that empiricism and gain a scientific understanding of the underlying principles. With that new knowledge, the engineer can then improve the product on the basis of a more solid understanding.

An engineer may never know the full scientific details, but by using informed judgement, technology can be applied with reasonable confidence.

2.2.2 What is an engineer?

Professional engineers are forever bemoaning the public's poor understanding of their role. Engineers are seen as people who drive engines or repair equipment or work with their hands – and get them dirty. Some recall colleagues who left school with good grades in mathematics, physics, and chemistry, and became engineers. This creates an image of professional engineers as scientists using mathematics to calculate the strength of components. Since the mathematics, physics, and chemistry taught at school are all precise, with closed-ended answers to every problem, it is assumed that engineering is similarly precise.

The 1929 edition of the Encyclopaedia Britannica gives an extensive definition of a professional engineer. It is reprinted with kind permission from *Encyclopaedia Britannica,* 14th Edition, ©1929:

> *Qualifications include intellectual and moral honesty, courage, independence of thought, fairness, good sense, sound judgement, perseverance, resourcefulness, ingenuity, orderliness, application, accuracy and endurance. An engineer should have ability to observe, deduce, apply, to correlate cause and effect, to co-operate, to organize, to analyse situations and conditions, to state problems, to direct the efforts of others. He should know how to inform, convince and win confidence by skilful and right use of facts. He should be alert, ready to learn, open-minded, but not credulous. He must be able to assemble facts, to investigate thoroughly, to discriminate clearly between assumptions and proven knowledge. He should be a man of faith, one who perceives both difficulties and ways to surmount them. He should not only know mathematics and mechanics but should be trained to methods of thought based on these fundamental branches of learning. Organised habits of memory and large capacity for information are necessary. He should have extensive knowledge of the sciences and other branches of learning and know intensively*

those things which concern his specialities. He must be a student throughout his career and keep abreast of human progress... the engineer is under obligation to consider the sociological, economic and spiritual effects of engineering.... The engineer's principal work is to discover and conserve natural resources of material and forces, including the human, and to create means of utilizing these resources with minimal cost and waste and with maximal useful results.

Alfred D. Flinn, an engineer from New York, wrote these words. He supervised the construction of the Catskill Aqueduct, the main water supply for Manhattan. His words are equally relevant today, three-quarters of a century later – but with one exception. Professional engineering is no longer the preserve of the male. The mathematical and scientific aspects of engineering only come in half way through his description, well after the need for judgement and courage. With the comprehensiveness of his description, and the wide distribution of the encyclopaedia, one wonders why the public has any confusion about the role of a professional engineer!

Psychologists and sociologists have found that organizations and their management are interesting to study. As a subsection of these studies, they have examined engineers – almost as a species. A prominent example is the study of engineers involved in the ***Challenger*** disaster written by Diane Vaughan (**1**). She saw engineering as a craft, and rule-following, bureaucratic profession. An engineer's first reaction is – here we go again, craft equates to getting one's hands dirty. As for rule following, there may be lots of standards and procedures but there is no rule book to tell you how to design. This requires knowledge, experience, and judgement that do not come in rule books. And bureaucracy is seen as governmental official-doms with inflexible and unimaginative routine. Surely engineering is not that negative? Diane Vaughan's study is well respected and well researched, and is worth deeper study.

Engineering as a craft

A craft is generally seen as a manual activity, albeit requiring skill learnt from doing, and through apprenticeship. Several dictionaries give a broader definition and define craft as a special skill. Many professional engineers work in offices and spend time producing documents and analyses. In this climate, it is easy to overlook the fact that engineering is about making and operating equipment. Craft and craftsmen underpin engineering. And in the broader definition of craft, professional engineers have special skills. So both the broader and narrower definitions can apply to engineering. Vaughan sees the production of engineering information as a craft – needing discipline and skill.

Engineering as rule following

Rule following is one of three levels of human performance identified in the research of Jens Rasmussen, Professor of Cognitive Ergonomics, Risø National Laboratory, Denmark, in the 1980s (2). The three levels are skill-based, rule-based, and knowledge-based. Skill-based actions are carried out almost automatically with learnt skills without the need for deep-thinking activities like walking and playing tennis and the piano. Rule-based activities require conscious thinking to assess and then apply the rules from our storehouse of experience. This applies to driving – knowing the rules of the road and the driving limits. Knowledge-based action requires time to consciously analyse unusual situations. This is used as a last resort, when it is necessary to resort to digging deep into one's brain to find a solution.

How do these definitions apply to engineering? Vaughan assesses that most of the activities of a professional engineer are rule-based. But the definition of rules covers a wide spectrum. During training, engineers learn the rules to the *engineering way* of thinking and working; a scientific approach to problem solving, a use of words more narrowly defined than in general use, like stress and strain; and the use of drawings with all the symbols and annotations for communicating ideas and requirements to other engineers and to the craftsmen who make the items. Then there are the written rules, standards, and regulations that incorporate experience.

Sociologists have described engineering as an *unruly technology* (3). There are not enough rules to cover all engineering work, there is no consistent set of rules, and the rules are continually changing. Engineers also bend the rules – often consciously when a knowledge-based assessment shows a better route.

On the basis of the sociologists' definitions, engineering is a rule-based profession, with occasional excursions into knowledge-based work as new problems arise and new concepts are invented and analysed.

Engineering as a bureaucratic profession

While the general use of the expression bureaucracy has negative connotations, the sociologists have their own definition. It is not a precise definition but a collection of characteristics (**4**). The most common are a hierarchy of authority, division of labour into narrowly defined tasks, having technically competent participants and the use of fixed procedures and rules for work. Other characteristics sometimes included are limited delegated authority, differential rewards, impersonal relationships, and the emphasis on written communication. Bureaucracies can be successful when the rules are used to assist work, rather than hinder or become ends in themselves.

When measured against these characteristics, engineering teams are bureaucratic. Engineering teamwork is handled by a hierarchy and by technically competent staff. Procedures and rules are needed for control; discipline is essential, as is a paper trail. The challenge for engineers is to use these bureaucratic tools while avoiding a descent into the abyss of negative bureaucracy.

The sociologists' view of engineering

Given the definitions used by sociologists, engineering is a craft, rule-based, bureaucratic profession. Some sociologists suggest that engineers are *servants of power*, that engineers are trained to be monopolized by large corporations, to further the principles of capitalism, and by governments to undertake their projects. As a result, engineers are just *tools of the system*.

Thcy also believe that the public is deceived by the myth that engineering is precise. Investigations of failures consistently show that engineering is characterized by *ambiguity, disagreement, deviation from design specification and operating standards, and ad hoc rule making* (**5**).

Educating engineers

An engineer is the product of the education system. Mathematics, science, and technology dominate the formal undergraduate education. This theoretical and numeric basis of engineering courses plays a small role in most engineers' future career, but it is still a necessary foundation. There is debate on whether engineers' education should include other aspects – economics, management, ethics, and social studies.

COMMENTS

However much thou art read in theory, if thou hast no practice thou art ignorant – but a beast of burden with a load of books. Gulistan of Sa'di c1256.

A total engineering education should comprise the formal under-graduate studies together with an apprenticeship to give practice of the craft skills that go into engineering. Following graduation, a period spent in different engineering tasks such as design, research, shop-floor management, site construction, project work, commercial and sales offices would give a deeper understanding. If a couple of years are spent in say five of these areas, then by mid-30s, an engineer could be considered a rounded and *compleat engineer* and ready to tackle a major engineering task. Education would still be a continuing process to keep up-to-date with technical developments, as well as learning more about the interaction of engineering with society. This type of complete education needs the support of institutions, businesses, and governments. It needs a belief that it is effective and justifies the addi-tional costs, costs that cannot be associated directly with a company's short-term financial performance, but will almost certainly produce better performance and fewer failures in the long term.

The views from the sociologists can make depressing reading for an engineer. Before discounting their views, it is worth understanding what they are saying and why. Maybe there is a grain of truth in their analysis, and maybe the public's view of engineering stems from similar thinking. Several sociological studies have been initiated by failures. Consequently, the studies see the worst side of engineering. A typical caricature of an engineer is a serious person who through numeric skills can assure the public that structures, components, and situations are safe. When a failure occurs, there is then a feeling of broken trust.

These comments relate to today's climate. In the halcyon days of supersonic aircraft, nuclear power, and space exploration, they were seen as engineering triumphs. Today there is cynicism, as well as groups dedicated to protesting against certain engineering activ-ities – nuclear power, environmental effects of hydroelectric dams, health effects of chemical plants, and so on. Is this a cycle that will eventually change? In Victorian times, the early euphoria for engineering advances changed to criticism of industry. John Ruskin bemoaned that *industry was always looking to make things a little cheaper and worse just to make money from people who only con-sidered price.* Are we just working through another cycle or seeing a societal shift permanently away from engineering? It is hardly the latter because society is more than ever before surrounded by, and dependent on, engineering.

Engineering products are an essential part of human progress. To make progress, engineers require skill, resources, responsibility, and the courage to make decisions. All these are essential for creating something new. Criticizing this needs little knowledge and little responsibility. Engineers are human and can make mistakes. Engineering mistakes tend to be prominent for all to see. Critics' mistakes appear at most as a small apology hidden in the inside pages. Engineers have to understand this one-sided predicament and learn to handle it.

Another comment on today's engineers is that they are unknown to the public. Engineering has no human face. Half a century ago, the UK public knew the names of the aircraft pioneers and other prominent engineers. They were seen as heroes; names like de Havilland, Fairey, Hinton, Whittle, and Moulton. Today, the public is unaware of who leads the design and engineering of the latest aircraft, the designer of the space shuttle, or the windmill farm, the TGV, or the Severn Bridge. These projects are just as demanding and have equally commanding leaders but they never appear on the public horizon. Surely they are not that uninteresting? Does this affect the image of engineering? Does it matter? Maybe the cognitive psychologists can help.

The view that engineers are just tools of the system is biased. It could equally be said of doctors in hospitals, accountants in auditing companies, lawyers in legal companies – and the characterization in these cases would be equally wrong. The fault is in believing that professionals only follow instructions mindlessly. This view ignores professional practices, the fact that professionals do think for themselves rather than be told what to do, and have a code of ethics.

The partial answer to the question *What is an Engineer?* is that an engineer is an essential part of society – providing the products on which society depends. An engineer is a practitioner in applying human judgement to the use of technology for society. To meet this challenge, a professional engineer needs all the skills and attributes listed by Alfred Flinn.

References
(1) **Vaughan, D.** (1996) *The Challenger Launch Decision*, University of Chicago Press, Chicago and London, ISBN 0 22685 175 3.
(2) **Rasmussen, J.** (1986) *Information Processing and Human-Machine Interaction*, Elsevier Science, New York, ISBN 0 44400 987 6.
(3) **Wynne, B.** (1988) Unruly technology, *Soc. Stud. Sci.*, **18**, 147–167.
(4) **Graham, W.K.** and **Roberts, K.H.** (Eds) (1972) *Comparative Studies in Organisational Behaviour*, Holt, Rinehart & Winston, ISBN 0 03084 392 8.

(5) **Turner, B.A.** and **Pidgeon, N.F.** (1978) *Man Made Disasters*, Second edition, 1997, Butterworth, ISBN 0 75062 087 0.

2.2.3   Cycles in engineering

An engineer's reaction to the word 'cycle' is to think of two-wheeled vehicles, or the Otto and Carnot Cycles, or fatigue cycles. These are important, but there are other more subtle cycles influencing engineering. Failures appear to be cyclical. Sibly and Walker noted that landmark bridge failures occurred approximately every 30 years. They did not arrive at a firm conclusion, but suggested that there was a cycle of success followed by complacency. Alternatively, there may be a communication gap between one generation of engineers and the next, and experience and judgement is lost.

Alan Graham and Peter Senge at MIT saw that the attitude of society to innovation becomes highly receptive in cycles that peak every 40–60 years. In between the peaks, society tolerates less radical improvements. They found a correlation between these cycles and the economic climate. Collapse of capital-producing sectors of the economy spurs innovation. Each new capital expansion uses new technologies. Time is taken to consolidate each new technology before the next wave takes over. The eras of the fastest development of automobiles, aircraft, gas turbines, nuclear power, and electronics are all examples of this effect. What will be next – nanotechnology?

Cycles to failure

Exploring the suggestions of Sibly and Walker shows a possible pattern that may apply to many engineering products. Three phases take place over time – gaining confidence, looking for improvements, and then extrapolating till it breaks (Fig. 2.2.1).

A product has a distribution of strength and load. The safety factor might be quoted as nominally B/A. A and B would be chosen conservatively, as more than the average load and less than the average strength. After a period of consolidation, while experience is gained and confidence grows, both the designer and operator look for improvements. The designer might remove conservatism in the assumptions. The designer might also believe that the distribution of strength is as curve 'A', because some failure modes have not been considered. This is a myth as the true strength boundaries are still the dashed lines. The strength might be reduced, on the basis of the myth, believing that this is a more economical use of materials.

The operator might believe that higher loads could be handled. In a climate of success, overconfidence grows and changes are made that

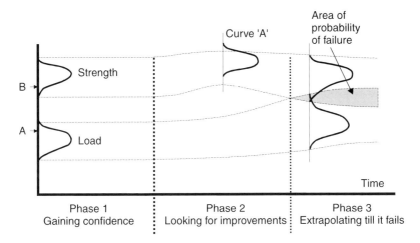

Fig. 2.2.1 **Speculative development cycle**

lead to phase 3, where the parties blindly take increased risk. The overlapping distribution of strength and load gives a probability of failure – a failure waiting to happen.

This pattern appears so common that one might believe that engineers have a Darwinian drive to evolve to failure! Instead of the species 'engineer' becoming extinct, it is the product that dies. The engineers, maybe with clipped wings, survive to produce the next product. Learning from failures is one means of breaking this cycle.

The timing of this type of failure cycle is variable. It depends on the pressure for 'improvements', which will most likely come from the competitive commercial climate.

Lessons from cycles

Cycles can lead to failures in several ways. There can be:

- a failure to spot trends from social and economic cycles. While there are no examples in Part 1, there are many examples of bankruptcies from this type of failure;
- a failure from lack of memory or knowledge, from previous generations;
- a product-development cycle failure, due to overconfident extrapolation until it fails;
- a belief that effects can be extrapolated linearly – but most effects have feedback loops that produce oscillating results. Once an effect becomes too pronounced, additional factors will

enter, providing compensation, and bring the effect back within reasonable limits.

2.2.4   Does history matter?

Pugsley and Petroski have argued the case for the importance of history for engineers. And CAIB (Columbia Accident Advisory Board) said – *History is not just a backdrop or a scene-setter. History is cause.*

History can help us recall and remember past failures so that they are not repeated. And history can tell us the how and why of process and technological evolution.

Lessons from case studies of past failures can help our understanding, even in the most advanced areas of engineering. Can more be learned? Engineering errors stem from human mistakes and failings. If the only history we read is a list of technological facts set in a time frame, then we miss the human aspects. There are a few biographies of prominent engineers, and these can add another dimension to our knowledge. They can help us understand how engineering decisions were made and how experience and judgement contributed to their decisions. However, the evolution of engineering is strongly influenced by its craft origins. Mechanics, technologists, and skilled workers all had a significant role. To fully appreciate the history of engineering, their contributions should be included.

History can help us comprehend the complex interaction between the developments in science and mathematics, and their application in engineering activities. For example, academic studies and advancements in chemistry during the nineteenth century in Germany led first to applications in commerce and then to the establishment of a strong chemical industry.

The examples in Part 1 show that the climate in which the work and decisions took place has had an enormous impact. The evolution of the UK rail industry started with a huge contribution from mechanics. It then progressed through the aggressive and laissez-faire attitude of the Victorian era. Later, it was nationalized in a climate of socialism and privatized in an era of free-market promotion. Did these changing climates have more impact than the technical evolution of the engineering?

The UK railways are compared with the European high-speed systems. To what extent has the culture and networking capabilities of the French polytechnicien ensured a co-ordinated approach across industry, government, and academia that keeps the French vision on track? They have managed to generate a national commitment among all players, including the public, that the TGV approach is right for France. There is nothing to show that French railway engineers are any different from those in the United Kingdom, yet the British railway industry is fragmented and subject to divisive and everlasting debate. In another context, British and French engineers, coming from different cultures, were able to successfully produce Concorde. The Japanese have quietly got on with the development of their rail system. Problems in technology, organization, finance, and management have all been resolved. What can we learn from these histories? Unfortunately, history shows us that engineers as well as industry leaders are reluctant to learn from the experience in other cultures and countries.

The history of engineering is multifaceted. So many disciplines are involved that it is hard to suggest who should be writing the histories. Should it be historians, economists, psychologists, or even engineers? An understanding of, and sympathy with, all the disciplines is needed. There are few examples of engineering history texts. One notable exception is the book by Stephen Timoshenko's *History of Strength of Materials* written in 1952 (1). He traces the progress from Galileo onwards. And he shows how the developments in mathematics and from different nations and industries have all contributed. He adds a human element by including a brief biography and small picture of all the leading players.

Maybe engineers should be lobbying to have engineering history studied and written about. It could then take its place with general history and political and military history, rather than being just a footnote in the history of science. Elevating the profile of engineering history may not only help engineers understanding but might also provide a larger audience with a better comprehension of the place of engineering in society.

Reference

(1) **Timoshenko, S.P.** (1953) *History of Strength of Materials*, McGraw-Hill, reprinted by Dover in 1983, ISBN 0 486 61187 6.

2.2.5 Learning from the military

It became popular in management studies to look for ideas from military experience. One example, for developing goals and strategies, is *Sun Tzu, War and Management* (1). Can engineers learn any lessons by studying military failures? Some answers may be found in the excellent book *On the Psychology of Military Incompetence* (2) written by a former officer in the Royal Engineers, Norman Dixon. The summary of Dr Dixon's research, with the possible engineering management equivalents in brackets, is that military incompetence can be characterized by:

- *an underestimation, sometimes bordering on the arrogant, of the enemy (competition, customer);*
- *an inability to profit from past experience;*
- *a resistance to reconnaissance (finding out relevant information) coupled with a dislike for intelligence, in both senses of the word;*
- *a tendency to reject or ignore information;*
- *a clinging to outworn tradition;*
- *a lack of creativity, inventiveness, and open-mindedness;*
- *a serious wastage of human resources;*
- *passivity, procrastination, indecisiveness, and a tendency to abdicate from the role of decision-maker;*
- *a tendency to lay blame on others;*
- *a belief in mystical powers – fate, bad luck, and so on.*

Most of these points have parallels for engineering management. The weaknesses to watch out for, and correct, are similar.

Dr Dixon found that, in most cases where military leaders failed, they had been chosen not for their prowess or intelligence, but because they conformed. He also found that the personality trait of authoritarianism often contributed to incompetence. Many social studies have shown authoritarians are more dishonest, more irresponsible, more untrustworthy, more socially conforming, and unlikely to make successful leaders.

Douglas McGregor (1906–1964), Professor in the School of Industrial Management at MIT in the 1950s, thought that authoritarianism would die out in management. Unfortunately, personality traits are not so easily killed off, and the announcement of its demise in management was premature.

References
(1) **Wee, C.H., Lee, K.S.** and **Bambang, W.H.** (1991) *Sun Tzu: War and Management, Application to Strategic Management and Thinking*, Addison-Wesley, London, ISBN 0 201 50965 2.
(2) **Dixon, N.F.** (1976) *On the Psychology of Military Incompetence*, Jonathon Cape, ISBN 0 7126 5889 0.

2.2.6   Maintenance holiday – a familiar story

The scene:

Annual budget meeting between a maintenance engineer and manager.

Manager: *I've read your business plan – you've done some great work. However business is very competitive, and the company needs to conserve cash. I'm sure you can find ways to meet your goals on a smaller budget.*

Maintenance engineer: *I presented a realistic plan. . . .*

Manager: *Yes, but I'm sure you could delay some work and reduce this year's cash flow needs.*

Maintenance engineer: *mmm. . . (reluctantly) OK, but there will be some risk.*

Same office, one year later.

Manager: *You did a great job last year. We had no major problems and you saved the company much needed cash. Unfortunately, as you know, we are still in a tough business environment, and we again need to minimize budgets.*

Maintenance engineer: *I know the business scene, but some of the maintenance is becoming critical. We cannot keep postponing the work.*

Manager: *Will it fail if we delay another year?*

Maintenance engineer: *I can't be precise, but you are taking an increasing risk.*

Manager: *I hear you, but that is what we are up against these days.*

COMMENTS

Does this story sound familiar? How often is it repeated in both public and private sector? Not only is the maintenance delayed but also the maintenance engineer becomes de-motivated. As this scenario continues, the responsibility moves from the maintenance engineer to the manager – although there is usually no formal recognition of the transfer. The responsibility probably falls in limbo.

Another aspect of this *maintenance holiday* syndrome is the lack of understanding of the real maintenance requirements. The designer may just specify the best guess at the maintenance needs. Does the designer have enough feedback from operation and maintenance experience to make recommendations? Should test work back up the recommendations?

Failures due to delayed maintenance are statistical in nature. One answer to the budget pressure is to be able to quantify the probability of failure. Then this risk can be firmly placed on the shoulders of the person cutting the budget.

While this story is told from the maintenance perspective, a similar conversation can take place for other aspects of engineering. Again, the danger is to leave the responsibility for performance unclear.

As a young engineer, I was struggling to keep complex fuel handling machinery in operation. An experienced engineer told me

> *Treat your machinery like a lady. Never turn your back on her. Listen for her every need and complaint. Respond immediately. Never think of penny-pinching – it will cost you more in the end – and give you hours of torture. Be sympathetic, learn to watch and listen for her signals. Never ignore her, don't get frustrated – just look after her.*

Good advice, but not politically correct today! Many engineers have coaxed machinery to perform with this approach. Unfortunately, this approach is not easy to use in a discussion with a financial officer.

2.3.1 Peer reviews

A common view in business is that if you open your operation to external inspection, you will reveal your dark secrets and give away your best ideas. Experience with peer reviews of nuclear plant operations gives a different perspective. The impetus came from two dramatic accidents, Three Mile Island and Chernobyl. Following these accidents, the Institute of Nuclear Power Operations (INPO) and the World Association of Nuclear Operators (WANO) were formed.

Institute of Nuclear Power Operations (INPO)

President Carter appointed the Kemeny Commission to investigate the accident at Three Mile Island, Unit 2. In parallel with the Commission's investigation, the nuclear electric utility industry founded INPO in 1979. The mission of INPO is *to promote the highest standards of safety and reliability – to promote excellence – in the operation of nuclear electric generating plant.*

What is INPO?

INPO is an independent technical organization funded by its members. All 28 US utilities that operate nuclear plants have become members. An international program brings in nuclear organizations from 11 other countries. A supplier participation program involves 11 nuclear suppliers, construction and engineering companies. INPO is a non-profit organization and does not lobby or promote nuclear power. INPO maintains a private web site accessible only by its members.

INPO's main programs

The main programs of INPO are:

- evaluations;
- training and accreditation;
- event analysis and information exchange;
- assistance;
- peer reviews through on-site evaluations by experts.

A set of eight performance-based indicators is used to objectively check performance. Industry targets are set every five years for each indicator.

What have these programs achieved?

There is a significant improvement in the availability of information and in the capability of staff through training. Symbiotic with this is the growth of a culture where operators are committed to safety and reliability. This has been achieved through peer pressure, openness, knowledge, and discipline.

All the performance indicators show great and continuing improvement. At the same time, the availability of the power plants has steadily increased.

World Association of Nuclear Operators (WANO)

WANO was formed in 1989 following the Chernobyl accident. It provides a similar role worldwide that INPO provides in the USA. Every company in the world that operates a nuclear electricity generating plant has decided to become a member of WANO. Like INPO, WANO is independent of commercial and government ties and is not for profit. Its only interest is safety. WANO has centres in London, Paris, Tokyo, Moscow, and Atlanta. The Atlanta centre links with INPO.

WANO's Mission is *to maximize the safety and reliability of the operations of nuclear power plants by exchanging information and encouraging communications, comparisons, and emulation amongst its members.*

Peer reviews

On-site peer reviews are conducted by a team of 12 to 15 specialists over a period of two weeks. Teams, formed for each review, comprise operating experts from other utilities. *Performance Objectives and Criteria* have been established. Review teams are briefed on these before carrying out a review so that there is consistency.

Peer reviews focus on ten functional areas that correspond with the areas needed to safely and reliably operate a nuclear plant. The functions are:

- organizational effectiveness;
- operations;
- maintenance;
- engineering support;

- radiological protection;
- operating experience;
- chemistry;
- training;
- qualification; and
- emergency preparedness.

A second group of performance objectives relates to the characteristics of an organization that cross organizational boundaries and apply to the whole workforce. They include:

- safety culture;
- human performance;
- self-evaluation (learning organization);
- industrial safety;
- plant status and configuration control;
- work management; and
- equipment performance and condition.

Following the on-site visit, the review teams produce a confidential report that identifies both strengths and weaknesses. It is then up to the utility being reviewed to take action.

Peer reviews are initiated by the utility by making a request for a review. In the United States, INPO performs inspections on each member's plants every 18 to 24 months. By 2003, WANO had conducted reviews on 90 per cent of all stations. This included all power stations assigned to the Moscow Centre. WANO aims to have every station reviewed by 2005.

COMMENTS
Can good performance hurt?

In a 2003 edition of 'Inside WANO' (**1**), it is noted that after over 10 years of good operation on a station there are increased risks of problems. Maybe teams have started to believe in their own success – leading to overconfidence. They found that warning signs were present, but were missed or overlooked. WANO is addressing this issue and aims to improve operational decision-making and find out how to maintain good performance over the long term.

Confidentiality versus public availability of peer reports

Anti-nuclear groups have criticized the INPO and WANO Peer Review processes because the results are confidential. They see this as secretive, and as hiding the truth from the public. On the other hand, INPO and WANO claim that confidentiality assures full and open discussion between the participants. The reports can be highly technical and use the language and buzzwords of the specialists. Making them available to a broad non-technical audience would require major editing to avoid misunderstandings. However, the public can be well-informed on the safety of a nuclear plant by reading the Regulator's reports, which are readily available.

Opportunities for team members
Participation in a peer review gives the opportunity to stand back from day-to-day activities and think about the fundamentals involved. It can give team members the chance to improve their own performance by seeing at first hand good ideas and practices, as well as helping others to improve.

Peer pressure
It is a great credit to both INPO and WANO that every nuclear utility in the world has voluntarily joined their organizations and are prepared to submit themselves to objective reviews of their performance against the best worldwide practice. INPO and WANO have created a peer relationship between themselves and their members as well as a peer relationship between team members and the staff on the plant being reviewed.

Peer reviews can create feelings of unease at being *audited* and concern at having one's faults exposed. Discussion between peers can also become competitive – *Who knows best?* or an aggressive – *I'll show you.* INPO and WANO have successfully created an environment where these psychological blocks are minimized.

Sharing information and knowledge
Many industries have regular conferences that present developments and experiences in that industry. While the papers at these conferences are promoted as sharing experiences, they are usually selective in the problems described. INPO and WANO appear unique in having created a culture of openness, and the full sharing of knowledge and learning between operating staff.

Could the nuclear peer review experience be used elsewhere?
The nuclear experience is based on a full commitment to in-depth reviews and exchange of information. It is based on a very disciplined, independent, and non-bureaucratic approach. These factors appear to have led to its success. It would be difficult to implement this process

with competing product manufacturers. However, there are industries where it could be applied, and not just for safety improvement but also for performance. Examples are railways, municipal waste management and water works. A valid question to ask would be – *Is it necessary to adopt such a rigorous program?* Any watering down of the discipline applied would probably significantly reduce the benefits. Hence, anyone thinking of using peer reviews should be prepared to go *the whole hog*.

2.3.2   Lesson learned

Peer pressure, peer recognition, peer sharing and learning from peers can be powerful tools to improve safety as well as to improve performance.

References

Information on INPO and WANO is available from their Annual Reports. Some information is available on the open portions of the WANO web site www.wano.info.

(1) Harrison, M. (2003) Lessons learned from recent events. Can good performance hurt? *Inside WANO*, **11**(3), 3.

2.3.3   Standing Committee on Structural Safety (SCOSS)

Following a few, but prominent, structural failures in the United Kingdom during the 1960s and 1970s, a Study Committee on Structural Safety was set up under the chairmanship of Sir Alfred Pugsley. Their report proposed the establishment of a permanent committee. The Standing Committee on Structural Safety (SCOSS) was formed in 1976. Today, SCOSS is an independent body reporting to the Presidents of the Institutions of Structural Engineers (IStructE) and Civil Engineers (ICE). SCOSS is supported by IStructE, ICE, and HSE.

The prime function of SCOSS is to be a watchdog – to identify in advance the trends and developments that might increase risk, and to make recommendations. The Committee has typically around 20 senior members of the profession drawn from industry, design, and academia. Membership is rotated. Topics for consideration arise from various sources – the Committee members, discussions the Committee has with the industry and government, and from confidential feedback from industry workers.

SCOSS publishes a report biennially and from time to time issues papers and bulletins. The Committee communicates its concern to those it believes can take action. Through its reports, papers, and bulletins it aims to communicate with a wide audience. While SCOSS is primarily concerned with UK activities, it believes that worldwide trends can be instructive. It has formed links to find out what is happening internationally.

SCOSS (**1**) publications are on the web site www.scoss.org.uk.

Some of the information, recommendations, and views relevant to a wider engineering audience are summarised below. The numbers in brackets [ ] refer to the SCOSS report that was the source.

Assessment of safety and risk at the design stage [10, 11, 14]
The control of risk depends primarily on the competence and integrity of individuals and organizations. SCOSS promotes a systematic approach to safety starting at the design stage to:

- identify hazards and define performance standards;
- quantify risks;
- use design audits;
- where appropriate, bring in independent assessment.

Formal risk-assessment and goal-setting procedures should be introduced. Designers should consider risks from a holistic, whole-life view. While this has been a requirement for some time, it is often not well done. *Life-care plans* should be started at the initial design stage. They tell the owner the basis of the design and the assumptions made in the design for inspection and maintenance.

Codes of practice, standards, guidelines, and regulations [1, 5, 8, 10–14]
These play an important role in communicating information on the best and safe practices. One issue is that the plethora and cost of these documents make it very difficult for a busy engineer to keep himself up-to-date.

There is often no clear definition of the purpose and role of the documents. They can be multi-purpose – for design guidance, contractual purposes, specification, provision of data, regulation, litigation, and so on. They speak with many different voices to many different users.

A new generation of codes is emerging that are more complex and comprehensive than those they replace. Should codes be written as aide mémoires highlighting pitfalls while design recommendations and formulae, which are updated more often, could be in handbooks? There is a need to converge on a single set of codes that clearly distinguish between performance requirements, principles, and rules.

Slavish adherence to codes can lead to hazards, which are not covered by the code, being overlooked. Codes are written for knowledgeable individuals. Despite using codes, designers still have the final responsibility for their designs and for using the right data. Codes are not intended to be quasi-legal documents to find the least onerous solution.

Shortcomings in codes should be addressed with urgency and obsolete codes promptly withdrawn. Users must recognize that there may be gaps in codes and that they may not cover recent knowledge and innovation.

Codes and Standards are gravitating from National to International. This brings in a broader base of input and knowledge but may create risks due to the variations in the understanding, training, and practices in different countries. Structural Eurocodes will be mandatory from 2007. UK engineers need to be prepared for this fundamental change. A co-ordinated implementation plan is needed – involving all concerned parties – with strong leadership. ICE has created a helpful web site, www.eurocodes.co.uk.

SCOSS is concerned at the declining representation on code committees from practising engineers and staff from government agencies. This becomes more critical as international standards are mandated. Without representation, local and national knowledge and requirements will not be heard.

Governments should support initiatives aimed at simplifying building regulations. Legislation has grown over many years in a piecemeal fashion – in uncoordinated statutes.

Robustness [10–12]
Robustness is one of the basic tenets of good structural design, which is less amenable to a simple set of rules. The aim is to provide robustness to protect against any conceivable problem.

Disproportionate damage [4, 10–12]
Structures should be designed to avoid subsequent and disproportionate harm following damage due to accidents, misuse, or exceptional circumstances. Examples are:

- the collapse of a high-rise flat due to a gas explosion caused by occupants replacing expensive electric heating with bottled gas heaters;
- the potential for progressive collapse of multi-storey car park structures and stadia;

- bridge bashing – the potential for a serious accident after an earlier accident that damaged a road or a rail bridge;
- protection against terrorist actions.

SCOSS recognizes that economic pressure can lead to reducing redundancy. It recommends that steps be taken to:

- avoid, eliminate, or reduce the hazards;
- select a structural form that has a low sensitivity to the hazards considered;
- tie the structure together.

SCOSS recommends that the fundamental property of resistance to disproportionate damage be required by regulation for all building structures.

Use of computers [6, 9, 10–14]

Computer hardware and software now dominate the analysis of structures and have given great benefit, but there are some concerns and questions:

- are the programmes and results adequately proven and checked?
- is the programme used in context?
- do the structural engineer, analyst, and software supplier all understand each others requirements and limits?
- is there a trend to rely on computer structural analysis generated by people with inadequate structural engineering knowledge or training? Powerful analysis software needs skilled users.

Software now makes it easy to spot the over-designed areas and see redundancies. 'Optimizing' the design under commercial pressure can result in traditional conservatism and redundancy being removed, which may result in greater risk.

SCOSS recommends the production of guidelines for the use of computers in structural engineering to help engineers, managers, and educators. IStructE has published *Guidelines for the use of computers for engineering calculations.*

Planning for inspection [1–3, 7–12]

Two areas of inspection have raised concern – the attachment of cladding on buildings and the inspection of tension members in bridges.

An architect often specifies cladding, but there is need for a structural engineer to get involved to ensure that the fasteners are adequate and that inspection is identified. Designers and builders often fail to inform a building owner of the need to inspect and maintain cladding. There is reluctance amongst all parties to take any action until failures occur. The structural engineer should make provision for inspection and repair, particularly for fasteners.

Tension members, holding-down bolts, cables, tie rods, and ground anchors in both buildings and bridges are too often in areas difficult to inspect. Designers should recognize the need for inspection, for the provision of tools to inspect hidden components, and for telling users about potential corrosion issues.

Falsework: full circle? SCOSS topic paper October 2002

The rush of activity and innovation in the UK construction industry in the late 1960s and early 1970s broadened the boundaries of design, materials, and scale of projects. These developments pushed the complexity of temporary support or falsework, and some failures occurred during construction. The concern was that falsework was not well regulated, and a committee was established under Dr S.L. Bragg to investigate. Subsequent to the Bragg Report, the British Standard BS 5975, the Code of Practice for Falsework was produced in 1982.

Changes since the 1970s

Some fundamental changes have occurred in the industry. The principle changes are:

- in the 1970s, most main contractors designed their falsework in-house. Now, most of the work is subcontracted;
- management contractors have largely replaced main contractors;
- proprietary falsework systems now dominate the market and design skills lie with the specialist organizations;
- a gradual loss of the traditional skills of the site foreman with a lifetime experience of what works;
- procurement maximizes commercial benefits with little regard for the flow of information;
- a harsher commercial climate exists.

Superimposed on these changes is the phenomenon known as *collective amnesia* whereby one generation often forgets, or is unaware of, the lessons learned by the preceding generation.

Current concerns

The HSE commissioned an Investigation into aspects of falsework, report 394/2001. They found:

- a lack of understanding of the basic principles and stability, at all levels;
- that wind loads are rarely considered and, if considered, tend not to be to the relevant BS;
- a lack of clarity in design briefs;
- a lack of adequate checking and a worrying lack of design expertise;
- a lack of accuracy in erection.

SCOSS conclusions

SCOSS concluded that without change it is only a matter of time before a serious accident occurs. Its key concerns are:

- competency of the designer;
- sufficiency of data;
- adequacy of supervision;
- competency of erection staff and supervision.

Recommendations

- Manufacturers of proprietary systems should ensure that clear and verifiable information is available for designers.
- Design staff has to be competent, well experienced, and should use CPD to attain contemporary knowledge.
- No design should be undertaken without sufficient data
- A chartered engineer with the requisite experience should review the design.
- Contractual documents should make it clear as to who is responsible for providing or determining the data needed for design.
- Training courses should address the concerns.
- HSE should review the provision of information and should consider producing an Information Sheet. It should also consider whether a more formal 'Guidance' is desirable.
- Those involved in falsework should re-read the earlier Bragg Report.

Despite these concerns, SCOSS recognizes that proprietary falsework systems have made invaluable contributions to the greater efficiency of the construction industry.

Summary comments from various SCOSS reports

- In many cases, failure results from the lack of application of existing knowledge, from divided responsibility, or poor communication. Wider publication of case histories on failures and risks would help towards a better understanding by all concerned. There is a continuing need to improve training and education at all levels [1].

- Increased expenditure on design and construction control might well be cost effective. Experience in bridge design suggests that the increase of about 10 per cent in total design costs for review and checking has been worthwhile in identifying mistakes. Independent checking might justify the use of lower safety factors [2].

- Problems arise where warnings must be issued to owners of structures at risk. It is very difficult to give the appropriate remedial measures sufficient publicity, without causing public concern through misrepresentation by some sections of the media [2].

- In 16 of the 24 topics considered in the first three years, there was a clear need for better communication, education, care or management. Such observations, even though they are most certainly accurate, are prone to inaction because of their generality [3].

- There is a tendency for contractors to criticize designers and vice versa, rather than working closely together [3].

- Improvement in safety requires constant publicity and education [4].

- A major cause of structural failure is human error. This leads to considering help from other fields such as psychology and the behavioural sciences [5].

- Legal and insurance considerations have prevented professional engineers from having technical discussions with their peers on matters of safety. This has resulted in difficulty in giving authoritative statements to the clients and the public. There should be no obstruction to communication or discussion between professional experts [6].

- The importance of regular and effective inspection of important structures cannot be overemphasized [7].

- Most failures are the result of more than one factor. One group occurs too frequently – the combination of the failure by the designer to appreciate the severity of actual climatic and aggressive conditions during the structure's lifetime and the use of inappropriate designs or choice of materials [8].

- Before assuming substantial responsibility, engineers need to assimilate the past experience of the profession. They need to develop an in-depth appreciation of safety and the hazards. A competent person will have an awareness of the limitations of their own experience and knowledge [10, 13].
- Complacency can preclude recognition of increasing risks. There is a strong tendency amongst those in government and others who are responsible for structural maintenance and procurement resources to make the comfortable assumption that all is well and will continue to be well even if resources are reduced [12].
- There is a natural but not inevitable tendency amongst engineers towards collective amnesia concerning structural failures of the previous generation and the lessons learnt from them. Over-confidence in more powerful design tools and over-optimistic extrapolation are constant temptations militating against caution [12].
- Where previously unknown structural behaviour is observed, whether failure has occurred or not, it is incumbent upon professional engineers to report the observation in the technical literature, so that others are alerted to the potential risks [13].
- The identification of dynamically sensitive structures, and understanding that their behaviour may not be well covered in an engineer's education [13].
- The consequences of climate change on structures should be regarded as a national and international issue. Changes should be quantified by continuous monitoring and analysis [13].
- Greater enforcement of safety regulations is probably unrealistic because of limited budgets and resources. The level set for insurance premiums could help the adoption of better practices [14].
- Who is responsible for reviewing buildings that were designed to now out-of-date codes [14]?
- The key to managing risk is to keep a clear vision [14].
- Risk assessments focus on the hazard to life and limb, but there are other risks to be considered – to image, profitability, integrity, and the sustainability of the industry [14].
- Most advice on risk management relates to large projects. Information and guidance should be made available for more routine engineering [14].

COMMENTS

SCOSS is effectively highlighting potential problems, not only for civil and structural engineering but also for other engineers. Its reports should be read by all engineers and kept close at hand for easy reference. Underlying most of the reports is the fact that failures are caused by human action, or inaction.

SCOSS has been able to steer a fine course between giving alerts for potential problems and at the same time not inhibiting developments in the industry. It is conscious of the need to avoid spreading undue panic while at the same time using its status and collective power to get the messages heard and hopefully acted upon at the appropriate level without the need for an accident to occur.

The review of falsework or scaffolding may appear to have little relevance to mechanical engineering. Closer reading shows many potential lessons for all engineers. The commercial practices for handling many areas of subcontracted design or components are changing. The impact on safety and risks might be missed.

The usefulness of the SCOSS concept raises the question of whether it should be used in other branches of engineering. If a similar committee had been reviewing mechanical engineering, would the problems of the UK railway industry have been highlighted earlier?

While SCOSS have been effective in raising issues, it notes that clear action has been taken on only 15 per cent of its recommendations. Half have variable to poor attention, and 35 per cent have had no action. The 14th SCOSS Report directed recommendations to either *practitioners* or *influencers*. To assist them in taking action the 14th Report has a checklist in the Appendix. Like SCOSS, most engineers will recognize the effort and struggle required to clear deficiency lists at the end of a project. Tidying up is a drag – but once completed, gives a sense of achievement.

The 14th (2003) SCOSS Report concludes that *much has yet to be done* – the same words used by Pugsley in his 1966 book on the safety of structures!

Lessons learned

- A senior peer group looking for potential problems in an industry can be very effective.
- Always know where the knowledge base is and that it is adequate for the task.
- Beware of industry changes that affect the knowledge base and methods of working.
- Changes in the industry or organization can slowly erode capability and knowledge without the impact being noticed.

- Despite advances, new risks can emerge. There is a need for constant vigilance to spot these trends in order to help avoid failures.
- Humans cause failures.

References

(1) *SCOSS*, 11 Upper Belgrave Street, London SW1X 8BH, United Kingdom.

2.3.4   The Hazards Forum

The Institution of Civil, Electrical, Mechanical, and Chemical Engineers founded the Hazards Forum in 1989. The Forum was established because of concerns about major disasters. Although originally focussed on engineering risks, the Hazard Forum's agenda has broadened to include other areas of risk. It aims to:

- *promote the public understanding of risk;*
- *promote the understanding of specific technological and natural hazards;*
- *identify key lessons from catastrophes; and*
- *work for the application of risk reduction and control strategies.*

It provides a multi-discipline interdisciplinary independent and impartial focus, promotes debate, and exchange of ideas at meetings, through its *Newsletter*, web site, and networking – and promulgates lessons learned.

The Hazards Forum has since grown and is now a registered charity and an independent organization with 65 members from a wide range of interests including engineering and other institutions, academia, regulators, companies, and private individuals.

A working group of the Hazards Forum produced a report in November 2000 entitled – *Public Understanding of Risk – An agenda for Action.* As a result, four evening events on risk-related topics are held each year. Past topics have included:

> *UK Government Handling of Risk; Regulation of Science; Risk and Inequality; Climate Change and the Private Sector; Social Amplification of Risk; The Board and Corporate Risk Culture; Risk Education and others.*

Three recognized experts speak briefly at each event and the remainder of the time is spent in debate and discussion. A summary of the event including the discussion is available on the Hazards Forum web site and is also published in the quarterly *Newsletter*.

Another recommendation from the working group was to establish a web-based database on Risk Analysis and Information. The *Newsletter* includes a useful reading list on risk and a calendar of related events.

COMMENTS

The Hazard Forum is a good source of information and provides an excellent opportunity to network with experts. Producing reports of evening events including summaries of the discussions allows the results to be known to a much wider audience than those able to attend in person. Elsewhere, reports on many good presentations to learned societies are unfortunately available only to the few people who attend. Even if a report covers the talk, it is seldom that the discussion is recorded.

Reference

(1) www.hazardsforum.co.uk.

2.4.1   The problem of probabilities

Chance favours the prepared mind – Louis Pasteur.

How do you predict chance? When tossing a coin, after a succession of heads, what is your guess at the next toss?

- heads again – banking on a winning streak;
- tails – because heads and tails need to balance out; or
- either – because the coin has no memory of the previous results.

The answers, including those from engineers, are distributed between these three views, yet the only logical answer is the last.

You might claim that there is another answer – the coin could stand on its edge. It sounds pedantic, but if the coin was being tossed onto soft mud, then this answer is possible. This emphasizes how easy it is to jump to a conclusion without first establishing the boundary conditions.

In day-to-day conversation, we express surprise at coincidences. Most occur so often that statistically they cannot be considered coincidences. The danger for engineering judgement is in believing that coincidences are remote and that unlikely events are not worthy of serious review. With engineering failures being the conclusion of a series of latent faults, it is not so much coincidence as waiting until all the necessary events have accumulated.

Engineers traditionally place undue value on the result of a single test. If statistical variations are considered, the scatter is believed to be small – much less than usually exists. A logical answer is to work only with statistical data derived from multiple tests. The cost may not be justified – nor may there be enough time to carry out the tests. Hence, designers have to balance available data with judgement in assessing probabilities. This appreciation of the risks should define the testing required.

A common practice is to fit limited test data into a standard statistical distribution curve. Practical results rarely fit standard distribution curves accurately. Yet, designers often need to extrapolate the tails of the statistics to get the necessary high reliability.

Statistics can be misleading. For example, it is commonly believed in North America that large heavy cars are safer than small light cars – and the statistics appear to support that argument. However, a review conducted in Texas delved into the details. Large cars tend to be more expensive and owned by older drivers. Younger drivers mainly drive the smaller cars, and statistically younger drivers have a higher accident rate than older drivers. When this fact is taken into account, there is no difference in the accident rates for large heavy cars and small light ones. The lesson is to always question the basis of the statistics.

Probabilistic risk assessments (PRA) estimate the total risk from all failure modes. The difficulty is in knowing all the failure modes – *what you do not know you cannot predict*. Despite this limitation, PRAs give valuable insight into potential risks.

In the half century since Pugsley and Freudenthal expressed their views on safety and reliability, there has been continuous interest and development in probability methods. A US Department of Transport (DOT) Report (**1**), although written in 1999 for aircraft structures, provides a good summary for general engineering use. The Report points out that *the legal implications of probabilistic methods are unclear at this time.*

NASA has had a mixed approach to PRAs. NASA managers have said *Statistics don't count for anything – they have no place in engineering anywhere. Risk is minimised by attention to design... PRA runs against all the traditions of engineering, where you handle reliability by safety factors* (**2**). On the other hand, NASA has a 323 page manual on Risk Assessment (**3**). NASA's target for 2002 was to have the loss of a space vehicle less than 1 in 250 missions, reducing to 1 in 500 in the next decade. In their 2004 budget, the PRA target for the Crew Rescue Vehicle is 1 in 800, significantly better than the Shuttle experience but orders of magnitude less than the figures that were ridiculed by Richard Feynman.

The corollary of Pasteur's quotation is that an unprepared mind will increase the chance of failure. A sound understanding of statistics and probability methods is essential for preventing accidents. This knowledge helps avoid the quick and casual judgement that *it will never happen.*

References

(1) *Probabilistic Design Methodology for Composite Aircraft Structures*, DOT/FAA/AA-99/2, Federal Aviation Administration, 1999.

(2) **Bell, T.E.** and **Esch, K.** (1989) The Space shuttle: a case of subjective engineering, *IEEE Spectrum*, **26**(6).

(3) *Probability Risk Assessment Procedure Guide for NASA Managers and Practitioners*, Version 1.1, August 2002, on www.hq.nasa.gov/office/codeq/risk/risk.htm.

2.4.2 Robustness

The benefits of robustness in structures were highlighted half a century ago by Pugsley as well as more recently by SCOSS. The type of failure to avoid is shown in Fig. 2.4.1.

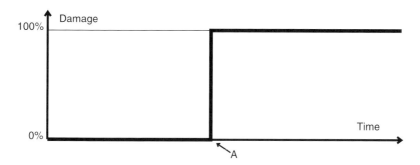

Fig. 2.4.1   Sudden failure

A sudden failure at point 'A' leaves no time for evacuation. An example of this type of failure is the bridge at Ynysygwas. What is needed is some warning signals. Figure 2.4.2 shows a more progressive type of failure.

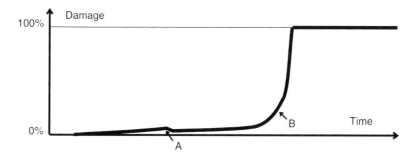

Fig. 2.4.2   Slowly developing failure

The slow start to the failure could be due to deterioration of materials or wear. Point 'A' reflects an improvement due to maintenance. If maintenance is delayed, then rapid deterioration may occur as shown at point 'B'. This type of failure may follow from a maintenance holiday. The growth of fatigue cracks follows this pattern. Most of the fatigue life is consumed with only small crack growth. Rapid growth occurs just before failure. The swamping of ro–ro ships follows the same pattern with a slow increase of roll or heel until a critical amount of water is on-board and the ship quickly capsizes.

The preferred mode of failure is shown in Fig. 2.4.3.

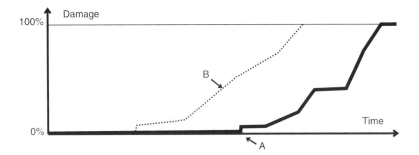

Fig. 2.4.3   Progressive failure with warning

The component or structure is robust so that the failure of a single feature, as at point 'A', does not result in the failure of the whole structure. There is sufficient time from point 'A' to the next failure for the structure to be evacuated. Preferably there are warning signs before any hazardous failure. Curve 'B' shows the occurrence of warning signs or signals. These signals may give time for preventative action, for repairs, and extra time for evacuation. But people have to take note of these signals. The collapse of the first bridge at Québec had plenty of warning signs as chords slowly distorted over days – yet they were ignored and many died.

While the emphasis on robustness comes from structural engineering, it is a concept worth adopting by a wider area of engineering. SCOSS recommends that the aim should be to provide robustness, or resistance to disproportional damage, for any conceivable eventuality. This is a much more onerous and difficult task than just providing robustness for the hazards considered in the design. Nevertheless, it is one way of giving extra protection for accidents that are unknown or have not been considered in a conventional risk analysis.

2.4.3   From fatigue to structural integrity

Fatigue is the major cause of failure in mechanical systems. Wear and fatigue compete to see which can dominate the life of components. Fatigue is still causing failures earlier than expected. This subsection looks briefly at fatigue. The simplest summary is the title of Professor Rod Smith's paper at a Fatigue Seminar (1), *Fatigue: A continuing problem despite its long history.*

History of fatigue studies

Studies of failures (2–4) in highway coaches and rail axles in the nineteenth century showed that repeated cycles of load could result in breakage. The first major accident caused by fatigue occurred on the Versailles Railway in 1842. The axle of the leading locomotive of a two-locomotive excursion train broke. Over 60 died as the train derailed and caught fire.

The first clear understanding that fatigue failures could occur at much lower loads than static overload failures came from tests organized by a Commission appointed in 1848. It conducted tests at Portsmouth, which showed that iron bars under repeated loading might carry only one-third of their static-breaking load without failure.

The evolution of engineering design has resulted in a more efficient use of materials and higher stresses. This keeps fatigue as the major life-limiting mechanism. Prominent fatigue accidents have been the crashes of Comet aircraft in the 1950s and, more recently, the structural failure of the North Sea oil platform, Alexander Kielland, and rail fractures.

Fatigue failures start from the details

From the early studies of fatigue, Wöhler (1819–1914) noticed the effect of large stress variations at sharp corners. Mechanical components are littered with stress concentrations in details such as bolt-holes, fillets, keyways, and welds. By the mid-twentieth century, R.E. Peterson had produced a reference book of stress concentration factors (5). But to understand the mechanism of fatigue, it is necessary to look in even more detail – right into the microstructure of the material.

Cracks start fatigue failures

Fatigue failures start from small cracks, often from defects occurring at the surface. Hence the importance of surface finish from manufacture and from surface corrosion. It is now accepted that cracks develop at the microstructure level because of local shear stresses and then grow in tensile stress fields. Cracks can start at material defects, like inclusions. Cracks can start in one grain and slow down when they bump up against the next grain boundary. Cracks can coalesce. While the question of how the cracks start is still in doubt, it is best to assume that polycrystalline materials have natural cracks of the order of the grain size. Given the right conditions, these can grow into fatigue cracks.

Fracture mechanics

The development of fracture mechanics has provided a means of predicting how cracks grow. As the technology has matured, it has looked in even more detail into the structure of the material. As a result, fracture mechanics has evolved into three regimes: linear elastic (LEFM), elastic plastic (EPFM), and microstructural (MFM) fracture mechanics. The typical crack size covered by these regimes is shown in Fig. 2.4.4 (**6**). LEFM has been around the longest. The larger the crack, the easier it is to inspect. The greater body of experience with LEFM helps assess and quantify the potential growth of large cracks. At the other extreme, MFM is in its infancy and its application probably requires university–industry research collaboration.

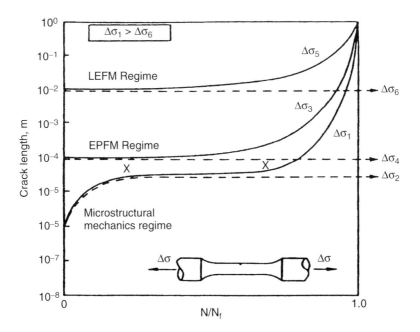

Fig. 2.4.4   Fracture mechanics regimes

In Fig. 2.4.4 N is the number of cycles and N_f is the number of cycles to failure. Crack growth accelerates towards the end of life, leaving less time for action once a crack has been detected.

It is essential to know the local stress at the crack tip in order to estimate fatigue life using fracture mechanics. This information can come from finite element analysis together with stress intensity factors. Stress intensity factors have been catalogued for a wide range of crack geometries (7).

Finite element analysis (FEA)

At first sight, finite element analysis is a mathematically precise calculation. But the nature of fatigue failure drives the need to know the stress distribution in fine detail, and in three dimensions. Hence FEA has become the art of determining the best node distribution. The drive towards finer and finer detail puts pressure on both the analyst and the developer of the computer programmes. But more detail is useless if the boundary conditions are not well defined. The application of FEA to fatigue predictions is the subject of debate. The prudent approach is not to assume FEA is the panacea but to use it as a complementary means for analysing and assessing experimental data and designs.

Fatigue design approaches

There are two main approaches to fatigue design:

- design for no crack growth. In other words keep the stresses, even in the fine details, below the level at which cracks will propagate; or
- design for cracks. Accept that cracks will grow, but keep the crack growth within bounds so that they will not lead to failure before the expected life of the product. This approach is often taken while simultaneously ensuring a *damage tolerant* design, where the component will not fail if a crack develops more rapidly than estimated. An inspection scheme is usually part of this approach so that the crack growth can be monitored during service.

Fatigue design needs to know the stress in detail. This in turn requires that the loads and duty cycle of the complete product or structure are known. This sounds easy, but in practice it is not. One design of aircraft may be used in vastly different locations, with different patterns of utilization. Careful assessment of possible mission mixes is essential to setting realistic design goals and to avoid over-design (8). Similarly, oil rigs designed for use in the Gulf of Mexico have been used in very different conditions in the North Atlantic, off the Canadian coast. Power stations designed for base load are used to load-follow. On top of these variations is the question – does the designer really know all the loads and vibrations that a component will be subjected to in-service.

In addition to the design loads, the fatigue life is impacted by residual stresses from manufacture, or high loading beyond yield from an accident. These are not easy to predict or quantify.

Fatigue issues for some common design features
Gears

The design of gears for routine use is well covered by a variety of standards. For more highly loaded and state-of-the-art gearing, the technology is less well established.

The fatigue failure modes for gears result from bending stresses and contact stresses. There are four common modes of fatigue failure:

- root bending, which results in the breaking off of the tooth;
- macro-pitting close to the pitch line and initiated subsurface;
- micro-pitting with very fine pits; and
- flank bending where the failure starts near the point of highest contact stress. It is often associated with micro-pitting.

Gears generally operate at high stresses and so the time from crack initiation to failure is short. Hence, the design approach is to keep stresses below the level for crack initiation. The difficulty with this approach is that the service loads are not often known accurately enough. Experience and judgement are needed to define the likely service conditions. The calculation of stresses can be complex. It has to be in three dimensions, and needs to include an assessment of geometric variables like manufacturing tolerances and any misalignment.

The current situation is summed up by Dieter Hofmann (9):

the fatigue strength of gears was fairly well understood when failures were predominantly root bending and macro-pitting. The improvements in gear steel and modern EP oils have resulted in micro-pitting becoming the most common mode of fatigue failure. This is poorly understood and no fatigue analysis is yet possible. The current design codes for gears are inadequate in terms of the accuracy of stress analysis for bending and contact stress, the failure to address the problem of micro-pitting, and in the poor definition of gear fatigue strength. In short, considerable research is still required to put the fatigue design of gears on a sound footing.

Splines

The evolution from keyways to splines allows higher loads to be transmitted through shafts. A high-performance application of splines is in the connecting of the shafts in aero-engines. A paper on this application by Dr Leen (**10**) shows how a good understanding of the failure mechanisms and factors affecting lifetime predictions can be obtained from a well-planned research programme.

The programme started by using small-scale tests to gain knowledge of the fretting mechanism. This allowed an initial analysis of fretting initiation models and the impact of slip amplitude, coefficient of friction, and contact pressure. Detailed FEA was made of the spline geometry being investigated as well as of the simplified 'representative' test samples. This ensured that both would have similar pressure distributions and slip amplitude. Various fatigue life prediction methods were assessed to see which would best fit the conditions on the spline. All this preliminary work gave a good background so that the test results from the 'representative' samples could be understood with good confidence. Finally, spline couplings were fatigue-tested.

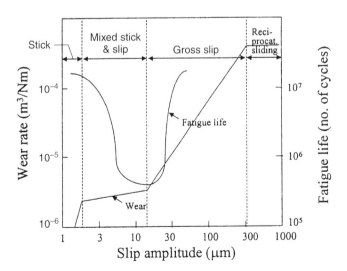

Fig. 2.4.5   Representation of typical fretting fatigue and wear (from Vingsbo & Soderberg, Ref. (11) ©, reproduced with permission from Elsevier)

The results from the whole programme give a sound basis for the designers to understand and be able to predict the fatigue failure boundaries of the splines under different loading conditions. The correlation between the spline tests and 'representative' samples allows for the evaluation of different materials and surface treatments to be made quickly on 'representative' samples.

Figure 2.4.5 shows how wear and fatigue life is affected by slip amplitude. At the change from mixed stick and slip to gross slip, the fatigue life is reduced by nearly two orders of magnitude. This is probably because there is insufficient wear to remove the embryonic cracks.

This example shows how the failure boundaries of an apparently simple engineering feature like a spline can be extremely complex to predict. The answer lies in getting sound empirical data from a well-structured R&D programme.

Welds

Welding is an attractive means of joining metal components but has complex and potentially poor fatigue properties. The very nature of welded joints tends to locate them at changes of section, which introduces stress raisers. In addition, the process of welding introduces crack-like imperfections and residual stresses. Design Codes based on empirical test data give a reasonable probability of avoiding fatigue failures under typical conditions.

The Welding Institute, (TWI), has an extensive store of practical information and advice on welding **(12)**. A paper by Maddox and Razmjoo **(13)** starts by saying *current fatigue design methods for assessing welded steel structures under complex combined or multi-axial loadings are known to be potentially unsafe*. At least that warning is clear, and there are recommendations in the paper on how to be more conservative in design while research continues. TWI has an ongoing programme to provide comprehensive design guidance.

Predicting weld fatigue life is influenced by the welding process and weld quality. Hence the need for designers to work closely with both welding engineers and inspectors in order to get a total solution.

Another problem is handling random loading.

Assessing cumulative damage

Few engineering applications have a simple cyclic loading pattern. Various tools that appeared to have the ability to handle mixed and random loading, such as the Palgren–Miner equation, are now seen to be optimistic in some situations. The simple Miner approach to sum fatigue damage at failure from various cycles is

$$\sum n(x)/N(x) = 1$$

where n is the number of cycles at loading x and N is the number of cycles to failure at loading x. Failures in welds and other features have been experienced down to $\Sigma = \frac{1}{2}$ **(14)**. Furthermore, components of random loading that produces stresses below the constant amplitude fatigue limit have been shown to add to the damage. This has occurred at stresses down to around a quarter of the fatigue limit. TWI have been exploring using Miner's rule while assuming the $s-n$ curve has no fatigue limit but continues with the same slope as at higher stresses.

Another way of looking at the probability of failure for multiaxial loading in the low-cycle range is shown in Fig. 2.4.6 from a paper by Professor Mike Brown of Sheffield University **(15)**. This clearly shows that most failures are within a factor of two of Miner's rule, either way.

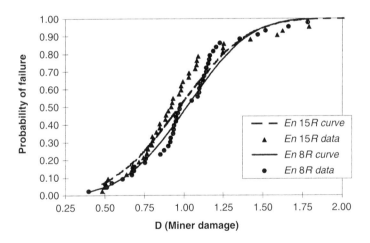

Fig. 2.4.6   Multiaxial fatigue failure probability with random loading

In the same paper, Brown explores the possibility of predicting fatigue life by using crack propagation models as an alternative to Miner's rule. He concludes that both methods have advantages, depending on the application. He points out that fatigue life predictions are influenced by the statistical nature of many factors such as:

- *the randomness of a crack meeting a microstructural barrier;*
- *the probability that initial cracks can coalesce;*

- *the uncertainty and variations of the applied loads;*
- *the difficulty of assessing and counting random loads in different planes, particularly when there is no correlation in their timing;*
- *determining the effect of mixed mode branching in different crack planes;*
- *variations in material properties; and*
- *variations in manufacture.*

While designers may use laboratory research and testing, many of these variables will have different statistical distributions on the full-scale structures or components.

The limitation of tests

Testing large structures is usually prohibitive. Small-scale fatigue tests are the main source of data on which to base a design. The small specimens are cycled at high speed to produce data for millions of cycles in a practical time. The tests are usually at constant load on artificially notched specimens. Under these conditions, an $s-n$ curve that shows an endurance limit at around 10^6 cycles for steels can be produced. Can these small-scale test results be used reliably in the design of full-sized components – only in those situations where the stress pattern is similar to that in the small-scale tests, and that is unusual. Another issue is the large number of tests needed to get a statistically meaningful answer when all the variables are considered.

From fatigue to structural integrity (SI)

Predicting service life is often more complex than assumed by design practice. How can other time and loading effects such as corrosion, temperature effects, and creep be dealt with? Instead of just concentrating on fatigue, a broader approach has to be taken and the term *Structural Integrity* (SI) has been coined to encompass all the effects and disciplines leading to failure. SI has been defined as the science and technology of the margin between safety and disaster.

Developments in many industries require a comprehensive understanding of SI to determine safe working life. Examples are highly stressed blades in gas and steam turbines, helicopter blades and gearboxes, highly rated rotating components such as pumps, generators, and compressors, cyclically loaded structures in variable conditions such as offshore oil rigs, chemical plant components and pipe work, and safety critical components in nuclear power plants.

Environmental effects

Temperature, creep, and corrosion can all affect the growth of cracks. It is not only the impact these have on the surface but also the effect at the crack tip. In the example of wheels on rails, liquids can change the mode of crack development from shear to tensile when they get trapped and pressurized at the crack tip.

These environmental effects are all time-dependent and time as well as cycles needs to/has to be taken as factors in determining failure.

Does a fatigue limit exist?

Professor Miller, Emeritus Professor at Sheffield University, has raised concern that:

- *cycles over 10^6 need an understanding of microstructural fracture mechanics;*
- *potentially dangerous situations exist at high mean stress, together with variations in low cyclic stress;*
- *fatigue tests do not take account of time-dependent effects;*
- *better experimental methods are needed.*

Professor Miller has been advocating that engineers should stop believing in the existence of a fatigue limit and look in more detail at a *Structural Integrity* approach.

COMMENTS

Many of the analytical tools for predicting the fatigue life of simple components are now mature. However, the ability to confidently estimate the life of complex components under high stress and variable loading, using only design office tools, is still elusive. Understanding and evaluating fatigue problems will probably remain an area of engineering where answers come from a judicious combination of calculation and empirical data. Answers will not be precise; there will always be a probability associated with fatigue life estimates. Inspection will be the prime way of preventing fatigue failures.

The broad nature of SI is a sound approach to understanding and avoiding failures. Professor Miller asks, *Structural Integrity – Whose responsibility?* (**6**) The answer is unlikely to be found in one person – there are too many disciplines involved such as:

- metallurgy;
- corrosion experts;
- statisticians;

- experts in FEA;
- experts in the fundamentals of strength of materials;
- researchers in fatigue studies;
- experts in fatigue testing;
- inspectors; and
- designers.

The aim is to harness all these skills, to provide sound products, and to keep Standards and Codes relevant and up-to-date. Judgement will be needed to steer a course between fatigue failures and being too cautious and over-designing. Then the user has to recognize the fatigue limitations and not overload the equipment.

References
(1) **Smith, R.A.** (2003) Fatigue: a continuing problem, despite its long history, *Fatigue, Real Life Solutions Seminar*, IMechE, 27 February 2003.
(2) **Morin, A.** (1853) *Résistance des Matériaux*, Librairie Hachette, Paris.
(3) **Rankine, W.J.M.** (1843) *Proc. ICE*, **2**.
(4) **McConnell, J.E.** *Proc. IMechE*, 1847–1849.
(5) **Pilkey, W.D.** (Ed) (1974) *Peterson's Stress Concentration Factors*, Second edition, John Wiley, ISBN 0 47153 849 3.
(6) **Miller, K.J.** (2001) Structural Integrity – Whose Responsibility? *36th John Player Lecture*, IMechE, 2001.
(7) **Murakami, Y.** (Ed) (1987–2001) *Stress Intensity Factors Handbooks*, Vol. 5, Pergamon Press/Elsevier Science Ltd.
(8) **Jones, M.R.** (2003) Fatigue in aircraft structures – an airbus perspective, *Fatigue, Real Life Solutions Seminar*, IMechE, 27 February 2003.
(9) **Hofmann, D.** (2003) Fatigue of gears, *Fatigue, Real Life Solutions Seminar*, IMechE, 27 February 2003.
(10) **Leen, S.B., Hyde, T.R.** and **McColl, I.R.** (2003) Fretting and fatigue of splined couplings, *Fatigue, Real Life Solutions Seminar*, IMechE, 27 February 2003.
(11) **Vingsbo, O.** and **Soderberg, D.** (1988) On fretting maps, *Wear*, **126**, 131–147.
(12) TWI, Granta Park, Gt. Abington, Cambridge CB1 6AL, UK and at www.twi.co.uk.
(13) **Maddox, S.J.** and **Razmjoo, G.R.** (2001) TWI, Interim fatigue design recommendations for fillet welded joints under complex

loading, *Fatigue and Fracture of Engineering Materials and Structures*, **24**(5).

(14) **Maddox, S.J.** (2003) Welded joints – fatigue design developments, *Fatigue, Real Life Solutions Seminar,* IMechE, 27 February 2003.

(15) **Brown, M.W.** (2003) Two methods of multiaxial life assessment, *Fatigue, Real Life Solutions Seminar,* IMechE, 27 February 2003.

(16) **Miller, K.J.** and **O'Donnell, W.J.** (1999) The fatigue limit and its elimination, *Fatigue and Fracture of Engineering Materials and Structures*, **22**(7), 545–557.

The Human Approach to Risk, Decisions, and Errors

2.5.1 Dealing with risk

Failure is when a risk becomes a reality. But risk is seen to mean different things. The first thought is usually 'hurt' – physical injury. The next is that risk is a gamble. Hence, the more scientific definition

$$\text{risk} = \text{probability} \times \text{hazard}$$

Hazards exist, and there is always a probability, no matter how small, that there will always be some risk. The question is how, and to what extent, should we manage and control the hazard and the probability? These questions have technical, societal, and ethical aspects.

Categorizing risks

Risks fall into two broad categories: those that are immediate and those that happen slowly. Immediate cause and effect risks occur, for example, when a tool, either by use or misuse, injures your hand. Slower risks occur when pollutants cause long-term health problems, such as exposure to carcinogens.

Decisions on risk-taking

Decisions on risk-taking are a balance of risk and benefit. But benefits to whom? Many decisions can result in risks and benefits being disproportionate on different sections of society. Chemical plants and power stations may benefit society as a whole and investors specifically, but may give additional risk to people living downwind of the plants. Governments are being increasingly pressed, on behalf of society, to take decisions on risk – especially after a disaster.

Government management or regulation of risk-taking raises the question *Should I have freedom to take my own decisions on risk?* Governments respond to pressure groups, and safety regulations increase. Litigation follows, and someone is found to blame. The trend in society is to be more risk averse. Yet at the same time, there is an increase in the popularity of extreme sports such as snowboarding, hang gliding, rock climbing, sky diving, and so on. The *benefit* seems to be the thrill of living near the edge.

The difficulty for the engineer is that the balance of risk and benefit that society will accept is a fickle and moving target.

Attitude to risk

Prior to an accident, there tends to be a more personal judgement of risk. This is not a new phenomenon. James Thomson, Lord Kelvin's elder brother articulated this in a lecture in 1873. He said people want to know *Is that plank over the stream safe for **me** to walk over?* The answer they are looking for is *Yes* or *No*. An engineer could describe the strength assessment made on the information available, list the uncertainties, and quote the probability of the loads being less than the strength. Such an answer would be seen as pontificating – ducking the question. It is likely to be answered by a rebuke – questioning the courage of the engineer to make decisions. Faced with this dilemma, a simplistic *Yes* or *No* answer is usually given without qualification. Then, when something goes wrong, the engineer is blamed and accused of misleading by not telling the whole story – a *catch-22* scenario.

People's attitude to safety is influenced by whether the risk is perceived to be under one's own control or imposed by others. In the former case, larger risks are taken, such as jaywalking across a busy street.

Countries and cultures have different attitudes to risk. What appears to be an acceptable level of road accident deaths varies from country to country. But road accident deaths everywhere are tolerated at a much higher level than risks in other forms of transport. The outcry at the Hatfield rail crash in which four were killed was out of all proportion to a road accident, where four are killed in a car crash. And this type of car crash happens every day.

The public accepts natural events – acts of God – much more readily than man-made failures. Man-made failures are accepted if there have been benefits and if those involved did their work responsibly. Failures where people or organizations are perceived to have performed inadequately generate a demand for action. The tendency is to press for tighter rules, regardless of whether they are logical in terms of the cost per life saved. Familiarity with repeated failures with small consequences is much more acceptable than with large failures, even when the total toll over time is similar.

Does the media influence public attitude?

The public get information on major accidents from the media – usually television. Does the media reporting influence public attitude? A report produced for HSE gives some interesting answers (**1**). They found that the public gets their information in short snatches and fragments. As a result, stark images and headlines are more easily recalled than extensive engineering explanations. Television is believed to be more controlled than the press and therefore more likely to be accurate. Known reporters and newsreaders can generate a feeling of trust.

The public reacts with greater concern to issues that they feel might affect themselves. They try to rationalize the diverse facts and views. Part of the rationalization process is trying to relate the facts into narrative stories of experiences. These are refined through everyday conversation, discussion, and argument. The HSE study concluded that the public are not passive absorbers of media information but make up their own minds. It notes a couple of additional points – there is usually a dearth of sound technical reporting, and some sections of the public are instinctively suspicious of people in suits trying to explain away blood on the ground.

While the HSE study reviews the media's influence on public attitude, another aspect is the media's role in the debate following an accident. Lobby groups use media data, particularly powerful images from pictures. They court the media to promote their own views. A good dispute makes interesting media. The challenge is to see who can be made to take the blame or to make changes. The other side of this is that remote accidents get little attention if there is no impressive image or media report, regardless of how many people died.

Information on risk

As well as data from the Hazard Forum, there are key papers on risk from the Royal Society (**2**), the Royal Academy of Engineering (**3**), the HSE (**4**), and the UK Government (**5**). Common themes run through these. The main one being the need to establish rational debate based on a clear understanding of the known facts and the uncertainties. The Royal Society sees the need to acknowledge the uncertainties in science as well as getting a better understanding of the available science. While most of the interest has come from issues such as BSE, GM foods, and climate change, there are lessons for engineering.

The main lesson is that risk assessment is an important part of all engineering. To get the whole picture, it is essential to consider the societal and ethical aspects. These aspects are evolving and need to be closely monitored – or for preference, get directly involved in the debate.

References

(1) **Petts, J., Horlick-Jones, T.** and **Murdock, G.** (2001) *Social Amplification of Risk: The Media and the Public*, HSE Contract Research Report 329/2001, HSE Books, Sudbury, Suffolk.

(2) Royal Society, Science in Society Programme. See www. royalsociety.ac.uk.

(3) Royal Academy of Engineering (2003) Paper 1, *The Societal Aspects of Risk*; Paper 2, *Common Methodologies for Risk Assessment and Management*; Paper 3, *Risks Posed by Humans in the Control Loop*. Engineering in Society accessible at www.raeng. org.uk/news/publications/reports/.

(4) Reducing Risk Protecting People, HSE's Decision-Making Process, 2001, HSE Books, Sudbury, Suffolk, ISBN 0 71762 151 0.

(5) UK Government, PM's Strategy Unit (2002) *Improving Governments Capability to Handle Risk and Uncertainty*. Accessible at www.number-10.gov.uk/SU/RISK/REPORT/0.1.htm.

2.5.2   Human decisions and errors

Reviews of failures put *human error* high on the list of causes. The top rate has risen from 30 per cent in the 1960s to nearly 90 per cent now **(1)**. Is this because society likes to have someone to blame or because accidents are now studied in more detail? The increased interest in this area has led to research in several fields.

Research on errors

> *The truth belongs to everyone, but errors are ours alone.*
> Ken Alder – The Measure of All Things, 2002.

The analysis of accidents and failures after WWII highlighted the problems that confront people who operate equipment. Poor instrumentation and poor man–machine interfaces caused failures. So ergonomic principles were applied, and better instrumentation was provided. The scope of studies broadened, and *human factors engineering* emerged.

There were parallel studies to understand how humans make decisions. This involved psychologists, sociologists, and cognitive scientists. Their work has covered a wide spectrum from trying to understand how the human mind works to observing behaviour **(2)**.

Errors are caused by decision-making that has gone wrong. Hence, a study of errors should start with studying the decision-making process. Hollnagel **(1)** argues that human errors are not a subject in their own right but are part of a continuum of human decision-making or problem-solving.

Decision-making

James Reason, Professor of Psychology at Manchester University, has observed eight steps in decision making. The first step is *activation* – recognizing that there is a problem to solve. The second is *observation* of the issue, then *identifying* and *interpreting* it within a context. The fifth is *evaluation* followed by *goal setting* and *deciding the process* for solving the problem. Lastly, there is the *action* to solve the problem. Not all problem-solving goes through each of these stages – the mind, consciously or subconsciously, takes shortcuts.

The three levels of human activity identified by Jens Rasmussen – skill-based, rule-based, and knowledge-based – apply to decision-making. The three levels impact the speed of decision-making, skill-based being the fastest as it is automatic. Rule-based takes a little longer to go through the steps to find an appropriate rule from experience. Knowledge-based takes the longest – it requires mental analysis.

Errors

Errors can stem from each level. For skill-based work, the errors are slips or lapses – like typing errors, Freudian slips, and slips of the tongue. For both rule-based and knowledge-based activity, the errors stem from thinking – mistakes or violations.

Skill-based errors

Skill-based errors take a number of forms. To help identify the most prominent types, consider two similar patterns of activity. The first going through steps ABCDE to F, and the second through steps one to six. Some of the error patterns are:

- doing the wrong task, or switching from an incomplete activity – going from ABC to 456 – possibly because of inattention or being distracted after completing task C. An example in design work is looking up a code for welding (step D) and in the process finding the code on lifting tackle. Checking the welding code may be dropped as attention is turned to studying the lifting tackle code (*when you stopped to think – did you ever start again?*);
- omitting a task – going through ABCEF, omitting task D. This may or may not be important or lead to a failure. An example is failing to carry out a check. Reason has noted that this is probably the most common human error;
- repeating a task – going through ABCCDEF. In most cases, this just consumes more time and may even reduce the chance of another error;

- delaying tasks – going through ABC delay DEF. Failures may occur if it is a time-sensitive process;
- intrusion or replacement of another task – going through ABC4EF. An example is isolating the wrong exciter at Uskmouth.

Rule-based errors

Some rule-based errors are:

- not recognizing that there is a problem to be solved. For example, not recognizing that static electricity and the doped skin on the Hindenburg were an explosive combination;
- working to the wrong rules. This was one of the contributing causes of the Gimli accident;
- misreading the signals;
- ignoring the redundancy of information. Deciding after too quick an analysis. Redundant information that does not fit the solution is quietly ignored;
- information overload or incomprehensible information. This can lead to a person being out of their depth. Alternatively, it may take more time than is available to sort out the facts. Trying to decipher the manuals while gliding into Gimli is an example of running out of time;
- getting lost in the complexity of the situation. Again, this can lead to a person being out of their depth. They can be blind to the inter-relationship of the problems and become fixated on one aspect or on their own hypothesis. Knowing when to call for help is one answer. People working outside their competence may have good intentions, but can easily walk into danger (cf. Homer Simpson's – *No problem!*);
- rule strength. Vice-Admiral Grace Hopper, a computer expert with the US Navy Reserve and co-inventor of COBOL said the most depressing words she ever heard were *we've always done it that way*. Said without understanding why, and without knowing the limitations. Rules that gain undue strength in people's minds cause unforeseen errors;
- dealing with problems on an ad hoc basis. Using only linear thinking and not considering how the issues link together nor what the consequences are.

Knowledge-based errors

These can be similar to rule-based errors but without the benefit of rules to help with the solution. As a result, they can be more complex. The psychology and physiology of decision-making plays a larger role. Thinking involves emotion. We get angry and upset when we cannot solve a problem in the way we expect, and this further exacerbates the finding of a solution.

One danger is oversimplifying the problem. Taking a *don't confuse me with the details* attitude and using a too limited or simplistic model of the problem. Another potential problem is overconfidence, and coming to a hasty judgement. Associated with this is persistence, which in some cases can be good but not when a change is needed. For knowledge-based problems, there is more need to use all of Reason's eight steps.

The studies of Tversky and Kahneman (3) have shown three types of biases in decision-making. The first is an illusion that the problem is similar to a previous problem. The bias leads to instant classification – if 'a' and 'b' are similar, then what worked for 'a' should work for 'b'. But are they the same? The second bias is that it is easier to recall more common phenomena than uncommon. They are more likely to be adopted as part of the solution to a problem. This might be a benefit, but if wrong can lead to misdiagnosis. The third bias is anchoring – when there is a reluctance to move from the familiar. This can be due to being too conservative, having too much mental inertia, or lacking the commitment to do something different. The perception from biases becomes reality in people's minds.

People can be hypnotized by the way they receive information. For example, an overheard stock tip is more likely to be acted on than a detailed financial analysis. A recommendation for a car heard at a social gathering is likely to figure more strongly than a review of test data.

Thinking can be divergent or convergent, linear or lateral. Assessing a problem with divergent thinking is likely to encompass more possible causes. Then convergent thinking and analysis can home in on the true cause. If convergent thinking is used too soon, then causes can be missed. Similarly, linear thinking can miss problems and solutions that Edward de Bono's lateral thinking might expose.

Errors and violations
The previous sections viewed errors as slips or mistakes. What is the chance of someone deliberately making a mistake – *committing a violation?* Most people work on the basis that their colleagues and company are law abiding and ethical. At the same time, they find it difficult to answer the question *Why should your group be more law abiding than a similar group in society?*

Most people will violate some rules. It is argued that overspeeding on the roads is not a violation but just using an accepted tolerance. Similar attitudes can cause errors in the decision-making process by going over the boundary.

Dr Bob Helmreich of the University of Texas human factors research project studied flight crew errors (4). He showed that more than half of the observed errors were non-compliance or violations, but only two per cent of these led to undesirable aircraft states. One could conclude that violations created few problems. However, his analysis showed that crews who made the most violations were almost twice as likely to make other errors.

Whose errors are they anyway?

An obvious target for blame is the operator. Recent work is questioning this quick assessment. Dr Simon Bennett (5) has studied air accidents and concluded that many that have been categorized as pilot error have other causes, causes such as poor maintenance, poor design, and organizational failings.

The studies of nuclear power accidents and incidents by INPO (Institute of Nuclear Power Operations) have shown a large percentage due to human error. But the largest classification of errors is omissions by maintenance workers rather than errors made by the operators on the desk. Similarly, Transport Canada (6) found that for both aircraft and rail transport, the most common maintenance error is omission – forgetting to do a step.

It is too easy to jump to the conclusion that the last in the chain caused the accident – and stop there. It is more revealing to ask why the operator or maintenance worker made an error – and why a simple error had to lead to an accident.

Organizational errors

Many errors are rooted in organizational failures. A study of 23 serious aircraft accidents from 1973 to 1982 showed that only 13 per cent were due to crew errors alone (7). Most were due to a combination of crew and organization errors. The organizations that had some responsibility were listed as the manufacturers, airlines, professional organizations, airports, and regulatory agencies.

Blaming the pilot or driver shows a bias towards accusing the individual at the controls – an active failure. Organizational errors tend to be made over a longer period and are passive failures. They may not be immediately apparent and can be overlooked. They become latent errors.

Reason's Swiss-cheese model

Organizations try to protect themselves against failures by adding layers of procedures or checking. Managers are expected to check their staff's work. Quality Assurance procedures add another layer of checking. These are attempts to achieve defence-in-depth. In reality, each barrier is dependent on humans, and humans make mistakes. So, instead of multiple solid and protective barriers, Reason sees the situation as a series of Swiss cheese slices – the holes in each slice representing the human flaws (8). So there is some possibility that a problem can escape through the holes in each layer and cause an accident. Some holes may be active errors and others latent – just waiting for the opportunity to surface.

One answer to the fallibility of the layers is to add even more layers. This may not create the expected results. More layers can create more confusion, less clarity, and confused responsibility.

Checking – man versus machine

With electrical control equipment, safety can be achieved with multiple parallel circuits. For example, three identical circuits can be used to check the state of equipment. If all three give the same answer, fine. If one gives a different answer, then it is reasonable to assume the other two are correct. On paper, higher reliability could be obtained by a three out of four system. But this adds complexity, which might be self-defeating in trying to get higher reliability.

Trying to use people to provide similar defence-in-depth is not so clear-cut. Individuals are influenced by their feelings towards, and assessment of, the people they are checking. The time allocated for checking is usually far less than the time taken to generate the idea in the first place. While this might surface obvious errors, the more entrenched ones will probably need a similar time to unearth as the original project took. If both the checker and originator come from the same background and culture, then both may be susceptible to missing the same error. Is it better for the originators to formally check their own work? It might require management supervision to ensure that the extra time is spent on checking rather than on continuing to evolve the original idea.

Impact of culture and environment

The culture of the organization and the attitude of the management and staff influence the likelihood of errors. An organization can have a variety of cultures and environments such as:

- a learning culture – where there is encouragement for staff to continue to learn. The converse is the type of organization that considers that the library is a place where staff waste company time!

- an introspective culture – where there is an inquisitive interest in the quality of performance. Past problems, near misses, and industry issues are studied in the quest for improvement;
- a culture of competence – where the staff are recruited for their competence, rather than just acquiring pairs of hands;
- a professional culture – where the professionalism of the staff has a large influence on the way work is handled;
- a safety culture – where staff and management at all levels have safety high on their agendas;
- a production culture – where the emphasis is on production at all costs. Someone down the line – maybe even the customer – will deal with any errors;
- an entrenched culture – unable to recognize failures or the need to change;
- a careless culture – where little attention is paid to failures – they are in the past;
- a 'group think' culture – where all staff are encouraged to think along the same lines. Errors that are not obvious can easily be missed;
- a culture of openness and debate – where ideas are openly discussed and debated with the aim of selecting the best, rather than allowing ego and prejudice to rule;
- a distracting environment – where it is difficult to concentrate. This can be due to noise, heat, inadequate space, and poor tools.

Admiral Gehman defined organizational, or embedded, culture as *the process and procedures used when operating outside the rule books – and what you instinctively do when the boss is not looking over your shoulder*. This emphasizes how difficult it is to fully know the extant culture. If this is seriously different from what management expect, then there is plenty of opportunity for problems.

Safety on the shop floor
Organizations may have several of these cultures within different parts of the company. An effective safety culture will also require a learning and introspective culture to be present. These days, it is seen as good management to have a safety culture. But the definition and implementation of a safety culture can be elusive. In many companies, the safety culture starts with meeting the safety regulations – clearly identifying where safety hats and glasses have to be worn, and making

sure that they are readily available. Shop floor workers and visitors get constant reminders by notices and posters. The notice boards proudly present the minutes of the safety committee and the latest safety statistics.

Safety in the design office
Although emphasis is placed on operations and on the shop floor, one of the most important places to build in safety is through the design process. Yet the design offices do not usually have their walls covered with warning safety notices. They are more likely to have photographs of successful projects or inspirational posters. Does this mean that designers pay less attention to safety? Definitely not, what they have is an unstated and underlying commitment to safety. Because it is unstated, it is more difficult for outside observers to assess their commitment to safety.

Safety culture – an individual or organizational issue?
An interesting debate is whether a safety culture is a characteristic of the individual or the organization. Culture manifests itself as a feature of a group. Some argue that to be committed to a safety culture requires absolute dedication to safety all day every day – it cannot be turned on when coming through the design office door or factory gate. Does this mean that people fond of extreme sports should not be involved in hazardous plant design? Should someone with a motoring offence be a concern? This argument is suspect, and ignores the human ability to adapt to circumstances. Maybe the psychologists should study whether someone participating in extreme sports and exploring the boundaries of risk might have a better understanding of risk in the work environment.

Culture and the working environment can have a major impact on safety. In many cases, the relationship is subtle, but nevertheless important. There is a big difference if everyone up to and including the Board of Directors has a common understanding of safety. If the working level feels that the Board is just mimicking safety mantra without understanding it, then the culture will be greatly undermined. Another concern is that changes in culture and deterioration in safety standards can easily be missed.

The role of near-miss reporting
Not all errors lead to failures or accidents. This is good, not only in terms of fewer accidents but it also allows the study of near misses. Studying near misses has become standard practice in hazardous

industries. It gives a warning of possible failure modes. Monitoring near misses can measure the effectiveness of error management practices.

While near miss reporting is a sound management tool, it docs need a careful balance between openness in disclosure of human errors and the assigning of blame. Attempts to improve safety by Draconian punishments are certain to disrupt the delicate balance.

Error detection

There are obvious errors and hidden errors. Latent errors are mostly hidden – at least for some time. Error detection needs a skill and attitude to watch for the unexpected, to expand thinking from known problems to the suspicion of new problems. Yogi Berra, the US baseball player noted for his insightful sayings, recounted his problem at school. His teacher, in frustration, asked the rhetorical question *Don't you know anything?* His reply was *I don't even suspect anything.* In similar vein, the best players in team sports concentrate on the goal, but are also well aware of what is going on at the other side of the field or rink. They have developed good peripheral vision. Being suspicious and having good peripheral vision helps detect errors. Do not be left with a blinkered view of the world.

Many errors are spotted and resolved before they develop into failures or accidents. People tend not to record errors they have corrected – that is just part of the job. But knowing how errors have been spotted could help increase the number detected and hence reduce failures. There are few studies of design office errors. One survey of errors in a structural engineering design office was conducted by Andrzej Nowak (**9**). He found that three types of errors occurred most frequently and were:

- rounding errors – usually insignificant;
- calculation errors – often of significant size, but as a result, easy to spot and correct before they get out of the office; and
- the need for more explanation. This can be due to lack of knowledge or poor communication. Lack of knowledge often compounds with poor communication. These types of error can easily spread to the next stage and become latent errors.

Studies of errors (**8**) show the typical distribution for each type. The split is skill-based errors 61 per cent, rule-based 27 per cent, and knowledge-based 13 per cent. Detection rates are similar for the three types, going from 86 per cent for skill-based to 71 per cent for knowledge-based. But the chances of recovering from a skill-based slip were twice that for a rule-based mistake. Rule-based mistakes were recovered three times better than knowledge-based mistakes. One researcher found that skill-based errors of the type where a step was missed were most resistant to detection.

When humans try to resolve errors, there is a preference to try the easiest route first. Skill-based answers are the easiest. If that does not work, look for a rule-based answer. If all else fails, in desperation resort to knowledge-based thinking and use one's brain!

It's all in the mind

Both psychologists and neurophysiologists are studying how the mind works. The neurophysiologists are beginning to understand the working of the mind and brain.

The neurophysiologists' view of the mind

The operating mechanisms of the brain are the nerve ends – the neurons. There are about 10^{10} neurons in a human brain performing all the activities of data storage, data processing as well as initiating actions – thinking and moving. We are born with a full set of neurons and lose about 20 per cent over a lifetime. Research has shown that neurons are not hard-wired but have dynamic connections with each other. The connections evolve and develop with the learning process. Some connections are stronger than others. The more they are used, the stronger they get. The brain is quite different from a hard-wired computer.

The neurons are key to memory. The memory is like a library continually in transition and reorganization. This may explain why the memory is not foolproof. Accident investigators find that the ability of witnesses to recall what happened is often flawed. The false memory can be very strong and firmly believed to be accurate. This may be because the facts have been recatalogued and associated with other similar facts.

Modern medical tools can explore which part of the brain is associated with different functions. It appears that there is not just one part of the brain that is the 'library' of stored data. Knowledge is distributed all over the place.

The psychologists' view of the mind

Psychologists' understanding comes from observation and empirical hypothesis. The memory works in stages – an immediate sensory stage, a short-term memory, and a long-term memory. The immediate sensory stage acts like a camera – quickly capturing a perception. The short-term memory has a timescale of 20 to 30 s. It is a bottleneck in human processing and can only deal with about seven different items. The long-term memory is the final stage for storing knowledge.

The strength of the long-term memory depends on how long it is since the information was used and how powerful was the original image. Studies show that practice improves performance. The strength is also dependent on how the input information was processed – did the knowledge come from quick casual reading or from serious study combined with note taking? Did the input process use stark visual imagery? Did the knowledge get associated with a situation and emotions or was it recorded as a piece of general knowledge without context? Research by Roediger (10) at Washington University has confirmed the popular understanding that visual images are more powerful than words.

It is unclear whether the long-term memory can decay. Whether information is forgotten or is just inaccessible – either permanently or for a period of time. Information can surface, apparently subconsciously after some time, minutes to hours after consciously trying to recall data. This process can help resolve errors. Memory can recall that a slip or mistake was made in earlier work.

As well as the short-term and long-term memories, there is a working memory. This is relatively short-term and appears to be the means for extracting and looping in and out of the long-term memory to carry out a specific task.

How reliable are human beings?

The error rate of human activities is estimated as 1 error in 10^3 to 10^4 actions. Duffey and Saul (11) have studied a wide range of human activity. They concluded that all activity has a clear learning curve, and that the minimum error rate generally lies between 1 in 20 000 hours and 1 in 200 000 hours. The lower figure of 20 000 hours is typical of industrial injuries, while the higher figure is for situations with extensively trained personnel like aircraft pilots. The exception is auto accidents at 1 in 6000 hours.

Several studies have aimed at quantifying human errors. Bello and Columbari developed one example for the Italian petrochemical complex ENI (12). Their method, TESEO (Technica Empirica Stima Errori Operatori.), breaks the problem into five aspects that, when multiplied together, give an estimate of the probability of human error. They are:

1. type of activity – with probability ranging from 0.001 for simple operations to 0.1 for non-routine activities that require special attention;
2. temporary stress level due to the time available to take action – varying from 10 when the time available is of the order of a few seconds to 0.1 when around an hour is available for decision;
3. operator's qualities – varying from 3 for little knowledge and training to 0.5 for skilled operators;
4. activity anxiety factor – ranging from 3 for serious emergencies to 1 for normal situation;
5. the ergonomic factor – ranging from 10 for poor working conditions and poor ergonomics to 0.7 for an ideal working situation.

This process gives an error probability of 1 in 10^3 for routine activities. The factors show an extreme range covering six orders of magnitude. This is an extreme extrapolation, but several orders of magnitude are likely. So training, good ergonomics, and designing systems that do not require difficult or hasty decisions can have an enormous impact on the reliability or error rate of a human. The importance of the mind and memory highlights the benefit of memory aids – mnemonics, checklists, notices, notebooks, diaries, and post-it labels.

Error management
Reason has identified three approaches to safety management:

- the person, or individual, approach – where safety depends on the individual's psychology and competence;
- the engineering, or system, approach – where engineering techniques such as improved ergonomics, simulator training, and hazard assessment analyses are used; and
- an organizational approach – where human errors are considered more a consequence than the cause. Safer work can result from culture change, improved working conditions, making information more readily available, training, and developing procedures.

Reason **(13)** summarized the main issues of error management as:

1. *The best people can sometimes make the worst errors.*
2. *Shortlived mental states – preoccupation, distraction, forgetfulness, inattention – are the last and the least manageable part of an error sequence.*

3. *We cannot change the human condition. People will always make errors and commit violations. But we can change the conditions under which people work to make these unsafe acts less likely.*

4. *Blaming people for their errors – though emotionally satisfying – will have little or no effect on their future fallibility.*

5. *Errors are largely unintentional. It is difficult for management to control what people did not intend to do in the first place.*

6. *Errors arise from informational problems. They are best tackled by improving the available information – either in the person's head or in the workplace.*

7. *Violations, however, are social and motivational problems. They are best addressed by changing people's norms, beliefs, attitudes and culture, on the one hand, and by improving the credibility, applicability, availability and accuracy of the procedures, on the other.*

8. *Violations act in two ways. First, they make it more likely that the violators will commit subsequent errors and, second, it is also more likely that these errors will have damaging consequences.*

Quote reprinted with permission of Ashgate Publishing © Reason, J.

References
(1) **Hollnagel, E.** (1998) *Cognitive Reliability and Error Analysis Method*, Elsevier Science, ISBN 0 08042 848 7.
(2) **Reason. J.** (1998) *Managing the Risk in Organizational Accidents*, Ashgate Publishing, ISBN 1 84014 105 0.
  Reason, J. (1990) *Human Error*, Cambridge University Press, ISBN 0 52131 419 4.
  Perrow, C. (1984 & 1999) *Normal Accidents*, Princeton University Press, ISBN 0 69100 412 9.
  Rasmussen, J. (1986) *Information Processing and Human-Machine Interface*, Elsevier Science, ISBN 0 44400 987 6.
  Dörner, D. (1996) *The Logic of Failure*, Perseus Books, ISBN 0 20147 948 6.
  Weick, K.E. and **Sutcliffe, K.M.** (2001) *Managing the Unexpected*, Jossey-Bass, ISBN 0 79795 627 9.
  Weick, K.E. (1995) *Sense Making in Organizations*, SAGE Publications, ISBN 0 80397 177 X.

Turner, B.A. and **Pidgeon, N.F.** (1978) *Man-made Disaster*, Second edition, 1997, ISBN 0 75062 087 0.

Vicente, K.J. (2003) *The Human Factor*, Alfred A Knopf, Canada, ISBN 0 67697 489 9.

(3) **Tversky, A.** and **Kahneman, D.** (1974/82) *Judgement under Uncertainty*, Cambridge University Press, ISBN 0 52128 414 7.

(4) **Helmreich, B.** (2001) *Flight Safety Australia*, January/February 2001.

(5) **Bennett, S.** (2001) *Human Error – by Design?* Perpetuity Press, ISBN 89928 772 8.

(6) **McMenemy, J.** *Most Common Maintenance Error*, Transport Canada, www.tc.gc.ca.

(7) **Bruggink, G.M.** (1985) Uncovering the policy factor in accidents. Published in *Airline Pilot*, US ALPHA Magazine, May 1985.

(8) **Reason, J.** (1990) *Human Error*, Cambridge University Press, ISBN 0 52131 419 4.

(9) **Nowak, A.S.** and **Carr, R.I.** (1985) Classification of human error, *ASCE Proceedings of the Structural Safety Studies*, May 1985 ASCE, New York.

(10) **Roediger, H.L., Meacle, M.L.** and **Bergman, S.T.** (2001) Social contagion of memory, *Psychol. Bull. Rev.*; and **Roediger, H.L.** and **McDermott, K.B.** (1994) Effects of imagery on perceptual implicit memory, *J. Exp. Psychol.*, **20**, 1379–1390.

(11) **Duffey, R.B.** and **Saull, J.W.** (2003) *Know the Risks*, ISBN 0 75067 596 9.

(12) **Bello, G.C.** and **Columbari, V.** (1980) The human factor in risk analyses of process plant, *Reliab. Eng.*, **1**(1), 3–14.

(13) **Reason, J.** (1998) *Managing the Risk of Organizational Accidents*, Ashgate Publishing, ISBN 1 84014 105 0.

2.5.3 Normal accidents versus High Reliability Theory

The study of accidents in systems that have a high-risk potential has produced two schools of thought – normal accidents and the school that believes in High Reliability Organizations or High Reliability Theory.

Normal accidents

Charles Perrow, Professor of Sociology at Yale and an organizational theorist, coined the term *normal accidents* (1). This concept stemmed from studying the Three Mile Island nuclear accident. His background brought a social insight into the examination of technological risk.

Perrow's thesis is that the engineering approach to risk is flawed – analysing what might happen, adding back-up equipment, and special operating systems still does not remove the risk of an accident. The combination of human error and complex systems will make failure inevitable – failure will be *normal*.

Complexity and tight coupling

Complexity is where equipment and systems are in close proximity and where they do not follow a linear process. Complexity increases with interrelated control parameters, and when redundant equipment has been added, in an attempt to provide defence-in-depth.

A system will have tight coupling if the components and subsystems are interconnected with each other. Action in one system directly leads to action in others – maybe in a time-dependant process. And the subsequent activity cannot easily be turned off.

Examples of complex systems are nuclear power plants, nuclear weapons systems, chemical plant, aircraft, air traffic control, R&D firms, and universities. Examples of tightly coupled systems are nuclear power plants, electric power grids, dams, air traffic control, and rail transport. Where there is both complexity and tight coupling, there will be inevitable accidents – they cannot be called unexpected, they are *normal*. Safety becomes just one of several competing goals.

Operators and maintenance staff are not the cause of *normal accidents* – they are hostages to them. Complexity prevents designers and operators from foreseeing all the things that can go wrong. If you cannot foresee problems, you cannot prevent them. It is not one problem that leads to an accident; it is inevitably a chain of events. The events are often minor and unsuspecting for anyone to be concerned about. But a combination of such events can lead to a disaster. In a complex and tightly coupled system, the failure or even minor incident with one component can quickly interact with others and affect the performance of everything else. The bigger the system, the more likely that problems will occur. The nature of *normal accidents* is that they are generally unexpected and often incomprehensible – at least for a critical period of time.

Organizations in complex and tightly coupled systems

In Perrow's view, managing a complex and tightly coupled system requires intense discipline and a complexity to match the complexity of the system being managed. This presents a paradox between having a centralized management to monitor problems caused by the tight coupling and decentralizing management to break the complexity into manageable pieces. This intense discipline is only found in the military and would be socially unacceptable in commercial ventures.

Where do we find normal accidents?

Perrow has studied major accidents in a variety of industries and believes they confirm his thesis. His assessment is qualitative rather than quantitative. Perrow notes that major accidents are fortunately rare – but this makes them more difficult to prepare for and to comprehend.

Scott Sagan (2) has extensively studied nuclear weapons systems and their deployment. While no major accident has occurred, Sagan unearthed many incidents, which appears to reinforce the *normal accident* theory. He sees engineers believing that reliable systems can be built out of less reliable parts, and having undue faith in limited testing. Responsibility is avoided by claiming that the rules have been followed.

Can systems be made safer?

Perrow recognizes that the likelihood of disaster can be reduced by training and by attention to culture, and organizational and equipment design. In the 1999 edition of his book, 15 years after its first issue, he notes that progress has been made. But the design solutions for organizations and equipment are to add more layers. More layers add more complexity and coupling, and can be counterproductive. Fixes added to try and deal with a weak design or weak organization are rarely effective over the long term. Perrow acknowledges that some technological advances such as simple gas turbines replacing piston engines, and computers replacing pneumatic control systems have been beneficial. But he questions whether there is a net benefit in the rush for speed, all-weather operation, and having greater and greater concentrations of energy in plants.

Decision-making

In an attempt to predict swings in the stock market, researchers have tried to understand how people make financial decisions. Similar studies have stemmed from work on artificial intelligence. Perrow followed these studies and sees three types of decision making:

- absolute rationality – which is a characteristic he sees engineers striving for. The ability to calculate risks and benefits;
- limited or bounded rationality – where people will only use a limited amount of rational thinking. Either the line of thinking is taking too much effort to reach a conclusion, or the results of

rational analysis are uncomfortable to live with. Then all sorts of short cuts, educated guesses, and gut feeling are used to reach a conclusion and decision. Humans are not optimal – they accept an answer that is good enough. Herbert Simon, the father of artificial intelligence and a 1978 Nobel Prize winner in Economics, labelled this *satisficing*;

- social rationality – which is how most of us live. People are all different and think differently. People with certain skills and ways of thinking chose jobs that need these skills – they develop an expertise. The sum total of people with different skills forms a capable social mixture. Society evolves and makes progress as the different skills develop and their products are used by society as a whole. But people do not automatically rely on the skills of others – the experts. They use a mixture of tools – rational analysis, limited analysis, biases, emotion, the excitement of exploration, and the dread of the unknown to come to their own conclusion and decision. Their decisions will be influenced by the social context of the time. Non-technical people can find all sorts of non-technical rationale for making decisions on technical issues.

Perrow concludes that the natural trend towards social rationality in the decision-making process will ensure that accidents in complex and tightly coupled systems will be *normal*. Risk is in the system rather than human factors. Production pressures will aggravate the risks of accidents. Organizations aim to extend and increase their powers. This drive for power drowns out any concern for understanding the risk they are taking – both for themselves and society. Organizations conspire to have the right to define and explain away accidents – and hate unpalatable publicity.

High-Reliability Organizations and High-Reliability Theory
High-Reliability Theory has blossomed from studies at the University of California, Berkeley (**3**). Todd La Porte and Gene Rochlin, both political scientists, and Karlene Roberts, an organizational psychologist, have been trying to find out why some complex human activities appear to work so well – with fewer accidents than might be expected. They have studied air traffic control, electric power grid management, chemical plant, and US Navy aircraft carriers.

The study of US Navy nuclear powered aircraft carriers
Large aircraft carriers have multiple nuclear reactors for their power plant. They operate within the way of doing things in the Navy – a

long established tradition. Then there is the different tradition of the air operations and aircrew. Operating the aircraft requires handling skills to best utilize the limited deck space, and air traffic control to manage the aircraft in the air. Finally, there are the continuous defensive actions to avoid attack, and operation in various weather conditions, day and night. Clearly an example of complex and tightly coupled operations – yet carriers appear to work reliably.

The detailed study **(4)** quickly showed a number of paradoxes when compared with commercial land based industries, such as:

- crews are young and generally inexperienced;
- crews are continually being reassigned;
- officers (management) also have a high turnover, with frequent reassignment;
- shore-based training only covers basic instruction;
- detailed operating instructions are available on board, but take longer to read than the time taken for the operation being described;
- even in day-to-day activity, crew and officers are frequently regrouped or morphed into different teams, depending on the immediate task.

In other situations, the keys to success appeared to be stability, maintaining a fixed routine, providing easy control with the minimum of challenge, and automating wherever possible. The aircraft carrier operation overturned all these ideas, yet still is a highly reliable operation.

What they noticed was that there is both a formal and an informal use of rules and relationships between crew and their immediate supervisors. The work is broken down into small task-oriented groups. But there is overlap and redundancy. This provides a continuous back-up for all key operations. The back-up is a genuinely redundant approach, rather than just another layer of supervision. In the event of an incident, the authority is quickly taken and accepted from the local expert – regardless of rank. Communication is through continuous conversation and verification, working in parallel at different levels, cross-checking each other.

Their interim conclusions are:

- there is a remarkable degree of personal and organizational flexibility despite operations continually increasing in complexity – flexibility that is not easy to describe in procedures;

- there is a dynamic process of replication where, at all levels, there is a continuous trainer and trainee role. Most people are working in each role at different times of the day;
- the pressure from working *on the edge* appears to bring out the best of human capability. This seems to be enhanced by being observed by the back-up crew, rather than relaxing and assuming that they will catch any errors;
- high reliability organizations have very clear, well-agreed-upon operational goals.

They noted that the Royal Navy and French Navy also operate carriers but have *somewhat different style and objectives* – yet still work reliably.

Summary of the High Reliability Theory
The Berkeley group is trying to identify how high reliability has been achieved, rather than saying how to achieve high reliability. It sees that high reliability is achieved through a culture of safety. But maintaining an effective safety culture over the long term needs strong and dedicated effort.

High Reliability Organizations are able to handle the double goal of safety and delivery at the same time. While there is decentralized authority, there is obvious leadership at all levels of management for both safety and delivery. Discretion, redundancy, and learning are all key to high reliability.

Discretion
Discretion in decision-making allows the unexpected to be handled. Rules and procedures can handle the issues they were written for – generally based on past experience. When a task group is knowledgeable and well trained, the best way of handling the unexpected is to improvise and deviate from the established procedures with discretion. The organization and management should openly accept this as good practice, while at the same time keeping it within bounds. If everyone adopts improvisation as the norm, then anarchy reigns.

Redundancy
The aircraft carrier study brought out the effectiveness of redundancy. Redundancy is a well-proven and well-established technique for providing reliable engineering systems, but is less well understood as an organizational tool. Most organizations aim for efficiency, getting the most productivity from staff without duplication. Redundancy of

staff is seen as inefficient use of resources and reducing an individual's responsibility. But the error rate from an individual may be too high in a high-risk process. Redundancy has a clear cost in additional salaries, but the benefits are less tangible and not easily seen in the financial accounts. In many industries, it might appear easier to opt for insurance rather than to use redundancy of staff to reduce the risk.

While redundancy in commercial organizations is an uncomfortable concept, it is easier to appreciate in an example from sports. Successful teams have players observing and backing up their colleagues so that they are ready to pick up the ball. The player with the ball is more focused on the immediate task. The back-up can take a broader view – assessing the risks and challenges, looking for the opportunities. The balance between wide scanning back-up and focused player is very effective. The same approach can be effective in commercial ventures.

Mindfulness

Karl Weick, Professor of Organizational Behaviour and Psychology at Michigan University, (5) uses the expression *mindfulness* to describe a common feature of high-reliability organizations. Both individuals and the organization are *mindful* of the risks. It is a concept difficult to define. *Mindfulness*, or being aware, is built into the psyche – it does not need to be continually articulated. A Navy pilot is *mindful* of the risks – he does not need to be told *be careful when you land on the deck*. *Mindfulness* underpins alertness and anticipation – it counters closed-minded blindness and rigid thinking.

The concept of *mindfulness* can be broadened to include integrity and ethics – or even the wisdom to balance long- and short-term perspectives. *Mindfulness* is the way of being aware of what is going on. It keeps an eye open for signals – it filters out the routine of normal activity and throws up the unexpected for questioning.

Mindfulness provides warning signals. It signals when to engage the mind in more demanding tasks.

Weick has also promoted *sensemaking* (6) – the skill of making *sense* of situations that have occurred. It is the process of interpretation rather than the interpretation itself, the process of putting experience and knowledge into perspective.

Organization

High-reliability organizations aim to manage accidents – preferably by preventing them, or at least sidetracking them into less hazardous results. High reliability organizations deal with complexity by using small task groups, with each group dealing with a small and manage-able number of risks. Task groups are encouraged to opt for moving large risks to smaller ones – providing escape routes, like gravel traps on auto race courses, similarly used as escape routes for trucks with brake problems descending steep hills.

With task teams concentrating on their own risks, there is a danger that other teams will be unaware of their problems. Com-plexity results in problems interacting. High-reliability organizations have to, at times, bring people together to see the whole picture and then decentralize to get on with the work. This creates the need for a dynamic organization.

The high-reliability organization has to have a broadly based safety culture – both wide and deep. But organizations are home to many diverse cultures and are rarely homogeneous. The skill is to allow the different cultures to coexist within the essential safety culture, without weakening or diluting it. For example, encouraging and allowing discretion should not compromise safety.

Redundancy needs careful management to ensure that it is always effective, rather than just another cost burden. To manage all the dynamics and paradoxes, a high-reliability organization has to be a learning and continuously reinvented entity. Despite the paradoxes, contradictions have to be avoided. A high-reliability organization needs to encourage *mindfulness* and *sensemaking*.

Dealing with decisions and errors

A high-reliability organization will have to deal with different cir-cumstances such as:

- the mundane – for example, a nuclear power plant operating steadily at base load, or a carrier stood down with no air oper-ations. In these circumstances, the rules and procedures domi-nate. The issue for management is to keep the staff alert and mindful of potential problems;
- the times of high activity – typified by full air operations on a carrier. Here there is the need for redundancy, allowing the individual with greatest understanding to take the lead;
- emergency – where emergency teams take control. Emergency training improves effectiveness, but the emergency will probably have unforeseen aspects requiring new thinking.

The high-reliability organization has to be flexible and able to change quickly from one mode to another. All staff have to recognize which mode is in play.

Learning

Never believe on faith, see for yourself! What you yourself don't learn you don't know.   Bertolt Brecht – The Mother, 1932.

Learning is essential for high-reliability organizations – learning at all levels. High-risk organizations adopt trial-and-error learning where possible, and in potentially dangerous situations, use simulators.

Learning has traditionally been through codifying experience into rules and procedures and databases. Ed Schein, Professor of Management at MIT Sloan School, suggests that all learning is coercive. An individual is expected to learn, either by the decision of others as at school, by peer pressure or by fear or guilt at lack of knowledge. These emotions lead to anxiety, which seems necessary for learning but also tends to inhibit learning. One fact is clear – learning is easier in face-to-face situations.

Experience on Norwegian offshore oilrigs **(7)** shows that a *human inquiry* approach works better than traditional methods. The tools are experimentation, stories, collective case studies, collective brain storming, and developing mental models rather than lecturing. The process uses imagination and interaction that leads to understanding – followed by commitment. The approach is underpinned by encouraging *mindfulness* and *sensemaking*, with a minimum of coercion. All members of the organization have to participate, and be seen to be committed. While learning is primarily with the individual, the organization contributes by promoting free exchange of information, growing the culture, and accepting redundancy and discretion.

In the words of Professor Frank Blackler, Lancaster University, knowledge should be *embrained, embodied, encultured, embedded and encoded.*

Weick's five habits for high-reliability organizations (8).

1. *Don't be tricked by your success.*
2. *Defer to your experts on the front line.*
3. *Let the unexpected circumstances provide your solutions.*
4. *Embrace complexity and,*
5. *Anticipate – but always anticipate your limits.*

COMMENTS

It could be said that the *normal accident* view and High Reliability Theory view are the pessimistic and optimistic views of the same situation. This is oversimplistic. *Normal accident* theorists, while being qualitative, appear to assume the lower range of human reliability for predicting errors. Then the lower factors in the TESEO assessment are used – probably on the basis of limited rationality. This produces an error rate of 1 in 10^2 to 1 in 10^3. On the other hand, the High Reliability theorists see the benefit of using all the tools that give the higher TESEO factors. If a human error rate can be 1 in 10^3 and there is fully effective redundancy providing another 1 in 10^3, then the redundancy can provide a combined error rate of 1 in 10^6. But if the redundancy is limited by resources and cost, the effectiveness can be reduced to only detecting say 1 in 10 errors. Then the overall system performance would be 1 in 10^4. This is not much better than a well-trained operator with good equipment, and two orders of magnitude worse than the most effective redundancy. These figures illustrate the huge range of difference in risk that can be achieved by effective training, equipment, organization, and culture. With all the tools of the High Reliability Organization, the risks of accidents can be significantly reduced.

Can risks be quantified?

Perrow has discounted the quantification of risk. He calls risk assessors the new shamans – as equated to sham. He sees quantification of risk as a tool to undermine social and cultural values by mathematical analysis that places a price on avoiding a death. If you cannot perceive all the causes of accidents, how can you calculate the risk? He sees risk assessment as a tool used by politicians and industry leaders to support their pet projects and their power. Perrow's rejection of risk assessment places engineers in a no-win situation. And his rationale is commonly used by those lobbying to stop engineering ventures, so it needs to be understood and countered. Trying to counter with rational arguments does not seem to work. Adopting limited rationality goes against engineering practice – so what should be done? There is no easy answer, but it is not an issue to abandon. Maybe engineers need to get advice and understanding from the psychologists?

Who uses absolute rationality?

Perrow believes that engineers and economists strive for absolute rationality. Surely, every profession and most academics also strive for absolute rationality? Aircraft pilots, ship captains and navigators, and operators of nuclear power plants also strive for rational behaviour.

But do engineers strive for absolute rationality – or is it a pious target? All engineering tasks require a multitude of decisions. It would be both time consuming and taxing on the brain to make every decision on the basis of absolute rationality. Most are based on *satisficing*, while saving the brain for those tough problems that need knowledge-based solutions.

How complex are systems?

Normal Accident Theorists have decided which systems are complex and tightly coupled. Would an engineering assessment reach the same conclusion? Take the extreme example used by the Normal Accident Theorists – nuclear power. A non-technical person viewing a nuclear control room would immediately classify it as complex. Yet the nuclear power process is simple. The nuclear physics is well-established – you need a fuel with sufficient fissile material (u^{235}), a moderator to slow down the neutrons and a coolant to take the heat away and generate steam for the turbine. Control is achieved by inserting neutron-absorbing material to limit the chain reaction. The complicating issue for nuclear power is that the heat energy after a reactor trip drops off quickly but exponentially. The sheer size of nuclear plants means that there is still a lot of heat energy to get rid of. This is where duplication of equipment is needed to cover all circumstances, like loss of power from the electrical grid system. The control room also has to deal with the water side of the boiler, steam to the turbine and electrical dispatch as in any conventional power station. Consequently, control rooms appear complex and daunting to the inexperienced observer, yet the basic process is simple.

Take a more common example – how complex is a car? Simply put, it has fuel to power an engine, which drives the wheels. The steering wheel is coupled to the front wheels to manoeuvre the vehicle. But a car can also be viewed as complex and risky. A volatile fuel is stored and piped past passengers in order to be exploded in the engine. A complex chemical flame front generates huge pressures that have to be contained in the cylinders by piston rings operating at high temperatures. The engine can only produce efficient power at certain speeds, and so a clutch, gearbox, and differential are essential for propelling the vehicle. All these components have sharp changes of section on shafts, as well as gears and splines, all of which are susceptible to fatigue failures. Accelerating the vehicle can alter the steering charac-teristics and results in tight coupling between the drive, steering, and suspension. Then there is the computer controlling alarms – brake fluid, coolant, fuel, and so on.

Complexity can be in the eye of the beholder. An engineering approach is to try and create a simple visual or mental image of the system – sufficient to get a good understanding of the key character-istics. Then the model can be developed to understand the details.

Summary

The work of the High Reliability Theorists shows what can be achieved to avoid accidents. But they are at pains to warn that achieving high reliability is not easy – maintaining it over the long-term is even more difficult. The potential for high reliability can be seductive, but before being carried away with the promise, engineers should be *mindful* of the record of *normal accidents*.

References

(1) **Perrow, C.** (1984 and 1999) *Normal Accidents; Living with High-Risk Technologies*, Princeton University Press, ISBN 0 69100 412 9.

(2) **Sagan, S.D.** (1993) *The Limits of Safety*, ISBN 0 69102 101 5.

(3) **La Porte, T.R.** and **Consolini, P.M.** (1991) Working in practice but not in theory, *J. Public Admin. Res. Theory*, **1**, 19–47.

(4) **Rochlin, G.I., La Porte, T.R.** and **Roberts, K.H.** (1998) The self-designing high-reliability organization; aircraft carrier flight operations at sea, *Naval War College Rev.*, Autumn '87, 76–90, reprinted Summer 1998, 97–113. Available on www.nwc.navy.press/review/1988/summer/art7su98.htm.

(5) **Weick, K.E., Sutcliff, K.** and **Obstfeld, D.** Organizing for high reliability, *Res. Org. Behav.*, **21**.

(6) **Weick, K.E.** (1995) *Sense Making in Organizations*, p 362. ISBN 0 80397 177.

(7) **Aase, K.** and **Nybo, G.** (2002) Organizational knowledge in high-risk industries: What are the alternatives to model-based learning approaches? *3rd European Conference On Organisational Knowledge, Learning & Capabilities*, Athens, 5/6 April 2002.

(8) **Weick, K.E.** and **Sutcliffe, K.M.** (2001) *Managing the unex-pected: assuming high performance in an age of complexity*, Jossey-Bass, ISBN 0 78795 627 9.

An Engineer's Personal Story Worth Repeating

2.6.1 What does it feel like to be associated with a disaster?

Fortunately, few engineers experience a major disaster during their careers. In the 1970s, an anti-nuclear activist used a simple argument to whip up public feelings at a meeting in London. This was at the time when a lot of publicity was being given to studies of the probability of accidents. The argument was – *Listen to what the experts say. They tell you that the risk is minute – one in a million, one in 10 million. But there is always a one. Do you want to be that one?*

At first sight the argument seems preposterous. Nevertheless, it is mathematically correct. Having dutifully performed an engineering task, what is it like if the remote statistics or an unanticipated event leads to a catastrophe on your project? Makes you that one.

The answer to this question has most eloquently been expressed by Leslie E. Robertson. He was the lead structural engineer for the World Trade Center (WTC). The design of the WTC introduced a lot of new innovations. Many of these have led to awards and have been incorporated into structural engineering. With all the novel features, great care was taken during the design process to properly understand the risks and get expert assistance. But the design never anticipated the scale of the terrorist attack of 11 September 2001.

Leslie Robertson reflected on his experience in an article in *The Bridge*, a publication of the National Academy of Engineering (Vol. 32 No. 1 Spring 2002). He said:

> *In my mind, the loss of life and the loss of the buildings are somehow separated. Thoughts of the thousands who lost their lives as my structures crashed down upon them come to me at night, rousing me from sleep, and interrupting my thoughts at unexpected times throughout the day. Those who were trapped above the impact floors, those who endured the intense heat only to be crushed by falling structure, are merged with those who chose to take control of their own destinies by leaping from the towers.*
>
> *The loss of the buildings is more abstract. The buildings represented about 10 years of concerted effort both in design and in construction on the part of talented men and women from many disciplines. It just is not possible for me to take the posture that*

the towers were only buildings... that these material things are not worthy of grieving.

It would be good to conclude this journey in a positive mode. We have received almost a thousand letters, e-pistles, and telephone calls in support of our designs. The poignant letters from those who survived the event and from families of those who both did and did not survive cannot help but bring tears to one's eyes. They have taught me how little I know of my own skills and how fragile are the emotions that lie within me. Yes, I can laugh, I can compose a little story... but I cannot escape.

Do these communications help? In some way they do; in others, they are constant reminders of my own limitations. In essence, the overly laudatory comments only heighten my sense that, if I were as farseeing and talented as the letters would have me be, the buildings would surely have been even more stalwart, would have stood for even longer... would have allowed even more people to escape.

Yes, no doubt I could have made the towers braver, more stalwart. Indeed, the power to do so rested almost solely with me. The fine line between needless conservatism and appropriate increases in structural safety can only be defined after careful thought and consideration of all the alternatives. But these decisions are made in the heat of battle and in the quiet of one's dreams. Perhaps, if there had been more time for dreaming...

In conclusion, the events of September 11 have profoundly affected the lives of countless people. To the extent that the structural design of the World Trade Center contributed to the loss of life, the responsibility must surely rest with me. At the same time, the fact that the structure stood long enough for tens of thousands to escape is a tribute to the many talented men and women who spent endless hours toiling over the design and construction of the project... making us very proud of our profession. Surely, we have all learned the most important lesson – that the sanctity of human life rises far above all other values.

Reprinted with kind permission of Leslie E. Robertson and NAE.

COMMENTS

The respect and feelings of all engineers go to Leslie Robertson. The question *Could I have done better?* must always be in the mind of people reviewing failures. The answer is usually *Yes*. A more difficult question to answer is *Should I have done better?* Even if the chances had been improved from say 1 in 10^6 to 1 in 10^8, there is still a 1. In other words there is always a risk, no matter how small. How resistant to terrorist attacks should we make public buildings? Should we design more terrorist-resistant buildings and end up with bunkers? Or should we concentrate on preventing the attacks?

There will be lessons learned from the collapse of the WTC that will improve engineering detail, fire protection, and evacuation measures. That is the nature of engineering. Another aspect of the nature of engineering is reflected in Leslie Robertson's words – *innovation, risk taking, professionalism, pride in the profession, and the value of life*.

PART 3

Drawing the threads together – conclusions and final comments

3.1.1 Is there a pattern to the failures?

Standing back, the first comment has to be that most products and projects do not fail. Failures are the exception.

What causes the failures that do occur? Are all disasters unique? A review of Parts 1 and 2, summarized in Table 1, shows the main factors appearing in disasters. The conclusion is that many of these factors are involved in most disasters.

Table 1   Factors appearing in disasters and failures

	Technical	Local human error	Organization	One step too far	Dynamic	Lack of Robustness	Ego	Failure of Foresight	Trail of Signals	Lack of Information	$ & Time Pressures
HINDENBURG	✔	✔	✔	✔	✔	✔	✔	✔	✔	✔	✔
SPAD	✔	✔	✔				✔		✔	✔	✔
RAIL FAILURE	✔	✔	✔		✔	✔		✔	✔	✔	✔
USKMOUTH	✔	✔	✔		✔	✔		✔	✔	✔	
CHALLENGER	✔	✔	✔		✔	✔	✔	✔	✔	✔	✔
COLUMBIA	✔	✔	✔		✔	✔	✔	✔	✔	✔	✔
HERALD-OF-FREE-ENTERPRISE	✔	✔	✔			✔	✔		✔	✔	✔
DEE BRIDGE	✔	✔		✔	✔	✔	✔	✔			
TAY BRIDGE	✔	✔	✔	✔	✔	✔	✔	✔	✔	✔	✔
QUEBEC BRIDGE	✔	✔	✔	✔		✔	✔	✔	✔	✔	✔
TACOMA BRIDGE	✔	✔	✔	✔	✔	✔	✔	✔	✔	✔	✔
MILFORD HAVEN BRIDGE	✔	✔	✔	✔	✔	✔	✔	✔		✔	✔
MILLENNIUM BRIDGE	✔			?	✔	✔				✔	
COMET	✔	✔	✔	✔	✔	✔		✔		✔	✔
GIMLI	✔	✔	✔		✔	✔		✔	✔	✔	✔
AIR TRANSAT FLIGHT 236	✔	✔	✔		✔	✔		✔	✔	✔	✔
CHERNOBYL	✔	✔	✔		✔	✔	✔	✔	✔	✔	✔

Another general conclusion is that few failures occur because of the most obvious failure modes. These are usually well covered. Failures mainly occur in the less obvious detail.

3.1.2   The three spheres of failure initiation

Failures are initiated in three spheres:

- technical breakage that started the disaster;
- human error by the operator, pilot, or local management; and
- organizational and institutional failure.

The three are shown in Fig. 3.1.1. If there is a communication gap between the three spheres as in Fig. 3.1.2, then there is even less chance of the initiating source being spotted before a failure.

Fig. 3.1.1   The three spheres of failure

Fig. 3.1.2   The three spheres of failure with communication gaps

Rickover's Nuclear Navy successfully engineered all three spheres. It is commonly believed that the cost and discipline of his approach is too much for the commercial world. On the other hand, the commercial world often ignores the cost of future disasters – as in the case of the Shuttle and UK Rail. It is not just the cost of the disaster, but also the impact of the changes resulting from the disaster. The Comet failures lost De Havilland the initiative for jet transport – where are they today?

The technical failure
Almost by definition, engineering failures start with a breakage. That is the trigger point.

Operator, pilot, or local management human error
It was popular to blame the person at the controls for disasters. Today, there is more attention paid to exploring why the errors were made. In most cases, these errors occurred at the time and place of the disaster. The *Columbia* failures were a little different, spread over the 16-day flight and separated in space.

Organizational failures
Organizational weaknesses set the seeds for disasters. They are usually present over a long time-frame and lie in wait for a combination of circumstances to coalesce. The failures are less easy to see and define than in the other two spheres. The boundaries are also less clear; for example, how does economic climate and political pressure lead to disaster? Was the Hatfield crash solely the result of the persons charged, or did the broader political and bureaucratic decisions contribute?

3.1.3   The nature of disasters
After identifying the three spheres where failures are initiated, a number of causes of failures are presented in Table 1.

One step too far
This is a common cause of disaster. The hazard from taking a step too far is in not recognizing the risk – optimistically believing that it is not developmental, or that the development is easy. The question mark against the Millennium Bridge recognizes that a more conventional bridge with similar modal characteristics could also have suffered synchronous lateral excitation. The Millennium Bridge problem was not due to its novel design.

Dynamic factors

All but two of the examples failed because of dynamic effects. It is easier to run a static analysis, but in the real world, dynamic effects dominate and need to be understood and analysed.

Lack of redundancy and robustness

This cause was present in all of the examples. Redundancy in equipment is a well-established tool. Improving the reliability of operations with redundancy of people is more difficult to achieve and is resisted by management pressure to reduce staff numbers. Robustness is a more comprehensive concept but is also difficult to achieve.

Ego

There are a surprising number of examples where the ego of the dominant player contributed to the failure. A weakness to watch out for both in organizations and in one's own actions!

Failure of foresight

Not surprising is that all but one example showed a failure of foresight, a factor highlighted by the research of Barry Turner. In many cases, this feature went back to the design concept and details – and remained as a latent failure for years, if not decades. Synchronous lateral excitation was a new phenomenon that surfaced with the Millennium Bridge. Robustness included during the design stage might have added lateral stiffness and avoided the embarrassment. But robustness often adds costs that are difficult to justify for unknown effects.

Failure after a trail of signals

Most of the failures had a trail of signals – which went unnoticed. Many signals were present months, if not years, before the failures. The difficulty is in spotting the vital warning among a mass of signals. Constant attention and questioning together with trend analysis can help identify the important signals.

Lack of information and communication

Nearly all the examples resulted from a lack of information and communication between the players. The exception is the Dee Bridge, where Robert Stephenson failed to inform himself that adding ballast would overload the bridge! Unfortunately, in every instance with the exception of the Millennium Bridge, the necessary information or clues were available, but not known to those involved.

The ***Columbia*** enquiry noted the trend to use PowerPoint© presentations, rather than presenting a thoroughly reviewed technical paper. Edward Tufte, a Yale Professor, has provocatively titled an essay – *Power corrupts, PowerPoint corrupts absolutely.* He points out that the concept limits what can be written and sentences have to be shortened. Companies often insist that the company logo be included, which reduces the space further. Information is force-fitted onto the screen. It provides a quick and easy way to present some information, but is too much left out? Richard Feynman hated set presentations; he saw them as sales pitches, preventing a full debate that would bring the issues to the surface.

Charles Steinmetz, Thomas Edison's right hand man in his laboratory, had good advice for researchers nearly a century ago. He recommended that researchers present their work in three ways:

- first, a thorough technical paper for the record and to inform peers;
- second, a more general technical review so that application engineers can understand and apply the new research; and
- third, a report in layman's language for a wide audience including the executive.

The conclusion is that more time needs to be spent on searching for information and on communication between all the players. Remember that there are many other players as well as Professional Engineers who can have useful information and insight – talk to all of them.

Funding, schedule, and commercial pressures

Visions without resources are just dreams – Admiral Gehman

Most examples were affected by limited funding, tight schedules, and commercial pressures, or combinations of these pressures. The question to ask is – did those putting on the pressure know the risks they were taking, and did they take the responsibility for the result of their pressure?

3.1.4   What are the common reasons for failures?

The technical reasons for failures tend to fall into a few categories that occur frequently. In mechanical engineering, the dominant reason, quoted in most surveys, is fatigue. In routine bridge engineering, the reason is scour of the foundations. In chemical plant, Kletz has shown that failures come from rust, insulation, plugs, operating temperature variations, and drains. All these failure mechanisms are in the details.

These failures are all in the technical sphere of failure initiation, but failure in the other two spheres is equally in the details. In conclusion, we have to be continually *mindful* of the details. *The devil is in the details.*

3.1.5   Why do failures occur?

Investigations of failure quote high percentages attributable to human error. This raises the question – what are the rest due to? Surely we cannot blame inanimate material or machinery. Failures labelled natural phenomena are failures where someone did not design for the extremes of natural conditions. Remember Bill Clinton's election mantra, handed to him by James Carville – *It's the economy, stupid.* For engineering failures, the equivalent is – *It's the people, stupid.* So step one is to recognize, and accept, that all engineering failures are caused by human effort, or lack of effort. There is nothing else to blame; there are no other excuses. Failures are not generally caused by stupid people. All people are fallible; we can all make mistakes. The question is – can we control engineering so that human mistakes do not lead to undue risk?

One tool is to incorporate robustness. Another is to dive below the surface and explore what we can do to limit the effect of people failures. This inevitably comes back to the design – design of the product, design of operations, design of maintenance, design of the management processes, and design of the infrastructure.

SECTION 3.2

The Role of Design

The designer has a heavy responsibility when failures occur. Failures in operation and in maintenance usually have their roots in equipment-design problems. The designer should not rely on operators and maintenance staff for overcoming design faults. Designers should not leave traps for operators, as was partly the case at Chernobyl. Nevertheless, the designer often escapes the blame because the design work was so far removed in time and place from the failure. How can designs be improved?

Pilots and nuclear plant operators are trained for years in highly structured courses and have to pass formal examinations. Designers generally do not have similar training, despite designers having an equal or more prominent role in safety. Like all engineers, designers are expected by their professional organizations to adopt CPD. While attempts are made to formally plan CPD, it is usually ad-hoc. Companies run training courses on a variety of topics and send engineers to symposia. Senior engineers monitor the work of their juniors. Would more formal training be effective in improving design? Academic input could help, but probably the best assistance comes by learning from the practising expert. The Master/Apprentice formula has proved an effective learning tool since at least the Reformation. Musicians run Master Classes. Can this approach benefit designers?

In the Introduction, Fig. 1 shows the elements of the design process. Stage 1, the design requirements, is more than just a quick specification writing exercise. The requirements need to be well understood. Go talk to the customer, visit the site, and get fully familiar with the requirements.

Stage 2 is a divergent thinking process – be creative and table many possible designs. A team effort is effective, but with no critical or judgmental comments to dissuade ideas being tabled. Saying – *That's a crazy idea* is forbidden. Then draw a chart of requirements against designs and assess ratings for each as in Fig. 3.2.1.

This chart helps focus debate on specifics, rather than on an ego trip arguing that *my design is better that yours.* For example, designs C and D both have total scores of 87. It probably does not matter which of these two designs is chosen. If anyone wants to debate the choice, then the two areas where there is a difference in rating against the requirements 1 and 4 could be debated.

Stage 3 – converging on the design choice, is a detailed assessment leading to the final choice. Throughout stages 2 and 3, there should be an awareness of all the failure modes, with at least back-of-envelope calculations to quantify the limits. Keep continuous pressure on simplicity. It is tough to achieve, but what is not there cannot go wrong. Complexity is less reliable and usually less predictable because it is more difficult to be sure that all modes of failure are covered.

Stage 4 is the detailed checking mode – the stage where Pugsley recommended that all bridge and structural engineers recalculate the loads and deflections.

SCOSS reports noted that more time spent on design could be effective. This extra time should not be for extending the normal

Fig. 3.2.1   **Design assessment chart**

design process, but should focus on areas that often do not get the full attention. Maybe an extra 10 per cent of design costs could be spent as follows:

- Four per cent on checking that the design requirements have been fully met and that all the failure modes are understood, and the risks quantified.
- One per cent on making sure that the design concept and details are well recorded in writing so that everyone downstream, whether they are in manufacture, construction, commissioning, operations, maintenance, or decommissioning, has the information.
- One per cent to go visit the site, walk-the-walk, visit the plant. Yogi Berra said it well – *You can observe a lot by watching.*
- One per cent to visit the factory shop floor and understand the manufacturing process.
- One per cent to talk to operators and maintainers.
- Two per cent to get the CEO or project promoter to come and talk to the designers and understand both the benefits and the limits of the design – and recognize who is responsible for any risks.

SECTION 3.3

Organizational Weaknesses

Engineers would not think of trying to assemble a mechanism with a piece missing. Yet, Pugsley warned of organizations proceeding when a key player is unavailable for any reason. Surely, a complete organization is as important as a complete product.

Rickover believed that it is essential to choose the best staff. Yet, how often do design and project teams have to make do with the pairs of hands available from the last task? In company employment terms, it is not easy to select only the best – but at least a check should be made that there is a minimum adequacy in all staff.

Most management texts recommend selecting the handful of aspects that most impact the business, and concentrating on them. They say that, if managers try to solve all problems, then they will be bogged down in the details. Because engineering disasters start at the details, there appears to be a conflict with the management advice. One tool to address this conflict has surfaced in several places. Appoint someone to have the complete technical understanding of the product or project.

In the past, the Chief Engineer filled this role. With larger design teams, and more complex technology, it is difficult for one person to be both on top of the technical details and be managing a large team. The answer is to separate the roles: allow someone to manage the teams and someone else, with some modest help, to concentrate on understanding the whole technology – to be the technical integrator. This person must have broad technical knowledge and understanding and must be able to communicate well with any member of the design team. The person, and position, should have sufficient power and stature to be both advisor to the staff and a consultant to the executive.

Both the operations at the hands-on-level, as well as the broader organization, need to have the same care in their design as in the engineering design. Another problem to watch out for is that organizational change can often disrupt the database of company knowledge and experience. Those making the change may be unaware that they are sowing the seeds for future failures – and usually the process is irreversible.

SECTION 3.4
What Does the Public Want?

The answer probably is – they do not know, at least explicitly. This sounds arrogant, but the public does not know the features of the car they will buy a decade hence. Did the public know they wanted cellphones, DVDs, and nuclear medical imaging? Searching out the promising areas for progress is not easy and requires a keen understanding and sympathy with the public.

The public wants progress from the engineering industry, but preferably with fewer life-threatening disasters. The reaction following a disaster is largely one of public perception. The public may be very complacent and apathetic prior to a disaster, but enraged and demand retribution and action after the event.

Regulation and risk

The public's response to disaster is often to demand that the government acts. This usually leads to more regulations. Consequently, we now exist in a tangle of regulations that needs a Houdini to find a way through. A recent Canadian survey found an estimated 640 000 public sector workers employed full-time in drafting and enforcing regulations. Nearly three per cent of the private sector workforce is kept busy complying with more than a million active government regulations. Inevitably, there are wacky by-laws. In Montréal, a 1966 by-law explicitly forbids the making and storing of nuclear weapons, with a $300 fine for the first offence, $500 for the second, and $1000 thereafter – similar to the fines for keeping premises untidy.

While this creates some amusement, there are more serious questions. Who is defining the regulations, and how are they given stature? In Asia, they refer to politicians as lawmakers – an apt description. Politicians rarely define the regulations, nor even review them. They are more frequently created by unelected government employees. An interesting example is the new Eurocodes for structures. The limited state design philosophy is based on choosing a failure probability. The codes, which will soon become law, effectively set a probability of failure at 7×10^{-5}. While this is a great improvement on the failure statistics that concerned Pugsley 50 years ago, it does raise some interesting questions. Is this what the public want, and who takes responsibility for setting this failure rate – in effect taking a chance that there might be a failure?

The environment, an uncomfortable relationship between the public, science and engineering

Over recent decades, there has been a growth in concern for the environment. This has led to an ethical debate often followed by a polarized dialogue and pressure for action. This has led to increased legislation. Some environmental laws are based on sound science, some are created in a climate of dark foreboding, and others accept junk science. There are many inconsistencies. For example, a review of US laws shows that the value implicitly put on a life in various legislation varies from $1 million to over $6 million. Ironically, one of the highest levels is for pollution from vehicles, which is hardly consistent with deaths from collisions of vehicles.

The debate, and highlighting past environmental disasters, has led to significant improvements. The environmental aspects of the full life cycle of a product including wastes and decommissioning are now being considered. However, in many cases the environmental qualities are not easy to measure in traditional financial terms. Some aspects have become a political tug-of-war – based on the use of power and emotions rather than fact. Also, there has been a shift from immediate and visible issues to invisible problems – such as the safe levels of potentially toxic chemicals. These tend to be long-term effects that take time to research and understand. Should engineers wait for the results from the long-term research or proceed at risk? Being cautious is an obvious answer, but this might also delay the benefits of new and improved products

In many areas, the pressure for new regulation is preceding the science. This creates problems for both engineers and regulators. An example in Part 1 deals with nuclear radiation. Similar examples are global warming and concern over magnetic radiation from power lines and cellphones.

A UK poll taken in 2003 shows that the public have a strong belief in science – a stronger belief than science can deliver on some issues.

The difficulty of resolving the debate on risk

There is a gulf in the debate on risk between scientists and engineers on the one hand, and the public on the other. Scientists and engineers use a scientific approach based on facts. The public can often take the same facts and through emotion, fear, experience, and gut-feel have a reaction quite different from that through the logical scientific process. Neither side can claim to be right, and the differences are as hard to resolve as differences between religious beliefs. The only advice is to try to minimize the gulf by ongoing dialogue.

The conclusion is that public concern is a complex and an ever-changing scene. Engineers have to understand the changes and recognize how they might impact engineering products and projects. One answer is to get involved.

SECTION 3.5
Making Better Decisions

Poor decisions lie behind all disasters. How can we make better decisions? A scientific approach is to gain a better understanding of the decision-making process. Behavioural and cognitive scientists are working on this, and engineers should keep abreast of their research. The field is broad. Many studies examine decision making in medical diagnosis. Results from this area are probably equally applicable to engineering – both disciplines have to effectively review facts and then reach a decision for action.

Brain scanning techniques are enabling doctors to understand how the brain and mind work. Recent studies show that the brain sends signals for action before any conscious request. This applies to skill-based actions and raises the question as to how much free will is already pre-programmed by culture and training.

We have learned the value of visual images and mental models to help memory. Mnemonics and note taking have been traditional tools. Engineering should be a continually learning profession. Hence, it is important to understand how people learn, how facts become knowledge, and how they are retained in the memory.

Better knowledge makes better decisions. This is generally accepted, but little is done to help individuals. Companies restrict Internet access, often cut subscriptions to technical journals in libraries, and keep librarians on a limited budget. They frequently consider attendance at lectures and symposia as perks, as are site and factory visits for design staff. Engineering is based on economy of material and labour, and time spent on learning may appear to reduce productive time. This emphasis on today's budget may inhibit the learning that could reduce tomorrow's costs – and minimize future failures. Learned societies can help in disseminating new data and be a focal point for debate.

Another way of making better decisions is highlighted by the proverb, *Two heads are better than one.* The two heads make the most contribution if there is a challenging debate that brings to surface all the issues and views. Such a debate can be exhausting as well as exhilarating – and bring strong emotions to the surface. Feynman loved this type of debate. Unfortunately, the trend is to avoid challenge and to go-with-the-flow, using the argument that this is how team players should behave. Do not let challenging debate become a lost art.

SECTION 3.6
The Last Words!

In the wake of a disaster, there is often denial by those involved – *I could not do anything about it.*

Errors can coalesce over time and lie in wait to make a failure more likely. The data shows that the highest human effort together with the ideas of the high reliability organizations can reduce the probability of failure by one or two orders of magnitude. To achieve these results, engineers have to adopt the highest professional and ethical standards – and maintain them at all times. Do not allow sub-standard performance either in design, manufacture, operations, or management. There can be no temporary slips or acceptance of less than the highest performance. The shuttle accidents are a good example. NASA has had a string of outstanding achievements – the result of millions of good decisions in the details. The success is, however, overshadowed by just two examples of poor decisions – on the field joint and the foam.

It is always worth asking – *what have I done to reduce the chance of a failure?* be always *mindful* of the risks.

Continually learn: adopt a questioning approach – always ask *Why?* and *What if?* Throughout the chapters 'not knowing' has been a large contributor to failures.

We think we learn from teachers, and we sometimes do. But the teachers are not always to be found in school or in great laboratories. Sometimes what we learn depends on our own powers of insight.   Loren Eiseley, The Star Thrower (1978)

Ask – *what have I learned today?*

Finally, it is worth re-reading the words of Leslie Robertson. Remember that *the sanctity of human life rises above all other values.*